D1462491

Voice/Data Telecommunications Systems

An Introduction to Technology

MICHAEL L. GURRIE
Telecommunications Specialist

PATRICK J. O'CONNOR
DeVry Institute of Technology

PRENTICE-HALL, INC. Englewood Cliffs, New Jersey 07632

Library of Congress Cataloging in Publication Data

Gurrie, Michael.
  Voice/data telecommunications systems.

  Includes index.
  1. Telecommunication. I. O'Connor.
Patrick J. (Patrick Joseph), 1947- . II. Title.
TK5101.G83   1986   621.38   85-9299
ISBN 0-13-943283-3

Editorial/production supervision and interior design: Ronnie Frankel
Cover design: Edsal Enterprises
Manufacturing buyer: Gordon Osbourne

Printed in the United States of America

10 9 8 7

ISBN 0-13-943283-3   01

Prentice-Hall International, Inc., *London*
Prentice-Hall of Australia Pty. Limited, *Sydney*
Editora Prentice-Hall do Brasil, Ltda., *Rio de Janeiro*
Prentice-Hall Canada Inc., *Toronto*
Prentice-Hall Hispanoamericana, S.A., *Mexico*
Prentice-Hall of India Private Limited, *New Delhi*
Prentice-Hall of Japan, Inc., *Tokyo*
Prentice-Hall of Southeast Asia Pte. Ltd., *Singapore*
Whitehall Books Limited, *Wellington, New Zealand*

Contents

Preface

The telecommunications industry is changing at an increasing rate, and as a result, the demand for "telecommunications-literate" people has outstripped the supply. *You* may be one of the people who will fill that demand, or you may be someone who needs to read this book before you can deal with the new generation of telecommunications equipment.

"What's that?", you say, "Everyone just went crazy over computers, and they told me I had to be 'computer literate,' until I've got digital depression! I've used telephones all my life. How can I be a 'telephone illiterate'!"

Well, probably, you're not. Unlike computers, which were out of reach of most people until a few years ago, telephones have been within reach—often very close to hand—for almost everybody. That is why the new developments in telecommunications technology are likely to catch you napping; the telephone is an old, familiar acquaintance and hasn't changed much in your lifetime. The computer is *clearly* new and different, and everybody is pretty much aware that a "computer revolution" is going on. As a result, everyone's getting "hyper" about computers and computer literacy, while equally awesome changes are taking place behind the mouthpiece of their telephones. What is the "telecommunications revolution," and why are so many people unprepared for it?

Three major reasons for this exist. First, there are more complicated uses for telecommunications today than there ever were in the past; as the capabilities of these systems get better, the performance that is expected of them also increases. As business expects, and demands, "smarter" telecommunications equipment, the people working with the equipment must be "smarter," too.

Second, competition has been introduced into the telecommunications industry. Where previously only one telecommunications company served a particular locality, there now are a variety of choices. Business demands for high-quality telecommunications equipment and services have literally forced the organization of companies to manufacture, install, and service this equipment. This brings about diversity through competition, and means that as time goes by, a telecommunications expert must know about many more kinds of equipment with more complex features than ever before.

Third, integration of the computer into the telecommunications network has been responsible in large part for the increased capabilities of the "smart" telecommunications system. The increased speed and abilities of these systems requires a technician with increased knowledge of the computer technology they contain.

What problems face you as someone who works with telecommunications equipment? One problem is maintaining "telecommunications literacy"—that is, to keep pace with the industry, supply the information needed to make full use of these systems, or to have the skill to install and service them promptly. This book is written to help you understand all you will need to know, to overcome the problems you may encounter.

Technicians, communications managers, and other professionals will find this book useful because of its real-world applications. We have intentionally steered clear of highly technical theory so that the reader can quickly obtain the pertinent information. Little previous experience is assumed. The reader will be taken through a progression from early developments through the most recent technology. We avoid emphasizing old or obsolete material, except where it helps to make a point concerning current techniques. Both detailed and necessary technical information and general background information needed by the newcomer are included in this book.

Personal growth and advancement in this field will require you to keep pace with new developments. Making your business advance beyond its competition will also require that it employ telecommunications-literate people who know what advantages accrue to the latest developments in equipment and services. The communications industry has proven itself as highly profitable for all concerned. With increased competition, it may also serve as a source of bankruptcy to those businesses which *aren't* concerned. Through the application of current telecommunications technology, manufacturers and suppliers have increased sales and production, and users have cut costs while increasing the efficiency of their businesses. "Aware" readers of this book will have the background to "read between the lines" to see the implications future developments have for their business. Whether you work with telecommunications equipment or manage those people who do, after reading this book you will not have to rely completely on the supplier's staff for all information.

This book presents an overall view of the telecommunications industry with emphasis on the systems and methods used by business. A telecommunications system might be installed by AT&T or one of their competitors, even by yourself. Systems like those used by large corporations require personnel to manage and maintain their use. Too often, people working with these systems find that they have difficulties because they don't see "the big picture" of telecommunications. Both newcomers and persons experienced in a specific segment of the industry will find that the material in this book gives them a broader view of the entire field.

To the reader: End of chapter review questions provided in this book include a number of "discussion" type questions that do not have a single "correct" answer. In some cases, there may not even be a "best" answer. We have made no effort to tell the reader what all the answers are, here or in any other publication, because we feel that a concensus arrived at through discussion and debate is a better learning tool than a solutions manual.

1

An Overview of the Industry

If you use a telephone, you are a part of the global communications network. Every time you converse with someone else by phone, you link into the most complex switching network ever devised. Since you are reading this book, you must be interested in the system and how it works. This chapter will give you some of the background and origins of the telecommunications system, a description of the organizations that it contains, and a look at some of the parts you don't see when you're just talking to someone across town.

1.1 BACKGROUND

1.1.1 Origins

Few industries are as important to our lives and economy as telecommunications services. It isn't easy to imagine how our lives would be without them.

Back in 1876, when Alexander Bell was getting the first telephone company started, many people were completely against the idea. For instance, there were those who felt the telephone was an instrument of the devil (some still do!). These are probably the same people, who, a few years later, were saying, "If God had meant men to fly, he would have given them wings." Others felt that the telephone was an amusing toy but of little real importance, and expected that the "fad" would soon fade. Bell displayed his invention to the Western Union company, hoping to sell it as a new type of telegraph receiver/transmitter to be attached to existing telegraph lines. The telegraph company was thus presented with a golden opportunity to get into the telephone business as a logical extension of the telegraph system.

You would think that the Western Union company would have snapped up such an obvious improvement on its existing communications equipment—that's what you would think—but in fact the general manager to whom Bell proposed his idea scoffed at it and said, "No future in it, sorry!" Too bad. Bell had to go and start his own company, being unable to get backing from Western Union. Western Union's manager lived to regret his mistake, but it was too late, and the telegraph company, because of their tunnel vision, missed out on the business deal of the century—of several centuries, in fact. Were he alive today, that same manager would have been even less happy to see AT&T/Bell become one of the largest single corporate entities in the world. He would have been amused, perhaps, to see what the government did to it once they decided it was *too* large, and broke it up. That never happened to Western Union.

1.1.2 Later Developments

In the middle 1870s, Alexander Graham Bell was a special-education teacher working with the deaf. In seeking to understand sound and sound communication better, he learned enough about the theory of sound to get the idea for a "speaking telegraph" that could transmit sound over long distances. The actual invention involved a lot of false starts, but ultimately, it was ready for demonstration in time for the Philadelphia Centennial Exhibition in 1876. Bell first exhibited his working telephone in this exposition, and soon had orders and potential subscribers in sufficient numbers to form a company and start a communications system.

By 1880, the telephone system had hundreds of subscribers and a network (in New York alone) large enough to require some sort of switching system to sort out the circuits and make the connections between sender and receiver. Thus the switchboard, telephone operator, and switching system began.

By the turn of the century, telephone interchanges in major cities had spread to the extent that the number of circuits outnumbered the Western Union telegraph system's, and "forests" of telephone poles with hundreds of wires supported by each filled the air above street intersections (see Figure 1-1).

As early as the May 1884 issue of *Scientific American,* it was reported that "wires . . . like threads in a huge web, hold the New York streets in a mesh-like tangle" and that one branch of the legislature had declared that "both electric-light and telephone wires must be underground by June 1, 1885, in cities of 50,000 or more." It appears to have been an optimistic deadline.

It soon became evident that the human operators were needed in larger numbers than the work force could supply. This fact, coupled with the astronomically growing number of possible interconnections between circuits, forced the development of the automatic switching system and dial-telephone device.

Today, telephones with rotary dials are being replaced by new technology. The Touch-Tone® pad has taken over the function of the mechanical switch, but still performs the same function: addressing the other circuits in the system without human intervention.

Figure 1-1   New York street scene from 1880s depicting "forests" of telephone poles and wires. From James Martin, TELECOMMUNICATIONS AND THE COMPUTER, 2nd ed., ©1976, p. 425. Reprinted by permission of Prentice-Hall, Inc., Englewood Cliffs, N.J.

1.2 ORGANIZATIONS

1.2.1 Network Today

The world telecommunications network is a complex linkage of different communications media. Everything from wires to satellite microwave links is used to carry the word from one place to another. The information carried on the global telecommunications system is conveyed by the largest assemblage of transistors, switches, wires, waveguides, fiber-optic light pipes, and other signal conductors that ever existed. It could be said without exaggeration that the global telecommunications network is one vast machine—the largest machine of any kind ever built.

The functions of this vast machine are divided into two parts. One part is a **public** network, and the other consists of **private** systems. The public network links together all the private systems (with the possible exception of the *really* private ones, such as intercoms). The dividing line between what's a **public** and what's a **private** system is not all that clear.

Public network

The public network is composed of numerous organizations, called **common carriers**, that provide telecommunications services to the general public. These common carriers can be owned and operated by business or government. In the United States, the largest of these was the Bell System, which included AT&T and other subsidiaries, a publicly owned corporation. In most other countries the public telephone system is operated by the government—in England, it is a division of the General Post Office—so that the public network is just another one of the public utilities. The facilities of all these telecommunications organizations are interconnected, allowing communications between any two places in the world.

Private systems and networks

Private systems are generally owned and operated by individuals or companies for their own use. A private system can be of any size. It can be anything from a small office with a couple of keyphones, up to the IBM network, which is international but comprises lines and switching circuits dedicated solely to IBM's use. A private system can deny access to anyone except users the owner authorizes, while a public network cannot deny services to any person without good reason.

1.2.2 Common Carriers

Local telephone companies (we will refer to them hereafter as **telcos**) may be either Bell or some other company, but once established, have traditionally had an exclusive "lock" on telephone operations in their service area. Of the 1500 or so telcos, about 85% of the business is controlled by Bell companies. Having the lion's share of the telco business in this country, Bell had an effective, although regulated, monopoly on telephone communications in the United States.

1.2.3 Deregulation and Divestiture

In 1934, the Federal Communications Commission (FCC) was established by the federal government. Its goal was to provide communications service to all Americans, and to regulate that service—especially radio broadcast communications—but all interstate communications media came under its jurisdiction. There was a perception that a "radio trust" or "telephone trust" might arise that would stifle all competition and hinder development in this field (still in its infancy at the time).

The Communications Act of 1934 (as amended) still governs the interstate communications of the United States. One thing it does is regulate the common carriers to protect the consumer from unfair prices for communications services.

Descriptions of services and rates are called **tariffs** and are filed with the FCC. Any changes or additions to the tariffs must be approved by the FCC. Similarly, state commissions regulate the common carriers within each state. Because of the two regulatory bodies, state and federal, it could cost more to call long distance within a state than it costs to call across the country.

By government order, AT&T has been divested of its interest in the Bell operating companies. AT&T still provides long-distance services through the AT&T Communications division. Money made by AT&T can no longer be used to subsidize the local telcos, or money from the local telcos used to subsidize production of telephones by a manufacturer. Before deregulation, phones could only be *rented* from Bell. That's like your power company saying, "Don't buy bulbs from Jones or we will shut off your power." Now you can buy (or rent) your telephones from the vendor of your choice. The cost of these systems is limited only by the pressures of the marketplace.

The Bell operating companies across the country have been split into seven independent groups. A holding company has been formed for each of those seven groups. For instance, Illinois Bell, Wisconsin Bell, Michigan Bell, Indiana Bell, and Ohio Bell are affiliated with a Midwestern-region holding company called Ameritech. The AT&T system's public network, for instance, no longer has any connection with the holding company, Ameritech, and Ameritech isn't allowed to make its own telephone equipment. This split was necessary to allow fair competition between Bell and the smaller companies. As a result, profits from the public network (still a monopoly, and still regulated) could not be used to subsidize a private (nonmonopoly, nonregulated) equipment division. However, Ameritech is not regulated and can charge whatever the market will bear for products and services of its deregulated private equipment division. Since Ameritech does not manufacture its own equipment, they buy it from other companies, such as Nippon Electric (NEC) or any other company whose equipment and price is suitable to their needs.

1.2.4 Specialized Common Carriers and Other Common Carriers

Specialized common carriers (SCCs) provide services that are not offered by the local telco or duplicate its services at a lower cost. MCI and SPC (Microwave Communications, Inc. and Southern Pacific Communications) are two examples of specialized common carriers that offer attractive rates for long-distance calls to certain metropolitan areas. Many SCCs can reach any location in the continental United States. They also allow you to communicate with other companies offering similar services, or may provide data communication links from computer to computer. Specialized common carriers may produce their own network links like microwave or satellite, or they may lease lines from other common carriers, including telco services.

As an example, let's suppose that a long-distance voice communication SCC, the Com Company, has a fiber-optic link between Chicago and Milwaukee. There is a lot of business traffic between Chicago and Milwaukee, since they are both major Midwestern centers of commerce. If a call placed through AT&T, via the local telco, costs $0.52, and Com charges $0.34, they should easily be able to attract business calls.

Suppose that a business in Chicago, the Harts Company, has a branch office in Milwaukee and must stay in touch regularly. They make enough calls to Milwaukee to warrant using the services of a SCC such as Com. Harts could use three methods to access the Com services:

1. **Dial-up service.** Harts could call up Com Company through the local telco and be charged for a local call only. Com would then use their link to Milwaukee to connect the call to the Milwaukee local telco. If the local costs are $0.09 and this cost is added to the $0.34 charge from Com, the call costs $0.43, still less than the alternative charge of $0.52.

2. **Specified common carrier.** Harts could "specify," that is, stipulate, that their common carrier on long-distance calls is Com, not AT&T. This is okay if Com reaches all the places that AT&T reaches, and if all the calls are priced lower than AT&T charges. If this is *not* true for Com, Harts may not be able to specify this option.

3. **Tie trunk.** Harts may choose to have a tie trunk put in between themselves and Com. The tie trunk is a private link provided at a flat rate—let's say $100—per month. If Harts makes 1000 calls to its Milwaukee office per month, this would be $0.10 per call allocated for tie-trunk use. This brings the cost of each call up to $0.44, a penny more than the cost of dial-up service. Now that Harts has a tie trunk, though, they may not need as many local lines— perhaps they can take out two local lines—thus offsetting the penny difference. Harts has also simplified its dial-up procedure to the Milwaukee office. Instead of dialing Com through the local telco, then giving Com their account number, then dialing the Milwaukee number, they have a dedicated line to Com, and have only the Milwaukee number to dial, saving dialing time and reducing errors.[1] If the management at Harts believes the old saying that "time is money," the extra penny for tie-trunk service is no loss at all.

Some problems arise from the tie trunk. Since Harts has only one tie trunk, only one call can be made to Milwaukee at a time. A second caller either has to wait or must make the call over the telco at a cost of $0.52. The amount of traffic on the tie trunk will determine whether Harts will need one, or two, or more tie trunks. Let's look at Com Company's side of the picture. The fiber-optic link cost them $12 million to put in and $56 thousand a year to operate and maintain. Selling calls at $0.34 each requires that a lot of calls be made before Com can show a profit.

Actually, a SCC such as Com would charge $0.34 for the first minute and $0.03 for each additional tenth of a minute. This is similar to the way telcos charge. Since the SCC charges in fractional minutes, and some telcos charge by the whole minute

---

[1]By law, "equal access" will make the number of digits dialed by all common carriers the same by about 1986.

(no matter how little of it you've used), there could be an additional price break for using Com's service compared to that of some other telco.

AT&T offer special lines called WATS, which are used only for long-distance calls. The charges for using these lines is substantially less than calling over the regular local lines. A call over a WATS line may cost the Harts Company $0.43, the same as Com Company's charge plus the local call. The office bean-counters might be happier with the WATS arrangement, since there is only one bill to pay, not two.[2]

1.2.5 Interconnect Companies

The local telco was, at one time, the only place you could go for a phone or system. State and federal tariffs would not permit you to connect anything to telco equipment unless they provided it. Being the only vendor of equipment, and also the only vendor of services, the telcos could enforce their monopoly by withdrawing service to anyone who violated these tariffs. For some reason, this autocratic attitude was ignored until the Carterfone decision in 1968. The ruling that a Carterfone (a non-Bell piece of equipment) could be connected to Bell lines without Bell having a "right" to "pull the plug" on the Carterfone user was a landmark decision in telecommunications.

Equipment connected to telco lines must still fulfill FCC regulations, and if directly connected, must still be compatible with telco signals and voltage levels, and could be purchased from AT&T or any other supplier.

Following the Carterfone decision, new tariffs were filed establishing how non-telco equipment should be connected to the telcos' lines. The Bell System began to get alarmed that some of this nonstandard equipment might damage, degrade, or otherwise compromise their lines and equipment. Data connections to telephone company equipment, for instance, required an interface, or **direct-access arrangement** (DAA)—a piece of coupling equipment used to protect telco systems from possible harm due to "unorthodox" equipment. For voice connections, such as PBXs and telephones, a **protective coupling attachment** (PCA) was required. You leased the interface from the telco to connect "that weird stuff" to the telco's lines. These interfaces are still needed to connect equipment not approved by the FCC and for "foreign" equipment that uses incompatible signaling.

The term **interconnect company** was coined by the Bell System. Interconnect companies do not particularly like to be grouped together under this heading; they don't call *themselves* interconnect companies. But there is a lot of Bell and not very much of them, so in the trade language, companies that provide equipment to be connected to the telcos' lines are called interconnect companies. Let's refer to these companies as **intercos** for the balance of this discussion.

Intercos sell, lease, install, and service private systems. These systems are then interconnected to the public network through lines leased from the local telco. These lines are run by the telco to a terminal block inside the customer's premises. Then

[2]Soon, AT&T will be billing separately for interstate WATS.

the interco connects its equipment to this mutual tie point. Since this terminal block represents the line of demarcation where the telco's responsibility ends and the interco's responsibility begins, it is referred to as a **demarc**.

There are around 2000 intercos in the country today. Small companies in this business may be located in one city or area only. Larger intercos may serve areas that encompass many states. The largest may be not merely nationwide, but international in scope. ROLM Telecommunications, for example, has major operating divisions in most large metropolitan areas, and markets their systems overseas as well.

Most intercos do not manufacture the equipment that they install. Because of this, they can offer equipment made by many different manufacturers. This can result in a flexible system, where every product line is second-sourced by alternate suppliers, ensuring a steady supply to consumers. A company that deals in a variety of different types of equipment, however, can have shortages when alternate suppliers of the same device do not exist. Equipment for a specific purpose may come from many different manufacturers. Where the equipment from one manufacturer is not interchangeable with that from another supplier, small companies do have difficulty keeping a stock of all possible items on hand.

Most intercos have their own service departments and rarely make arrangements for other companies to do the servicing. The majority install their own equipment; a few also delegate some tasks, such as cabling, to subcontractors.

If you are seeking a telecommunications system, you have several choices: You can lease or purchase equipment from many intercos. These include former divisions of the Bell System which are now completely separate subsidiaries that do not provide telephone service, such as trunks, operators, and so on, and are, in fact, interconnect companies. Whereas in the past you could pick up a phone and order telephone equipment from "the phone company," you can no longer lease or purchase equipment from your local telco. The law now requires that there be two groups, one for basic services such as telephone numbers, and the other for the equipment. You can no longer rent telephone equipment under a tariff on a month-to-month basis, except for equipment already installed as of January 1, 1984 (which was taken over from the local telcos by AT&T Information Systems). You should base your decision not only on the prices and options available, but also on the supply capability and service reputation of the interconnect company.

1.3 HARDWARE

1.3.1 Telephones

That funny little critter that sits on your telephone stand is such a taken-for-granted item that you probably haven't thought very much about what's in it or what it's connected to. At the beginning, the telephone was an intercom without a dial, telephone numbers, or even a microphone in the true sense of the word. Bell's original 1876 equipment used the same mechanism for both microphone and speaker. The speaker mechanism was a moving permanent magnet pushed back and forth in the middle of a coil by a diaphragm. Pressure waves from the voice moved the diaphragm, and the changing magnetic field in the coil induced electric currents in the coil. At the other end of the circuit, the current reversed the process and pulled the magnet in the earpiece half back and forth, moving the diaphragm and producing sound waves. In 1877, Thomas Alva Edison invented the carbon microphone, permanently separating the functions of mouthpiece (transmitter) from earphone (receiver). The "carbon mike," still used today, is capable of producing much stronger current signals from a voice input (it *does* need a battery to run, which the "dynamic mike" system didn't). Early in the twentieth century, the telephone network had been extended to all parts of the nation, but you had to shout to be heard over a long-distance call. This was due to the fact that no effective amplifier existed for voice signals. (An amplifier for telegraph signals, called a relay or "repeater," had been invented in the 1840s—this is what made Morse's telegraph commercially successful.) The development of the vacuum-tube triode made it possible to "repeat" (amplify) audio signals (voice waves) as the relay had earlier made "repeating" of telegraph dots and dashes possible. Invented in 1908 by Lee De Forest, the "audion" tube was also the backbone of the radio telecommunications system later.

Recent modifications, such as the Touch-Tone® keypad and solid-state switching, have modified the appearance and action of the telephone, but inside each one, you'll still find the carbon microphone **transmitter,** electromagnetic speaker **receiver,** and the switchhook (even though the earpiece no longer hangs on a hook). These basic parts are still the same as in the 1880s, and newer telephone system equipment still retains compatibility with the older models. Early dial telephones (rotary dials) can still be attached to the present system.

The "electronic" telephones of today, incorporating digital logic and microprocessors into their composition, have a lot more "smarts" than did phones of the recent past. They can display the numbers being dialed, store and retrieve frequently used phone numbers (and dial them for you), and even display the cost of a call as it's in progress. The capabilities of microprocessor-based telephones are limited only by the imagination of designers.

1.3.2 Links

The first telecommunications links were used by the telegraph companies, which connected the key and sounder with wires. Most telephone communications links also use wires for at least part of the signal path, but the longer the distance covered, the less likely it becomes that the *total* signal path will be wires. Long distance may use such media as coaxial cable, waveguides, or fiber optics to communicate high-frequency "RF," microwaves, or light waves. Some links may be designed to convey voice signals, digital data communications, or both. **Multiplex** communication allows use of a link to carry more than one channel of information at a time. Each type of link is named according to its purpose and/or its physical characteristics. Your long-distance call may travel via any combination of methods that is available to reach your intended destination.

Rather than referring to the telephones attached to the link as telephones, we may call them **stations** to indicate anything at the end of a line where a transmission originates or is received. Telecommunications links carry transmissions for station equipment such as telephones or computer terminals. Many individual links may be connected together to make up the total path of a transmission. The "local line" that connects your telephone to the telco's switching equipment is the first link used when you place a call. Your local line is then connected or "switched" to other links, perhaps to the local line of the telephone you are calling. If you are calling a telephone that is not served by the same local office to which your phone is connected, other types of links (besides local lines) must be used. The greater the distance between you and the person you're calling, the more types of links your call is likely to use.

The lines that connect your phone to the local office, or a business phone to the switching equipment, are simple pairs of wires. A *pair* is two conductors that make a complete circuit containing your phone and the switching equipment at the other end. One conductor is designated *tip* and the other *ring*. These terms come from the appearance of the kind of plug used in manual switchboards. Each patch-cord used on traditional telephone switchboards has a plug that has two points of contact. As you can see in Figure 1-2, the tip of the plug is connected to one conductor, and farther back on the plug, the ring is connected to the other conductor. The conductors are named after the parts of the plug they're connected to.

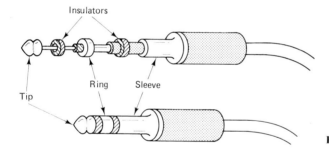

Figure 1-2   Telephone switchboard plug.

The local line to your telephone is being used only when you are making a call. The rest of the time, the wires connecting your telephone to the local office are doing nothing. There is nobody else who can use them. Most of the time your personal link is idle. Most of the day, the conductors that make up your link are no use to anybody. The phone company went to a lot of trouble to connect distant points with conductors. If all of the links out there were used as inefficiently as your local line, it would represent a great waste of conductors and manpower. To take better advantage of the conductors, they should be shared among several users. There are several ways to do this:

1. Used mostly in rural areas, a **party line** allows several people with telephones connected to the same local line to use the line alternately. One party uses it at a time (although the others are often in the habit of listening in). This is a little like having extensions of your home phone in other people's houses. Various methods

exist for telling whose call is incoming. In the old days, each party listened for "their" ring. One person's ring might be "two shorts and a long" and another's "two longs and a short," while a third's might ring steadily. Now phones on a party line can each be rung separately. The advantage of party lining is obvious. Only one pair has to be strung from the local office rather than a whole lot of pairs. This makes more efficient use of the lines, and less idle time, but the disadvantage is obvious, too. You can't make a call unless nobody else is on the line, and you haven't got very much privacy if someone else wants to listen in. Wherever feasible, party lines are being replaced by local lines for each subsciber.

2. **Trunk lines** are used between switching systems (for example, between a telco's local office and a PBX or the telco's toll office). They carry one conversation at a time but are switched between different local lines as needed to complete the connections. These links "concentrate" the many local lines to a few trunk lines. Since a telephone, or any other station for that matter, may be used for only a fraction of the day, only a few trunks are needed to handle a large number of local lines. Figure 1–3 shows a simplified layout of the connection between two switching systems.

With trunks carrying only one transmission at a time, the number of transmissions between switching systems is limited to the number of trunks. If it happens that all the trunks are in use and someone tries to place a call, the call cannot go through, and the person receives a "busy signal" indicating that the trunks are busy. This is commonly referred to as **blocking**, and the busy signal (which repeats more rapidly than an ordinary busy signal) is called a **reorder** or an **all trunks busy** signal.

3. A **multiplexed link** allows multiple transmission on the same link at the same time. This may be accomplished in several ways. The ultimate result is that the transmissions are mixed together in a way that permits them to be "taken apart" at the distant end of the link. **Multiplexing** and **demultiplexing** are the names given to the processes of "mixing" and "unmixing." The advantages are clear; even fewer conductors are needed to connect distant points. The circuits required to multiplex and demultiplex the various channels on a multiplexed link are an extra expense but are far outweighed by the savings in cost of running large numbers of conductors across the countryside. Your radio is an example of a multiplexed link. Multiple channels of radio transmission are broadcast from different antennas, and your radio picks them all up at once with its aerial. The tuner in your radio allows you to demultiplex one channel (the one you *want*) from among all the channels being transmitted at the same time. If the radio stations were transmitting over wires, (cables) instead of using "wireless" transmission, the cabling would work exactly like a mutliplexed link (in fact, it would *be* a multiplexed link). An illustration of this type of arrangement is shown in Figure 1–4.

These three types of multiuser connections are called **shared links.**

The term **line** is often mistakenly used in describing many types of links, from a handset cord to the trans-Atlantic cable. The specific name (if you can find one) should be used to avoid confusion. A telephone line could be a row of people waiting to use a pay phone. Another thing that causes confusion is having multiple names for the same type of link or equipment. Subscriber loop, local line, or user circuit

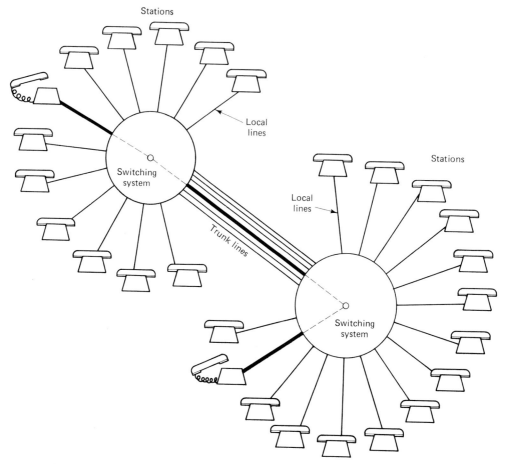

Figure 1-3 Switching systems connection. Local lines connect stations to switching systems; trunk lines connect switching systems to one another.

may all be used interchangeably in referring to the wire that connects a home telephone to the telco's switching system. Usually, there is one term that is used more commonly than others. Wherever possible, we'll use the most popular term, but will include the others together with the definition. A glossary of terms is included at the end of the book.

1.3.3 Switching Systems

The purpose of "making a call" is to connect one telephone to another. These connections now include the connection of computer terminals, facsimile machines, and video. Equipment used to make the connections are called **switching systems**, or simply **switches**.

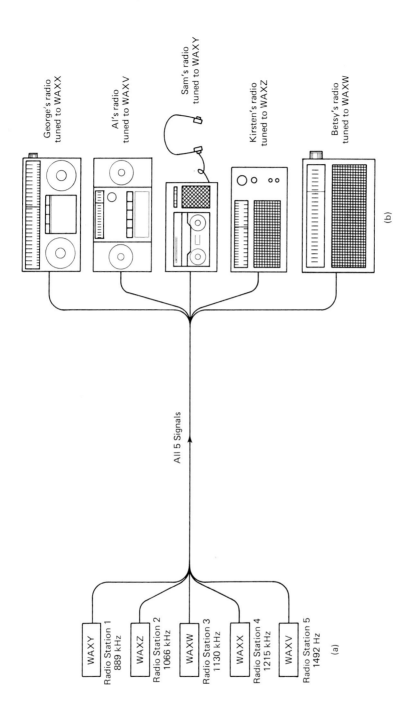

Figure 1-4 Multiplex linkages. (a) Multiplexing (left), each channel is a different frequency. (b) Demultiplexing (right), each receiver tunes in a different frequency (channel).

George's radio tuned to WAXX

Al's radio tuned to WAXV

Sam's radio tuned to WAXY

Kirsten's radio tuned to WAXZ

Betsy's radio tuned to WAXW

(b)

All 5 Signals

WAXY
Radio Station 1
889 kHz

WAXZ
Radio Station 2
1066 kHz

WAXW
Radio Station 3
1130 kHz

WAXX
Radio Station 4
1215 kHz

WAXV
Radio Station 5
1492 Hz

(a)

13

The first switches were manual. The earliest of those was put into service in 1878; it had 21 phones connected to it. Subscribers were connected to one another by operators using patchcord plugs and jacks on a primitive switchboard. As more phones were tied to these manual switching systems, more operators were needed. In fact, as more phones are added to a system like this, the number of operators you need goes up faster than the number of phones. When the system gets big enough, you reach a point where you need more operators than there are subscribers! Before this happens, something's got to "give." In this case, the problem can be solved by developing an automatic switching system that doesn't require a human being's attention to make and break each connection.

In 1892, Almon B. Strowger's **stepping relay**, an electromechanically driven rotary switch, was first commercially installed in a switching system, although it didn't come into general use until the 1920s. This made the addition of a dial possible; now the telephone user could make a connection with another telephone automatically without "bothering" the operator. The dial was merely a pulse-generator that "stepped" the switch around, making the connection with the other telephone. Circuits like this are still used by some of the intercos and Bell operating companies.

Today, switching systems are not only automatic, but computer-controlled. This powerful improvement has given switching circuitry far greater speed and smaller physical size than earlier electromechanical switching equipment.

The switches used by the telcos are enormous. They have the capacity for connecting thousands of lines. Metropolitan areas have dozens of switches which may be located in different sections of the area. These switches require a building to house the equipment and staff needed to maintain them. The switch used in a private system may be small enough to fit on a desktop and support only a few telephones. Switches used by the local telco are powered by the local power company, but also keep batteries "floating on the line" to keep the system going in the event of a power failure. This is why your phone still works to call the power company when your electricity goes out. Even in a major power blackout (residents of New York take notice!) the telephone service will continue to function (at least for a while). Diesel generators may also provide additional backup for the battery system.

The **private branch exchange** (PBX) is a telephone system switch whose basic function is the same as the switches used by the public network, but is part of a private system. These basic functions include connecting the phones of a business to one another or to a trunk of the public network. In this private system, the PBX is located on the user's premises and corresponds to a **central office** (CO) in the public network. It forms the "hub" of the private system in the same way that the CO forms the hub of the public system in a neighborhood.

Manufacturers of PBXs have used different names to describe their equipment, but the basic concepts of use are the same. Names such as **computerized branch exchange** (CBX), **private automatic branch exchange** (PABX), and **electronic branch exchange** (EPBX) are all used to describe different kinds of PBXs.

The first difference in using a single-line phone and a private system with a

PBX is the **access code**. The access code is a number dialed before making an outside call—a call outside to the public network. The digit 9 is a common access code to go to the public network. To make a long-distance call another access code may be used—perhaps 8 to connect to a WATS line. Internal calls—calls within the building—don't require an access code; you just dial the number of the extension. Each of the phones within the private system are called **extensions**. You may be thinking of extension phones in your house, where all the phones are "in parallel" on a single line—this is not the case with the phones in the private system (although they're still *called* extensions)—each phone can carry a separate conversation. Access codes are also used to access common equipment in a system. To "page," you might dial 77 to be connected into the public-address system. Some common equipment in a system, such as the paging arrangement just described, might not require an access code; instead, you just dial it like any other extension. Extension nmbers are usually three or four digits long; access codes are one or two digits. Different extension numbers are assigned to each line in the system, but several phones might share common numbers (called common "appearances"), all ringing when the number is dialed, although this is more typical of key systems than of PBXs.

As the user sees it, switching systems have some special abilities. Examples of these are second call waiting, three-way calling, and speed dialing. **Second call waiting**, for instance, is the feature that puts a beep on the phone while you are talking to someone, to let you know that there is a second caller trying to reach you. **Three-way calling** permits you to make a call, and then, when you establish a connection, permits you to make a second call, and link all three phones together. **Speed dialing** is a feature that permits you to use a short sequence of keystrokes, perhaps only two, to dial a frequently used number stored in a digital memory. You could dial 1-800-323-4500, by just dialing *7, where the 7 indicates that this is the seventh number stored in the digital memory device. These features may be obtained from the local telco or can be found in a PBX.

The switch contains programming that chooses what features a person may have access to. This is part of the "smarts" that are in the switch. If the switch is a PBX, the smarts are located in the PBX, locally. If the switch is at the telco, the smarts are located remotely from the user, but do the same things. A **smart phone** can contain a processor itself, capable of doing things like speed dialing without using an outside switch to do this at all.

For the **dumb phones** that don't contain their own processors, the smarts are at the switch. To add features, nothing has to be modified in your phone; the changes are all made at the switch.

Abilities that the user doesn't see include **routing**, the "thinking" that the switch must do to get a call through the best available choice of connections. For instance, you might have to call California from Chicago via Florida if a direct circuit were not available. These are part of the "invisible" abilities that are "transparent" to the user, because they don't change anything about the way the user operates the phone. In a PBX, alternate routing may be done to use a tie trunk to make a special call using an OCC, which costs less than calls placed through the telco. This is also one of those transparent features that doesn't have to involve the user in the mechanics of making the connection.

Since the PBX is the heart of a private telecommunications system—which is the main subject of this book—we will explain the features and abilities of PBXs in greater detail in a later chapter.

1.3.4 Key Systems

Key telephone systems are generally used in smaller businesses, such as a doctor's office or car dealership. They may also be used in a single division of a large business, where the people in the division share a common function.

Figure 1–5 shows two typical styles of keyphones. Unlike PBX systems, which commonly use single-line phones, the key system usually uses multiline phones. The multiline **keyphone** has several buttons on it, one for each line. These line buttons can be used to connect to trunks from the telco or to various extension lines from a PBX. When a **key system** is used with the lines of a PBX, the user has the same access to features and abilities as any other user of the PBX. When the key system is connected to trunks, each button is a separate telephone number. The telco, rather than the PBX, is the source of the different lines. Since the key system user isn't working through a PBX, the features of a PBX are not there, only the features of the telco and the keyphone itself.

Just in case this appears very clearcut, there are **hybrid systems** with some of the features of a PBX and a keyphone. Features are available for these systems from "nothing more than pushbuttons" to "everything in a PBX."

Operation

To place a call on a key system, you would press a "line" button (to select a line not already in use). A lamp associated with the line you selected goes on and shines steadily, indicating that you can now pick up the handset and dial your call. (Sometimes, you have to pick up the handset before the lamp will light and connect you to the line.) To put a call on **hold**, you push the "hold" button, and the lamp blinks on and off. To return to a call on hold, you press the line button for the line you want.

Incoming calls (**ringing in**) also cause the lamp to blink on and off, but at a slower rate. The audible ringing sound can be from a bell or from a small speaker in the keyphone. To answer the call, you press the line button that's blinking and pick up the phone. The line may not have an audible ringing-in signal (it may ring at another keyphone that also has that number appearing on it), but the lamp will always show the state of that line.

Let's define some terms here that are commonly used in telephone jargon to refer to the state of a line:

1. **Idle** (not in use). This state is indicated by a lamp that is OFF.

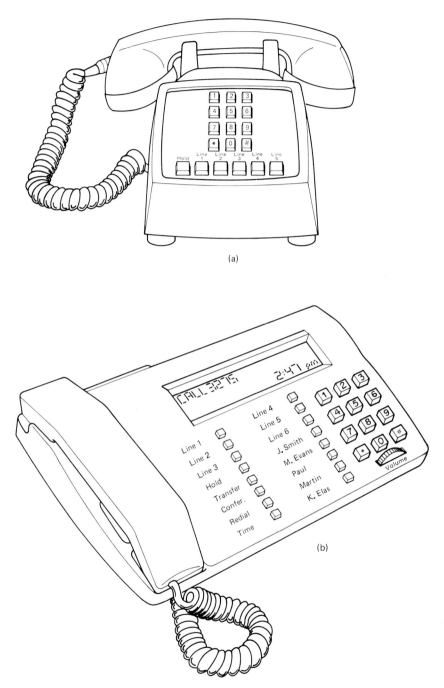

Figure 1-5 (a) Five-line keyphone. (b) Electronic keyphone with feature buttons and display.

2. **In use** (also **occupied** or **busy**). Somebody is engaged in a conversation on this line—shown by a lamp that is lighted steadily.

3. **Hold** (also **call holding**). Someone is waiting for a called party to return to a conversation—shown by a rapidly blinking (winking) light (blinks about twice a second).

4. **Ringing in** (someone is calling in). This situation is indicated by a slowly blinking light (about once a second).

Sometimes, you want to talk to someone else within the same key system. A suite of doctor's offices, for example, might have three trunk lines, 555-1123, 1124, and 1125. Suppose that one of the doctors picks up line 1123 and wants to talk to another doctor in another room of the suite. The other doctor could also punch-in line 1123 and pick up the handset, but the two doctors would be talking over the dial tone. This would make the conversation almost impossible. Doctor 1 could call up doctor 2, dialing up the entire phone number, 555-1123, from another line, but this would place the call through telco equipment, and the doctors would be charged for the call. The solution is an **intercom** line. Doctor 1 would call doctor 2 by pressing the "intercom" line button and dialing the extension number of doctor 2's intercom line. Doctor 2's intercom line would indicate a call ringing in.

This is called a **dial intercom**. Another type of intercom is called a **manual intercom**. To engage in a coversation on a manual intercom, dialing is not necessary. All the phones connected to a common manual intercom circuit are a "party line" and anyone picking up one keyphone whose intercom button is pushed can talk to anyone else on the same circuit without dialing or dial tones. With a manual intercom, you do not dial and ring in on the other phone as you would with an ordinary call. Usually, you push a separate button to buzz the person you're calling.

In both cases, the intercom line is totally independent of the outside telco and does not use their equipment, nor is there any charge from them for the call.

Line assignment

Imagine a keyphone system with 10 lines coming in and a keyphone with 10 buttons. All the other keyphones in the system could be 10-button phones, called a **square** arrangement. In a square system, the 10 buttons on any phone are arranged the same way on every other phone. Another arrangement, a **nonsquare** arrangement, would use several phones with a few buttons, each handling only a fraction of the 10 incoming lines. Or alternately, some phones might have all 10 lines while others had only a few of them.

A typical nonsquare arrangement for a suite of executive offices with nine lines coming in is shown in Figure 1-6. In this office, a receptionist at a 10-button phone receives incoming calls. Each of the executive suites has a five-button phone, and each phone has a dial intercom line on the last button. Incoming calls are answered by the receptionist, who would use the intercom to announce to the executive that there is a call holding, and on which line.

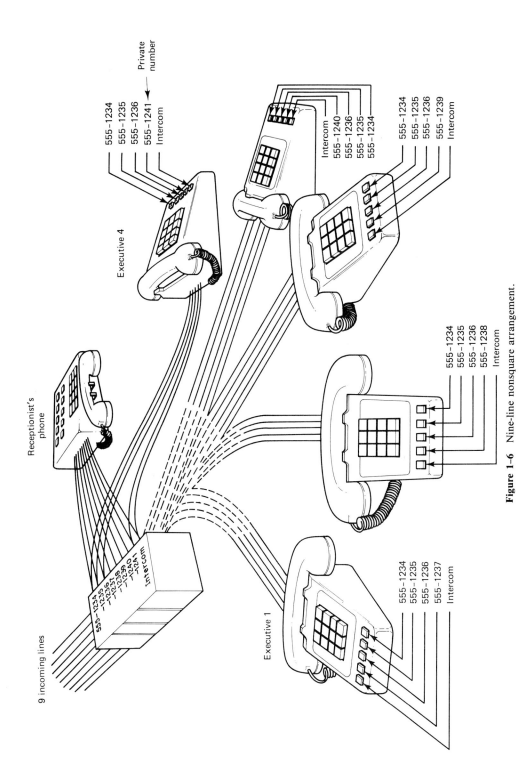

9 incoming lines

Receptionist's phone

555-1234
-1235
-1236
-1237
-1238
-1239
-1240
-1241
Intercom

Executive 4

555-1234
555-1235
555-1236
555-1241 ← Private number
Intercom

Intercom
555-1240
555-1236
555-1235
555-1234

555-1234
555-1235
555-1236
555-1239
Intercom

555-1234
555-1235
555-1236
555-1238
Intercom

Executive 1

555-1234
555-1235
555-1236
555-1237
Intercom

Figure 1-6 Nine-line nonsquare arrangement.

19

Note that each executive's phone has the same first three lines. The fourth is the executive's individual line, and the first three are "general." A general call to the company can be taken by any executive who's free. The company would list these three numbers in the telephone book. Each executive's fourth line would have the business-card number the executive uses.

The fourth and fifth lines on the executive's phone are the only ones that will ring the phone. The other three lines will ring only at the receptionist's desk, and the receptionist will decide to whom to direct the call.

In the example just discussed, the lines coming into the receptionist's keyphone were CO (central office) lines. In a larger office, instead of CO lines, the lines arriving at the keyphone could be extension lines from a PBX.

1.3.5 Data Communications: Digital

Digital data transmission is becoming the most widely used form of telecommunications. The amount of information communicated by digital means has already exceeded the amount sent by ordinary voice transmission, and as time goes by, the amount of digital information continues to increase compared to voice.

At the present time, voice signals and digital signals are considered different types of transmissions. When voice is used to produce an electrical signal, the signal is a varying voltage called an **analog waveform** that produces an electrical wave which matches the pressure pattern of sound waves. Digital information—numbers and letters—is transmitted in a code that uses only two kinds of signal. The two states—on and off—used to transmit **digital** information are simpler to use and transmit than the complex voice waveform. For a long time, digital techniques were used only to transmit data—such as binary codes—from one computer to another. These are still important uses, but the digital method of data transmission is now being used even to encode the voice signals mentioned earlier into digital binary numbers. Eventually, all data and voice transmissions may take place using signals with just two levels. The concept, which is explained completely in Chapter 4, is complex, but enables transmission of voice *AND* digital information to be done *more reliably* and with *less distortion* than in the *analog* method used since the days of Alexander Bell.

The telegraph actually was a digital system. The dots and dashes of the Morse code used two-level signals. A dot and a dash weren't actually different levels of electrical voltage, but were distinguished by how long the key was held down. The "key closed" and "key opened" conditions were the two levels of signal that the telegraph system used. One major advantage of using this kind of a signal was the fact that when the signal was amplified (boosted in power) as it traveled long distances, the **repeater** (signal booster) was also able to *reconstruct* the original signal without distortion. Digital repeaters in modern telephone systems have the same capability. Unlike these digital repeaters, analog voice transmissions must be repeated by circuits that amplify both the distortion and the noise picked up by the signal in its travels. This means that "purer" and "clearer" sound can be conveyed by digital means than by analog means.

A very large amount of information can be transmitted by digital signals in a very short time. Spoken words contain a lot of wasted, repetitive signals that don't have to be used to convey the same message digitally. For instance, a spoken letter W ("double-yoo"), transmitted by voice, would take about a second to transmit so that somebody listening to it could recognize it. To do this, several hundred cycles (waveforms) of sound would be repeated on the telephone circuit, with each waveform being much like the others. To transmit a W in the digital code called **ASCII**, the information would take only 10 waves (or a wave made of 10 pulses) and would not have to be repeated at all to be recognized. This would take about 37 thousandths of a second.

Within the time of a short telephone conversation, the Sunday newspaper could be transmitted from New York to Chicago. Printed matter is often transmitted digitally to distant locations where it is inconvenient to ship the paper copies directly. At the receiving end, the data is printed directly for immediate distribution. For example, by being transmitted digitally and printed locally, the *Wall Street Journal* appears in most cities at the same time that it appears on New York newsstands. The savings in time and cost are obvious.

A computerized **bulletin board** is data stored in the memory of a computer that you can "reach in" and "look at" through an interface that involves a telephone and a video screen. The ability to access ("look into") the contents of a computer's memory is not limited to bulletin-board services. Libraries permit investigation of "stack" material and listings of abstracts through computer/telephone data links. **Electronic transfer of funds** (ETF) permits you to do banking remotely. It also permits banks to do banking with other banks remotely. Dow Jones permits subscribers to access their stock market information files. **Electronic mail**, carried interoffice or around the country from one terminal to another, is presently in common use in business. Teletype, the news wire services, and telegrams are also instances where data transmission is used to "get the message across."

The nonspeaking deaf can now communicate with their local emergency services (fire, police, etc.) through *TTY* (actually an abbreviation of "Teletype," but currently used for any remote-printing unit). The deaf are not just using TTY for emergency services; they can "talk" with anybody else who has a TTY terminal. This application of data communications opens a whole new area to deaf people that had been closed to them previously.

To connect yourself to the "digital world," you need a TTY-style terminal, which consists of a keyboard similar to a typewriter and a printer or video display. The terminal cannot (usually) talk directly into your telephone. A "magic" box called a **modem** (modulator/demodulator) is the link that permits your terminal's digital codes to be communicated through your telephone.

To use the **TTY/modem** system, you just dial the TTY/modem to which you want to "talk." When the other end "answers," it will be with a tone that tells you that you've been answered by another machine. Then you activate your modem and the communication link is established between your terminal and the system at the other end of the line.

Two styles of modems exist, **direct-connect** and **acoustically coupled.** The *direct-connect* type is simply wired into the telephone line—you activate it after receiving the answer tone by flipping a switch. The **acoustic** type, an indirectly connected interface still widely used although sensitive to external sounds, is equipped with two rubber cups that hold the handset of a standard telephone. One contains a speaker and the other a microphone. The signals travel into the telephone through the air, rather than directly through wires—hence the name "acoustically coupled".

The use of a **modem** takes it for granted that the telephone uses **analog** communication between TTYs. This assumption was good in the past but is no longer necessarily true. Today's voice communication might be conveyed between transmitter and receiver by **digital** signals directly. In this case it would not be necessary to convert the data into sound for communication. Instead, the digital signals from the TTY might go directly into the circuit; it would be the *ordinary voice telephone* that would require a converter—from an *analog waveform* to some digital transmission form such as PCM or delta modulation. The circuit to do the **coding** and **decoding** of voice into digital code is called a **codec** (coder/decoder).

1.3.6 Equipment Manufacturers

A number of companies manufacture telephone equipment of all kinds—since deregulation, more than ever before. Some, such as Western Electric, you would be very familiar with, since they manufactured all the telephones and switching equipment used by the old Bell System. Others, especially foreign manufacturers, haven't received much exposure in this country. When the Bell System had the "lock" on telephones (rented, never sold), you got the phone your local operating company let you have, whether made by Western Electric, GTE, or whoever. Now that you can buy your own phones, and every hardware and department store seems to be selling telephones, you can become acquainted with a number of features and styles you wouldn't have seen before. The public network's switching equipment is still manufactured by Western Electric, Automatic Electric, Northern Telecom, and a number of other companies. Large manufacturers of PBXs (private branch exchanges) are ROLM, Northern Telecom, Mitel, NEC, and at least 30 other companies. Manufacturers of switching equipment are often affiliated with computer manufacturers for purposes of standardizing terminal/host computer and switching systems that will all be part of a common voice/data network. ROLM is affiliated with IBM, Intecom with Wang, and AT&T has gone into computer manufacturing on its own, now that they're not restricted from doing that sort of thing. NEC, being Japanese, has for a long time, been in the business of making both telephone switching equipment and computer systems, since they were not regulated by American monopoly and anti-trust laws.

1.4 TRENDS FOR THE FUTURE

1.4.1 Example Conversation on a Future Telecom System

Your phone rings and you answer from anywhere in the house by saying "identify caller." The voice-recognition capability built into your telephone interprets your request and tells you that the caller is Jack Smith. Jack, a friend and business associate of yours, is identified to you before you actually answer the incoming call. To answer, you say, "connect call," and your phone completes the connection that is ringing in. "Hi, Jack," you say, "What can I do for you?", and Jack tells you that he has some information about some property in Phoenix in which he is interested. He asks if you want to see what he has found. "Sure," you say, and you direct the system, "Telephone—switch call to video." Jack appears on the video with some papers in his hand, and for the next few minutes, your monitor displays graphs and pictures as Jack explains them to you.

You and Jack decide to get another friend involved in the discussion. You instruct the system to call Elaine, and the three of you continue the discussion. Each of you, during the conversation, is making other calls. Jack calls his broker's system and enters a sell order for some of his stock, then transfers the money into an account set up jointly with you and Elaine. From this account, Jack, you, and Elaine plan to purchase the property you have been talking about. Jack is already convinced that it is worth buying, but while you continue to discuss the details, Elaine is hunting up in financial database services information that is relevant to the price of similar parcels of land, to see if the land is overpriced. You want to see the property before making a commitment. Meanwhile, Jack is making sure his lawyer is provided with all the details so that the property can be transferred to your joint possession with minimum delay. While the details are completed and the lawyer initiates a title search, you are setting up airplane reservations for the three of you to go see the property. The airport limousine has already picked up Elaine, and will be arriving to pick you up in 15 minutes. You leave your home telecom system with instructions to watch the house, answer calls and take messages, water the plants, and feed the dog until you get back. The system already knows what callers should be connected to you when you call, so you have their calls transferred to the motel where you will be staying in Phoenix.

1.4.2 Voice/Data/Video System

If you think this sounds like the home life of a crew member on the Starship *Enterprise*, you're mistaken. All the capabilities displayed by the system in the example above can be found in systems today, or at least connected to them. The first place an integrated system like this will probably appear is in businesses. The trend is well established, however, for today's business phone features to become available on home phones tomorrow. "Call waiting" and "call transfer," for instance, have been part of private-network PBX systems for years, but are now being offered to the public in some areas by the local telco.

Of the items described in the example, the two that are probably furthest from realization, before a telecom system like this can become a household item, are the voice-recognition "smarts" in the system and the high-speed link for video. Progress in microprocessor programming should, within the next decade, eliminate the voice-recognition problems associated with today's systems, and if cable-TV services become incorporated into a more interactive sort of public network, the high-speed video-link problem will also go away.

Some experimental cable video systems (for example, *videotex*), already have some of these interactive features, and to a certain degree, these permit the viewer to control what's happening on the screen. For an educational program, this provides the viewers with the ability to answer questions, and perhaps, in the future, to see their scores on the screen, while maintaining privacy from everybody else. The integration of the home computer into the system makes it possible to sift through incoming information in a "smart" way that a telephone or television could never do. The telecommunications system of tomorrow may have voice, data, and video capability coordinated by the home computer.

Drawing an analogy to the phone lines used today, we can envision in the future a voice/data/video system linking home, business, and industry.

1.4.3 Applications of Tomorrow's Systems

Entertainment

One aspect of the entertainment available from the voice/data/video system of the future would be its flexibility. With the "dial-a-movie" feature, you could pick out for yourself those films in the TV station's library that you would like to see. Through the use of fiber-optic links rather than the coaxial cable currently in use for cable TV, the entire movie could be "downloaded" into the memory of your home video system in a few seconds, to be played back in "real time," at your leisure. Fiber optics can carry signals hundreds or even thousands of times as fast as the video signal that runs an ordinary television receiver. Sent in a fast "burst," like an audio tape whose sound is played back at a higher speed than you recorded it, the video signal could be transmitted at superfast speed and then played back at ordinary video speed from the computer's memory banks.

Government

Government by referendum—where every citizen is able to vote on every issue that is being debated in the houses of government—has been considered impossible for a large population like that of the United States. It was even considered impossible by the Founding Fathers of this country when they wrote the Constitution—although the population was much smaller then—and because of this, *representative democracy* was instituted here instead of *true democracy*. With the voice/data/video telecommunications system of the future, direct voting would be possible on any issue in which the voter is interested. Voice-recognition techniques used to "understand" spoken words over the phone could also identify the speaker, as distinct from other people calling on the same phone. People who tried to vote twice would not succeed in disguising their voices, since the LPC (linear predictive coding) methods used for voice recognition are identical to "voiceprinting" and could be used to identify a voice with the same reliability with which fingerprinting identifies a person. This system would be interactive, with the results of each referendum stored in a database that could be accessed by any voter who wanted to see how people voted on a particular issue. Functions of this type of interactive government channel would include making information that should be of public record—such as the amount spent by the county government on roads and the mileage of road work actually done—accessible to the public through a terminal.

Education

We're not talking "TV College" here. Instead, we're looking at interactive, two-way, station/viewer communication that is very much like CAI (computer-aided instruction). In this case, however, the interactive programming would work out of a vastly bigger database, with much more sophisticated programs than the home user's computer could hold by itself.

If you wanted an education in telecommunications, you could get a lot more from this system than out of a single "rigid" source—such as this book—if you just know the right questions to ask. Videotapes, text, spoken dialogue, and library research would all be at your fingertips, and you would never have to leave your front room to access this information. With a "help" command on each access feature, you could tailor your own learning experience to the information you needed, and channel the course of your studies into a new direction in midstream. Let's say that you've ordered from the New York Public Library, the chapter on data communications in this book. You find, as you read the chapter, that we are using the word "modem," and you have no idea what a modem is or whether it's an important thing to know. You could "jump out" of the reading you are doing and ask the system to "help" you with more information on modems. The system might provide you with titles of books and periodical articles, or might give you a "dictionary definition" of the word out of its own dictionary. Narrated films, graphs and charts, and live interaction with resource persons (academic experts on the subject) could all be a part of these interactive programs.

News

For news, you could scroll through the headlines on your video terminal, stop at any headline you'd like to see, and seek out in-depth detail. This would apply also to voice or newscast as well as text stories. A newscast could be recorded into the day's database as a "tree structure," where the commentators would be shown on the screen headlining the main stories of the day in "abstracts", and you could stop the recitation at any point, to "branch" to the report giving the in-depth story you're interested in, while skipping the details of the stories you don't want to see.

Also, the option of "going back" and looking at the headlines again to see details of an earlier story would be available. Try doing *that* with an ordinary newscast of which you missed the first half of!

In this sense, the news could become "page-able" in a way that is currently possible only with newspapers, and with live-action reporting that no newspaper could match. Like pulling out the sections of your Sunday newspaper today, you could "pull out" news in the categories that interest you. If you want to look at the comics page, you could see the cartoons as film animation rather than just as "stills," although still pictures would probably be available, too, if desired.

Marketing

As much as you hate television commercials today for interrupting your programs, this is nothing compared to the annoyance that can be visited on you by the minions of automatic marketing. You may have already had the opportunity to be annoyed by telephone advertising by the obnoxious marketing counterpart of the telephone answering machine—the automatic dialing machine. This delivers impersonal messages to you via your telephone, and then expects you to say nonobscene things into your telephone mouthpiece to suit the needs of the advertiser. You may be deluged with junk mail, but you can always throw it into the wastebasket without looking at it. "Junk telephone," and it future counterpart, "junk telecommunications," uses a channel that you just can't refuse to pay attention to—your ringing telephone. Since this might be an emergency call, you are compelled to answer it, whether it's going to be "junk" or not. *Or are you?*

Not if you have "screening" capability. If the public network voice/data/video system is developed in a thoughtful fashion, your telephone will have access to the calling number. Your home computer will be able to match the calling number against a list of "acceptable" callers, to identify if the call is coming from someone whom you are on talking terms with. Perhaps this type of advertising will be regulated by the FCC, and possibly not allowed (if you vote that way in the electronic referendum).

Of course, this type of advertising might not be totally unwelcome. You might put out the "welcome mat" for advertising from auto dealerships, for instance, if that old "beater" of yours has finally given up the ghost. Competitive advertising messages (including pictures of the cars in your price range) might be supplied by the advertisers based on your needs, as enunciated in your request to the public network's main computer. This arrangement could send you only the advertising you *need*, and help the advertisers, too, by making it possible for them to advertise to you only those things in which you are likely to be interested.

REVIEW QUESTIONS

1. What types of media are used in telecommunications systems? What types are most common? Where?

2. What is the difference between a public network and a private network?

3. Are common carriers part of a public or private network?

4. Who owns and operates private networks?
   Who owns and operates public networks?

5. What do the terms telco and interco refer to?

6. What was the significance of the 1968 Carterphone decision?

7. Give examples of services offered by:
   (a) AT&T
   (b) Regional holding companies
   (c) Bell operating companies (BOCs)
   (d) OCCs
   (e) Interconnect companies (intercos)

8. What is the present connection between AT&T and the Bell companies?

9. What common carrier offers WATS?

10. What is a demarc?

11. How many telecommunications links are used on a call between Chicago and Washington, D.C.? Why?

12. Do intercos manufacture the equipment they sell and install?

13. What is a repeater used for?

14. What are the three basic parts of a telecommunciations system?

15. What is a wire pair, and to what do tip and ring refer?

16. Name some types of station equipment.

17. Why are access codes needed?

18. Why is there no charge when one telephone calls another in private system?

19. Why *is* there a charge when one telephone calls another in a public system?

20. What do we mean by multiplexing?

21. What types of links are multiplexed? Why?

22. How is it possible to get a busy signal when calling an *idle* station?

23. How are local lines used differently from trunks?

24. How do PBXs differ from switches used in the public network?

25. Are the "smarts" for a telephone's station features located in the telephone, the switch, or both?

26. Name three different functions for which buttons on a keyphone might be used.

27. What are a square arrangement and a nonsquare arrangement in a keyphone system?

28. What type of communications uses analog waveforms?

29. How many *states* are used in the digital waveform?

30. Name a device that converts digital data into an analog waveform.

31. What device converts a voice signal into a digital signal?

32. Why is a telegraph considered digital in character?

33. What is the best method for *repeating* the signal without the noise on a long-distance transmission (digital or analog)?

34. Name an application of home telecommunications that is not presently available
but probably will be in the future.

35. Are the exotic home-telephone features mentioned at the end of this chapter
currently available to the commercial, business, or scientific user?

36. When the technology for a feature is available, why doesn't it automatically
become incorporated into the repertoire of every telephone user's station
equipment?

2

Basic Telecommunications Technology: Telegraphy

In this chapter we hope to prepare you for a better understanding of telecommunications technology by showing you the underlying electronics technology on which all telecommunications is based. It would be outside the scope of this book to try to explain all of electronics. Such electronics technology as is directly related to telecommunication will be brought forth and described with examples drawn from the history of telecommunications. Historically, telecommunications equipment began simple, and got complicated later. In the same spirit, we will cover the electronics used in telecommunications equipment by going back to the historical beginning, starting with the fundamentals and moving forward to the present.

2.1 BASIC ELECTRICITY AND MAGNETISM

2.1.1 Origins of Electrical Communications

In 1800, Alessandro Volta invented the electric battery and provided a source of steady, continuous electric current for the first time. Before this, American and European experimenters had attempted to communicate over long distances through wires using static electricity, but their attempts had not been very successful. Luigi Galvani had accidentally discovered current electricity (using a frog as a voltmeter!) by crossing two dissimilar pieces of metal with the frog's leg. The frog's leg jumped when it crossed the two pieces of metal, which was a pretty good trick since the frog was dead. The current that Galvani had discovered flows steadily, as opposed to static electricity, which is dissipated all at once in a "spark." Galvani didn't know what he had discovered. His countryman, Volta, proved that the electricity Galvani had discovered could be produced easily without frogs. Almost any porous material that conducts electricity could be substituted for the frog's leg (it wouldn't jump, though) and the current generated by the two dissimilar pieces of metal would still flow. Volta stacked up alternate pieces of metal and vinegar-soaked cardboard and produced a "voltaic pile" which generated current electricity. This was the first **battery**.

There was already a "boom" in electricity research going on at this time (using static electricity), but the development of a source of steady current electricity gave an impetus to a great burst of invention and discovery in the next few years.

2.1.2 Electromagnetism

Early in the nineteenth century, Hans Christian Oersted discovered that an electric current would deflect a magnetic compass needle placed near a wire carrying current from one end of the battery to the other. Immediately, the result of this electrical research was applied to communications. Wheatstone and others arranged complicated arrays of compass needles and wires to convey messages across long distances by electric current.

The basic idea of using electrical current to communicate information is shown in Figure 2-1. The ability to interrupt the flow of current makes communication possible. By turning the current in the wire on and off, it's possible to let somebody else at the other end of the wire know that you've got something to say. This is done by a switch or *key* in the circuit that connects and disconnects the conductor so that it provides a complete path for current from one side of the battery to the other or blocks that current. In one of Wheatstone's early designs, small compass needles were made to point at different letters of the alphabet on a board (see Figure 2-2). For 20 letters (no numbers), there were 5 wires. As each letter was selected, two needles deflected and pointed at a letter where the needles crossed. This was not a very efficient communications link, and it could not go very far, because of something called **resistance**.

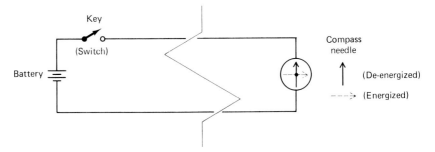

Figure 2-1 Using electrical current to communicate.

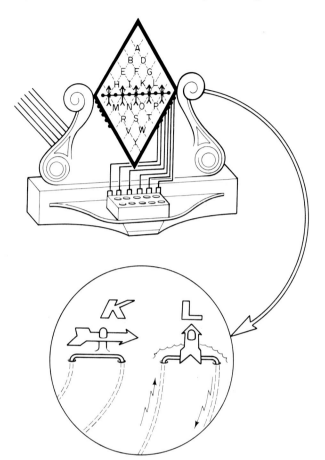

Figure 2–2   Wheatstone's five-needle
telegraph.

2.1.3 Resistance

Current in a wire depends on the battery and the wire (no big surprise there!). The
longer a piece of wire is, the more **resistance** there is to the passage of current through
it. Georg Simon Ohm discovered that the amount of current varied with the battery,
so that a battery twice as big (twice as many pieces of metal in the "pile") would
push a current twice as large through the same circuit. If the circuit were made longer
(its resistance increased), the current got smaller.

   This is known as **Ohm's law**, and it's useful to us because it tells us that the
longer the distance across which we want to communicate, the harder we'll have to
"push" with our battery to get the same current. Why is the current important? Because there must be a current to get **magnetism** around the wire. The more current
flows, the stronger the **magnetic field** around the wire. Oersted discovered this with

compass needles. Compasses point the way magnetic forces "push," but don't respond to electric force. The current in the wire produces a magnetic force, and it's this force that allows us to identify the current flow (without using frogs' legs). If we try to communicate using a long wire, the battery must be made larger, or the compass needle made more sensitive. The problem with this is that there is already a magnetic field hanging around to which compass needles *usually* point (the Earth's geomagnetic field), and if the current in the wire is too small, the compass won't "see" the magnetism produced by the wire because it's "covered up" by the earth's stronger field.

Joseph Henry of the United States and Michael Faraday of Britain discovered in the 1820's that a coil of wire wound around a piece of iron produced a much stronger magnetic field with the same current. This wire coil is called an **electromagnet**. It made possible electrical communications across long distances. "Long distances" were still limited to a few miles, and telegraph communications across country were not feasible (commercially practical *at a profit*) until Samuel F. B. Morse invented the **repeater** in the 1840s. The repeater, now called a **relay**, was simply a switch closed by an electromagnet, but it made long-distance communications practical, for reasons we will now discuss.

2.1.4 Simple Telegraph System

In Figure 2–3 we see the three basic parts of a telegraph system—the **battery**, the **key**, and the **sounder**—with a wire connecting them all into a complete current path. Current must flow through the key to the sounder to make a "click" at the receiving end of the telegraph.

This system isn't too much different from one of Wheatstone's early telegraphs. It was a technical success, able to reliably transmit messages across several miles. It was also a commercial failure. The reason has to do with what happens as the **conductor** gets longer. There are three things happening in this circuit.

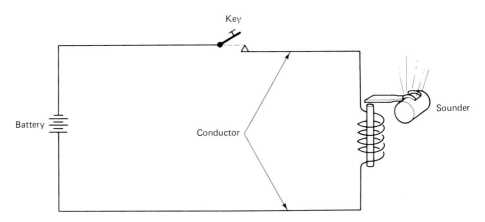

Figure 2–3 Basic parts of a telegraph system.

We measure *how much* of each thing is happening using three quantities. **Voltage** answers the question: *How big is the battery?* As more plates of dissimilar metals are "stacked up," the voltage (named after Volta, of course) gets larger. It is measured in units called **volts** and it turns out that each pair of dissimilar-metal plates added to the stack increases the "push" (the voltage) that pushes current through the circuit by the same amount every time. We use the letter V to stand for voltage. The **current** flowing around the circuit is measured in units called **amperes** (after André Marie Ampère, a French researcher also working in the early nineteenth century). This answers the question: *How much current* flows in the wire?" We use the letter I to represent current (Ampère, being French, got to name current with a French word that begins with "I"). Finally, the **resistance** of the current path is what limits the flow of current. It is the "friction" in the conductor that keeps the electrical charges from going faster and faster without limit. (Since objects are supposed to keep speeding up as long as they're pushed on, the reason the current doesn't keep getting faster and faster must be the friction in the wire.) We use the letter R for **resistance** and measure it in units called **ohms** after Georg Simon Ohm, who figured out the relationships among these three quantities:

$$V = I \times R$$

$$I = \frac{V}{R}$$

$$R = \frac{V}{I}$$

which we use to find one of these values when we know the other two.

2.1.5. Resistance and the Need for Repeaters

Ohm's law is useful to us because it is the law that will show us why Wheatstone's telegraph doesn't work (commercially) over long distances. We also need to know that resistances found in different places around a current path (resistances) in a **series circuit**) add up. The circuit works as if it has *one* resistance, and that resistance is the *sum* of all the small resistances at different points along the path:

$$R_{\text{total}} = R_1 + R_2 + R_3 + R_4 + \cdots$$

Armed with these two pieces of information, we'll look at three examples of the circuit in Figure 2–3, each with different numbers. We have to take some things for granted. First, we're using a wire **conductor** about the size of coat-hanger wire, but longer. It has a resistance of about six Ohms per thousand feet of length. The **sounder** is made of wire about as thick as a sewing needle, with a few hundred coils around its iron core. This coil has a resistance of about 100 Ohms. The sounder we are using needs 10 volts to go "click." The **battery** is 24 volts, which is a lot more than the minimum needed to click the sounder. Finally, the **key** is a switch whose contacts are made of heavy metal and has no measurable resistance when the contacts are closed.

In the first figure (Figure 2–4) there are 16,000 feet of wire conductor (about 3 miles) with a resistance of 96 Ohms (about equal to that of the sounder). We have this much resistance in the conductor because it is 16,000 feet of wire with a resistance of 6 Ohms per 1000 feet. With 16,000 feet, there are

$$16 \times 6 \text{ Ohms} = 96 \text{ Ohms}$$

The total resistance of the current path is 196 Ohms, because we can add the conductor's resistance (96 Ohms) to the resistance of the sounder coil (100 Ohms). We can probably throw in an additional 4 Ohms for the part of the current path that travels through the inside of the battery. That makes the whole current path's resistance

$$100 \text{ Ohms} + 96 \text{ Ohms} + 4 \text{ Ohms} = 200 \text{ Ohms}$$

The voltage of the battery is 24 volts, because that's the rating of the battery we used in all three cases. The current can be computed according to the Ohm's law calculation that says.

$$I = \frac{V}{R}$$

Thus

$$I = \frac{24 \text{ volts}}{200 \text{ Ohms}}$$

This is 24/200 of an Ampere, which is also 12 hundredths of an Ampere. That's not much current, but it's enough to **energize** (magnetize) the electromagnet of the sounder. How much of the voltage from the battery reaches the sounder?

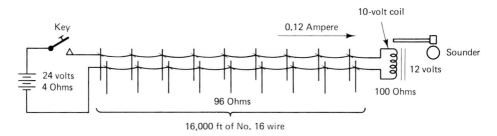

Figure 2-4 Resistive losses on a telegraph line.

You can figure out the voltage that the sounder gets from the size of its resistance and the current flowing in the circuit. Since there is only one current, and it's flowing everywhere, the current in the sounder is the same 12 hundredths of an Ampere (0.12 Ampere) that we calculated from Ohm's law. We can use Ohm's law to figure the voltage on the sounder, since we know its resistance and the current going through it, using the rule

$$V = I \times R$$

Thus

$$V = 0.12 \text{ Ampere} \times 100 \text{ Ohms}$$

This is 12 volts. It is exactly *half* the battery voltage. Since the sounder needs 10 volts to click, it will work, but there's not a lot of voltage to spare.

The second diagram (Figure 2–5) uses a conductor 30 miles (160,000 feet) long. Its resistance is

$$160 \times 6 \text{ Ohms} = 960 \text{ Ohms}$$

The total resistance in this circuit is

$$960 \text{ Ohms} + 100 \text{ Ohms} + 4 \text{ Ohms} = 1064 \text{ Ohms}$$

(Quick, Watson, the calculator!) According to Ohm's law, this makes the current

$$I = \frac{V}{R}$$

Thus

$$I = \frac{24 \text{ volts}}{1064 \text{ Ohms}}$$

This is 0.022556391 Ampere, roughly. (Is that what *you* got?)

How much of the 24-volt supply gets to the sounder? Not much. Ohm's law applied in the other direction (to find what voltage drop there is across the sounder is

$$V = I \times R$$

Thus

$$V = 0.0225564 \text{ Ampere} \times 100 \text{ Ohms}$$

This is 2.25564 volts. That's a lot less than the 10 volts we need to click the sounder.

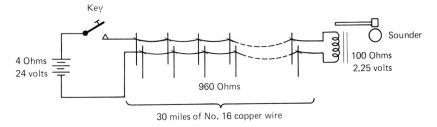

Figure 2–5   Effect of increasing line length.

The third diagram (Figure 2–6) shows one way to get the 30-mile circuit to work. The wire carries enough current to put 2.25 volts on the sounder, so we run a second wire alongside the first, and it carries some more current, so that the sounder gets some more voltage, and so on, until we have 10 wires running side by side. The 10 wires conduct 10 times as well as one wire does. The resistance of all these parallel

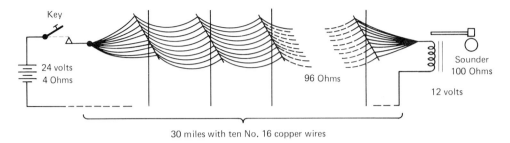

Figure 2-6 One method of reducing line losses.

wires is one-tenth that of the single wire. Since you can remember that the single 30-mile wire is 960 Ohms, you can guess that the 10 wires are 96 Ohms. This is exactly the resistance that we started with in Figure 2–4. The only difference is that we now have 10 conductors, each 10 times as long as in the first figure. This costs 100 times as much for wire conductors.

This is the problem that made Wheatstone's telegraph commercially impractical for the long haul. To go 10 times as far costs 100 times as much in conductors. To go 100 times as far (300 miles), it would cost 10,000 times as much (we'll leave it to your imagination why). To go 1000 times as far (3000 miles), which is about as far as it is across the North American continent, it would cost 1,000,000 times as much as the 3-mile wire. This is no way to run a business. How much copper is this? (We tried it, and it comes out to as much copper as there is in 1 billion copper pennies.) This is about 10% of all the copper minted in the United States in the year 1979.

The **repeater** is shown in the circuit of Figure 2–7. It is used to extend the length of the circuit with 30 miles of wire by using ten 3-mile lengths with repeaters and additional batteries. Let's assume that the **relays** (repeaters) have about the same resistance as the sounder coil. Then the 12 volts that arrived at the sounder in Figure 2–4 is used to close the contacts of a switch that is the key for a new telegraph circuit with a fresh battery. Every three miles, one of the telegraph poles has a battery and a repeater. A completely "fresh," reconstructed signal proceeds down the next three miles of wire, and so on. The cost of the system with repeaters is 10 times as much for a system 10 times as long, and 100 times as much for a system 100 times as long. Costs don't escalate the way they did in Figures 2–5 and 2–6.

It wasn't the Morse code that made Morse a rich man. Nor was it the key, or the sounder, or even the wire. It was the repeater. Without the repeater, long-distance communications would never have been commercially practical. The "bottom line" facts are that anybody who went into competition with the Morse system and *didn't* use repeaters would go broke.

2.1.6 Telegraph System with Repeaters

The **relay** (repeater) was the first **amplifying** device used for electrical signals. By making the sounder coil close a pair of switch contacts (the way a key does), our telegraph

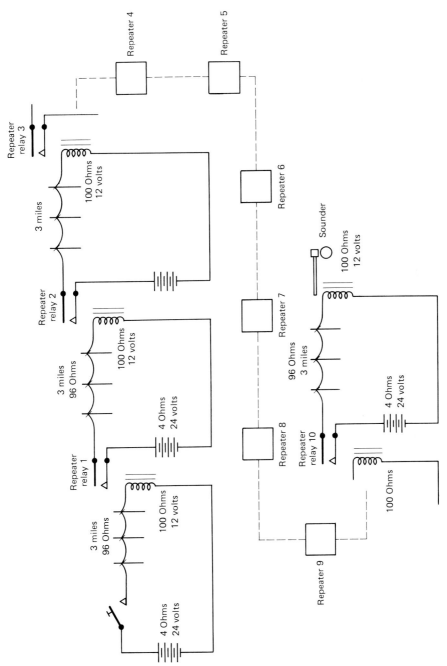

Figure 2-7 Using repeaters to compensate for line losses.

is able to use a small current and a small voltage to operate another circuit with much more voltage and current. For instance, in the circuit in Figure 2–7, each relay at 3-mile intervals got a voltage of 12 volts from the line to which it was attached. The 12-volt voltage did something. It closed a pair of contacts that put 24 volts onto the next section of wire. That means that the signals from the original key, which had lost half their strength, were recreated on the next length of wire at **full strength** again (24 volts).

Think about what would have happened in 1845 if there had been no relays to repeat the signal every 3 miles. There would be a man in a little booth copying Morse code as fast as he could listen to it, and then repeating it into *his* key for transmission down the next three miles of wire to the next relay station, and so on, down the line. Each man would get an opportunity to repeat the message he received with a whole new set of mistakes, and the reliability of a system like this after several thousand repeats between New York and San Francisco is easy to imagine. By using relays as repeaters, Morse automated these men out of a job before the jobs even existed!

2.2 TRANSMITTING INFORMATION BY TELEGRAPH

How is information transmitted on a telegraph? Anybody who's ever watched the old Western movies knows that there are "dots" and "dashes" and the operator makes them by tapping something called a "key." This type of information transmission is called **serial transmission**. The information that makes a letter S, for instance, isn't transmitted all at one time, but takes three separate dots with spaces between them, plus a longer space to separate the "S" from the next letter. Serial transmission means that the information that makes an A, or a Y, or an S is transmitted on one wire, but with different *timing*. A diagram that shows the current arriving at a sounder coil might look like Figure 2–8.

A figure like this is called a **timing diagram**, because it shows how much current there is at different *times*. In this figure you can also see that the difference between a dot and a dash is the *length of time* the current is on. This type of diagram is the standard way of representing all kinds of electrical events that vary with time. Usually, it's called a **waveform** or **trace** when it appears on a graph or an oscilloscope. The name for this kind of encoding is **pulse-width modulation**. The width of the pulse

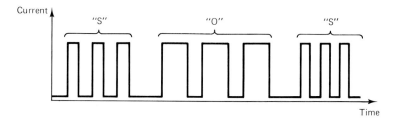

Figure 2-8   Morse code timing diagram.

(voltage ON) is different for a dot or a dash because the width on the diagram is equal to the *time* the pulse lasts. In Morse code, there are only two widths of pulses, dash width and dot width. In other systems, a great deal more information could be carried by pulse-width modulation by using many different widths of pulses.

2.2.1 Simplex, Half-Duplex, and Full-Duplex Communication

All the telegraph circuits that we've seen until now are one-way communications links. There is a definite **transmitting** end and **receiving** end in each diagram, and data goes only from the transmit side to the receive side, never in the other direction. Stringing miles of telegraph wire across the countryside requires a great expenditure of effort (and money). It seems terribly inefficient and wasteful to take all that wire, which can carry signals in either direction equally well, and use it in just one direction. There are several approaches to two-way transmissions. It's possible to transmit even more than two messages at a time on the same wire.

The simple-minded telegraph circuits introduced at the beginning of this chapter allow transmission in one direction only. This is called **simplex transmission**. (Just remember "simple-minded" and you'll remember "simplex.")

There are two varieties of two-way transmission on a single line: **half-duplex** and **full duplex**.

1. In **half-duplex transmission**, there is a transmitter at each end of the line. Transmitters "take turns" sending messages from each end of the line. This is a little like the reversible lanes on an expressway. In the morning, traffic may be inbound on these lanes, and in the afternoon, the lanes are used for outbound traffic only. Although the "reversible lanes" carry traffic in both directions, it is only *one direction at a time*. Traffic is never traveling in both directions in the same lane (thank goodness!). Half-duplex is "alternating simplex," really, but it uses the wire for transmission in both directions.

2. **Full-duplex transmission** uses a line to carry information in both directions at once. This works like the northbound and southbound lanes of the expressway we mentioned earlier. Traffic travels both ways at the same time in different lanes. Full-duplex telegraphy and telephony are most easily done on multiwire lines. Each wire is working individually in simplex.

Figure 2–9 shows half-duplex (a) and full-duplex (b) communication on telegraph lines. One of the most important things to note is that the relay repeaters don't work in reverse; that is, you can't put a current through the contacts and expect the coil to magnetize. Each relay has a definite *input end* and a definite *output end*. This means that the relay which sends data in a "northbound" direction cannot receive data and send it "southbound" at the same time. Another relay handles the return signal.

In the definition of a full-duplex system, we introduced the terms "telegraphy" and "telephony." It's pretty obvious what the difference is. Telegraphy is done in

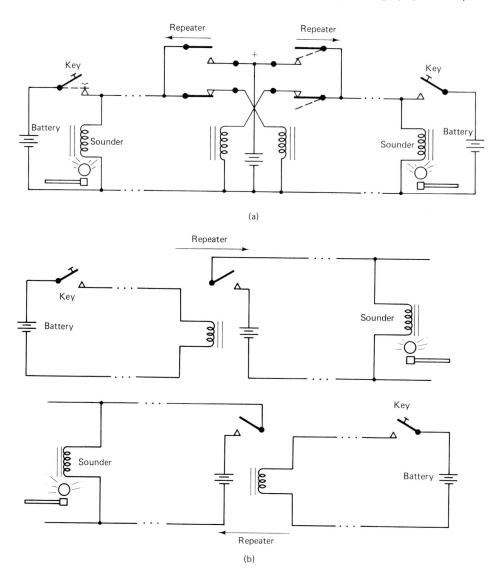

Figure 2-9 Half-duplex (a) and full-duplex (b) communication.

Morse code (or in other, more advanced, digital codes), and telephony is done with voice signals. Two-wire lines such as those used for the simplex transmission of telegraph signals are also used for telephone communication. Either simplex or half-duplex telegraph (or telephone) commnication can be carried out on two-wire lines. For full duplex, two separate current loops are sometimes used to carry telephone information both ways at once. Various ingenious methods were devised to let telegraph operators do full-duplex with one current loop, but these are not needed in telephony.

2.2.2 Multiplex Communication

Multiplex communication allows many simultaneous transmissions on the same line. In 1875, for example, Edison devised a "quadriplex" system with two key operators on each end of the same wire transmitting messages simultaneously. The main point with any multiplexing scheme is to make sure that the messages don't get mixed together. There are three primary ways of communicating multiple messages at the same time: SDM, TDM, and FDM. SDM stands for **space-division multiplex.** It's really the simplest way to get a lot of messages from one place to another at the same time. You use a separate current loop for each message. Cables with hundreds or thousands of individual wires can carry simultaneous messages, one on each pair of wires.

TDM stands for **time-division multiplex.** In this case, only a single current path is used. The current path is called the **common line,** and is similar in form to a round-robin conversation. Imagine three people sitting around a table; each is given one minute to "say his piece," then the next person "gets the floor." Each person gets a turn at contributing something, and the conversation keeps going round and round, but as long as only one person is speaking at a time, there is no confusion as to what is being said, and everybody can understand what's going on. The telephone equivalent of this would be a line carrying three conversations using TDM. Each conversation's electrical signal "gets a turn," and only one conversation is on the common line at any instant, but the turn "passes" around to all three conversations, so that nobody gets missed. In reality, each conversation gets its turn in a very short interval of time, and all three conversations are multiplexed onto the common line, one after another, and around to the first again, thousands of times a second. At these speeds it is possible to reconstruct what was said in each of the three conversations, even though each one gets its turn only one-third of the time. The common line is used to carry the multiplexed conversations to a receiving point where they can be **demultiplexed.** This involves figuring out "whose turn it is" at each instant. Figure 2–10 shows how each of three waveforms (which represent a sound made by a different telephone user in each of the three cases) is "chopped" into pieces, so that every third piece gets a turn. A **sample** is taken from waveform 1, then from waveform 2, then 3, and back to 1 again. The pieces segmented-out are put together into a wave, shown in the middle of Figure 2–10. This wave actually samples any one conversation only a third of the time. Finally, the wave on the common line is "taken apart" and each piece assigned to a different destination based on what time it is. In other words, *who* is receiving depends on *when* the segment arrives. We have taken out the pieces of the wave on the common line that arrive at "3-time." Every time the number 3 comes up in a sequence such as 1, 2, 3, 1, 2, 3, 1, 2, 3, 1, 2, 3, 1, 2, 3, . . . , we take out the piece of a wave that is present during that interval, and always send that 3-time piece to the same destination. The wave that arrives at destination 3 is shown at the bottom of Figure 2–10. Even though two-thirds of the time, nothing is sent to this destination, you can still reconstruct the wave pretty well from the remaining one-third pieces that are present. Separating out only the pieces that arrive

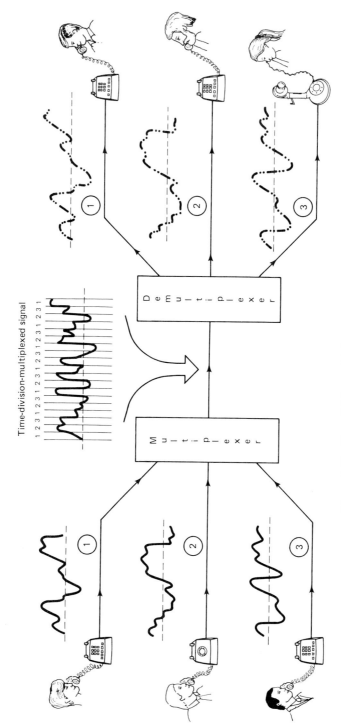

Figure 2-10 Time-division multiplexing of 3 voice signals.

Time-division-multiplexed signal

1 2 3 1 2 3 1 2 3 1 2 3 1 2 3 1 2 3 1 2 3 1

at 3-time, in our example, is **demultiplexing**. In this case, we're **demultiplexing channel 3**, which is one of three channels. By using more **time slots** and sampling the signals at a faster rate, we can put more than three channels on the same wire. It is possible to fill-in the gaps (between 3-times, for instance) using a circuit called a *sample-and-hold* and get a pretty good replica of the original wave in conversation 3 back at the top of the diagram.

FDM stands for **frequency-division multiplexing**. It is exactly the process used to separate channel 2 from channel 3 on a television receiver. The only difference between tuning different channels or stations on a TV or radio and demultiplexing voice channels in telecommunications is that the signals may travel in a conductor and are not always broadcast by "wireless" methods.

Statistical multiplexing is a sophisticated form of TDM in wide use. In STDM (**statistical time-division multiplexing**), a time slot is allocated to a channel only when it has something to transmit. When channels transmit data only in bursts, this can be very efficient, although it is more expensive and complex. If each channel transmits data in a uniform, steady fashion, the efficiency of STDM is much less, and ordinary TDM will probably be chosen instead.

REVIEW QUESTIONS

1. What is the difference between a voltage and a current?
2. What effect does resistance have on electric current?
3. What are some telecommunications uses for electromagnets?
4. What is the resistance of a relay coil if, when attached to a 10-volt battery, it has a 1/100-Ampere current? Would the resistance be larger or smaller if the current that the 10-volt battery put through the coil were 2/100 Ampere?
5. How many volts (battery voltage) would be required to produce a 1-Ampere current in a 10-Ohm coil?
6. Give an example of a system that uses serial transmission.
7. Is simplex communication in one or two directions?
8. Is half-duplex communication in one or two directions?
9. Is full-duplex communication in one or two directions?
10. Can two simplex links be used together to form a full-duplex link?
11. Can SDM (space-division multiplex) be used to put many conversations on a single wire?
12. Can TDM (time-division multiplex) be used to put many conversations on a single wire?
13. Can FDM (frequency-division multiplex) be used to put many conversations on a single wire?
14. Could a pulse-width-modulated signal have more than two widths of pulses? How many widths are used in telegraph transmission?
15. With wire that has a resistance of one Ohm per 100 feet, how far could a signal from a 24-volt battery be sent to a 100-Ohm 8-volt sounder, to operate it without a repeater? What distance would be possible (without repeaters) if a 48-volt battery is used?
16. Using **repeaters**, how far can a telegraph signal be transmitted?

3

Basic Telecommunications Technology: Telephony

3.1 RECEIVERS AND TRANSMITTERS

Telephony is not as simple as telegraphy. The voice signal is not simply formed by a current that's on or off. Instead it is formed by a varying current called a **waveform**. Air pressure on the telephone mouthpiece changes rapidly when a sound enters the mouthpiece. The changes in pressure (which *are* sound) are converted by the microphone in the mouthpiece into changes in electrical current. The changing electrical signal (which is *not* sound) is easier to transmit and to amplify than sound waves are.

The carbon microphone found in a standard telephone is shown in Figure 3–1. A metal foil cup mounted to the vibrating diaphragm converts voice vibrations into variations in the resistance of carbon granules packed between the cup and a larger cone-shaped metal chamber. One electrical contact in the telephone circuit connects to the metal foil cup and the other to a button in the center of the microphone assembly. The button is actually a cap that holds the carbon granules in place inside the conical chamber. The foil cup attaches through the metal diaphragm to the rest of the microphone housing, which forms the other electrical contact.

Variations in pressure on the diaphragm (and hence the cup that bears down on the carbon granules) change how tightly the carbon granules are packed together. This changes the electrical resistance between contacts, because tightly packed carbon granules conduct better than do loosely packed granules. These variations in resistance are not usable by themselves to generate an electrical signal, but when placed in a circuit that puts voltage across the two contacts, these variations result in a varying current. This current with varying value is called the **voice signal**. The electrical principle of the carbon microphone is the same as when it was originally designed

44

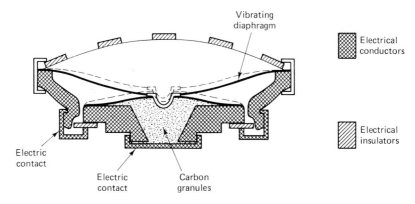

Figure 3-1   Carbon microphone.

by Edison in the 1880s. (Bell originally used a transmitter that was the same as the receiver.) In normal operation, the voltage across the transmitter (microphone) is about 5 volts, and with variations in resistance, this voltage varies about a volt either way when you speak into the transmitter.

The receiver (the earphone) is a scaled-down version of the speakers in a stereo system. It has two active parts, the **voice coil** and the **speaker cone**. There is also a large permanent magnet which interacts with currents in the voice coil, but stands still itself. When current flows in the voice coil, forces are generated that act to move the voice coil. The force is transferred to the speaker cone, and its motions back and forth produce pressure waves in the air. This is what you hear when you hear sound from the receiver. This receiver mechanism [Figure 3-2(a)] is almost the same one Bell used in his original patent in 1876. There are some variations in the mechanical design and shape of the magnet, but the principle of receiver operation hasn't changed at all.

Both of these devices are examples of **transducers**. Transducers convert one form of energy into another (usually one of these forms of energy is electrical). The transmitter converts acoustic (sound) energy into electrical signals, and the receiver converts electrical signals into acoustic energy.

The electrical signal produced when the resistance in the carbon microphone varies is a changing current that looks like the waveform shown in Figure 3-2(b). Compare this current wave with the waveform from the telegraph (Figure 2-8). You can see that the voice wave has a lot more levels of current (in the telegraph wave there are only two).

3.1.1 Simplex circuit

Very few simplex telephone circuits exist, but there is one example we could use. In Figure 3-3, we see the circuit(a) and the symbolic diagram(b) for a very simple public-address (P.A.) system. The P.A. circuit doesn't include "frills" like an amplifier,

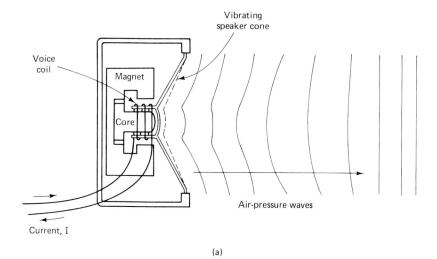

(a)

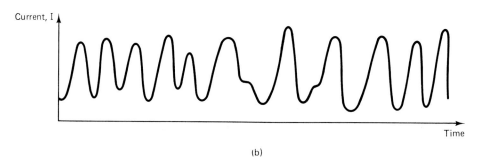

(b)

Figure 3-2   (a) Receiver mechanism. (b) Current wave.

but just has the "essentials" for carrying a voice message from a source to a destination.

The carbon microphone (transmitter) controls the current flow in the circuit. Acoustic energy changes the resistance of the carbon in the mike and causes current fluctuations. The changing value of current causes the speaker cone to move in and out, producing acoustic energy at the receiver end. Since this is a *one-way* communication link, it is **simplex**.

Most P.A. systems use an amplifier to boost signal strength before the electrical signal is applied to the speaker. With the addition of an amplifier, a very weak signal at the transmitter can become a very loud speaker output at the receiver. Without the amplifier, though, our carbon-mike circuit still works. The acoustic energy at the transmitter can control a much larger amount of electrical energy in the circuit, and the sound can be louder at the receiver than it was at the transmitter. The only limit on *how* loud the receiver's output becomes is the voltage applied by the battery

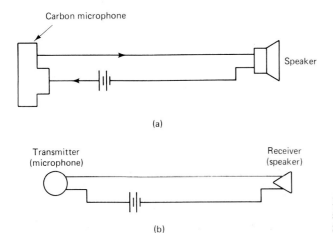

Figure 3-3 Simple public-address system. (a) Schematic diagram. (b) Simplified symbolic diagram.

and the heat generated in the microphone. Eventually, if the current becomes too large, the carbon becomes too hot and burns up.

3.1.2 Duplex Voice Communication

As with telegraph signals, duplex voice communication is two-way communication. Unlike telegraph communication, though, duplex telephony can be done on a single circuit without the complicated demultiplexing required on the telegraph. The problem on a telegraph is that you can't tell your ON voltage from the other guy's ON voltage. There is either a click or there isn't. You can't tell where the click came from. With voice signals, however, you can tell one voice from another, and you can hear both voices at once, because the electrical voice signals add up just like acoustic voice waves in the air when two people are talking in the same room. You can hear both people, because their voices *add* acoustically. The pressure wave is the *sum* of what's in both voice waves. In the electrical circuit, this is called **superposition** of voltages, and it works just the same way. Since you can hear both voices, you can do full-duplex voice communication on a telephone, provided that you can understand two people talking at the same time. Most human beings do not do this very well, and they have a habit of *taking turns* talking at each end of the telephone circuit (most of the time). Talking alternately at each end of a communication link is **half-duplex** communication. For human conversation, half-duplex capability is all that's needed. A simple intercom is an example of duplex voice communication. We can see how a simple intercom works in Figure 3-4. The transmitter and receiver are in series at both ends of the circuit. In a series loop, you can hear yourself talking on your own receiver because the current from your transmitter goes through the receiver in your handset as well. This condition is known as **sidetone**—it's an important way to tell if your phone is working, since a phone without sidetone feels "dead" when you listen to it. You're usually not consciously aware of it, but when there's no sidetone, you get a definite feeling that something is wrong. More important is the fact that

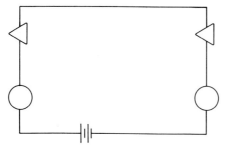

Figure 3-4 Simple intercom diagram.

the current from your transmitter goes through the other handset's receiver, and they can hear you on your receiver when you talk, which is the whole idea.

Since all the transmitters and receivers are in series, this is a **simple series circuit** with just one current loop. When the current varies anywhere in the circuit, it varies everywhere in the circuit at the same time. It is the simplest circuit that could be used for bidirectional communication between two places.

3.2 INDUCTANCE AND CAPACITANCE

In Figure 3–5, a multiple-phone circuit is shown. The phones have now got the capability of being "hung up," because each has a **hookswitch** (also called a *switchhook*). It is called this because on early phones, this switch was actually a hook where the receiver hung when not in use. The T-shaped switch in the diagram is closed (completes the connection) when the handset is lifted and is opened (disconnects the handset) when the handset is put down where it pushes on the switch. These two conditions are called **on-hook** and **off-hook**.

The circuit in Figure 3–5 contains the familiar battery as a source of current. The **handset** is the transmitter and receiver, which are together in a single unit. The **hookswitch** is the T-shaped switch that connects the handset into the circuit. Each handset is a separate current path to the **coil**. When a voice signal causes the current to vary, the coil presents a greater opposition to varying signals than to steady, direct current (DC). Thus the alternating (varying)-current (AC) part stays on the phone side of the coil, while the DC current from which the voice signal is developed passes freely around the circuit.

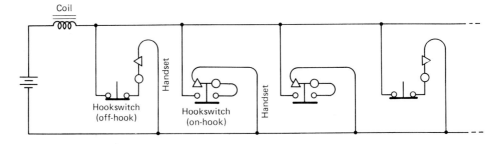

Figure 3-5 Multiple-phone circuit.

3.2.1 Inductance

If that seems a bit confusing, perhaps we can explain it a bit by describing what a coil does. The quality of the coil that is important here is the **inductance**. This has to do with the fact that current must deliver energy to a magnetic field in order to make it larger and receive energy from the field when it gets smaller. The size of the magnetic field around a coil depends directly on how much current is going through it. When the current through the coil increases, as when a surge of air pressure compacts the carbon granules together in the transmitter, the coil must use up some of the additional electrical energy available in the circuit to build a larger magnetic field. At the coil, while this field is expanding, electrical energy is being lost from the system just as it is at a resistor, where the electrical energy is lost in the form of heat. Thus the coil looks like a voltage drop, where energy is lost, and resembles a resistor. Remember, though, that the coil only does this while the magnetic field is *changing*. This requires a *varying* current. If the current is steady, and the magnetic field neither expands nor contracts, there is no energy entering or leaving the circuit at the coil. The coil just looks like another piece of wire in this case. So what happens? If the current in the circuit is steady and constant, as when nobody is talking, there is no voltage drop on the coil and the voltage on all the telephones is the same as the battery, with no variations. During the part of the telephone's operation where the current is varying, the variations produce a voltage on the coil. The remaining part of the circuit gets the voltage of the coil and battery together. Sometimes, the battery voltage and the coil voltage are opposite, as when the circuit is losing energy to the coil's magnetic field. Other times, the energy in the magnetic field is returning to the circuit as that field collapses back into the coil. This returning energy actually *adds* some voltage to that of the battery, and the voltage the telephones "see" is larger than the battery alone. Even though the voltage of the battery is constant, the voltage the phones receive is varying, because they receive the total of the coil and battery together. All of the telephone receivers are attached in a connection called a **parallel circuit**. They are all getting the same voltage, because they are all attached to the same two wires—the wires that get voltage from the combination of battery and coil.

It might occur to you to ask at this point why anybody would use a circuit like Figure 3-5, with the parallel telephones and the coil. Why not just put more telephones in a series loop like the one shown in Figure 3-4 until you have all the phones you want? After all, the wiring would be simpler—just one wire running all the way around—and there would be no need for the added expense of a coil.

The answer is—it won't work! Figure 3-6 shows how our series loop would look if we tried to use the same four phones in Figure 3-5. Two of the phones in Figure 3-5 were "hung up" and the remaining two phones were talking to each other. In Figure 3-6, if we hang up two of the four phones—or even one—the circuit is broken and no current can flow anywhere. Even the phones that are off-hook still have no current, because there isn't a complete circuit for current to circulate in. Nobody can talk to anybody else in the series circuit unless *all* the phones are picked

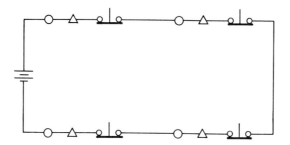

Figure 3-6 Series loop.

up. (If you hang up a phone in a parallel circuit, you hang up *one* phone, but if you hang up a phone in series, you hang up *all* the phones.)

In the parallel circuit, any number of phones can be picked up. This has its limits, however, because when too many phones are being "driven" by the signal from one transmitter, the signal energy is split up between all the different phones. When there are five phones on the circuit, for example, 80% of the volume has been lost. That's enough loss that the signal is hard to hear. If more phones are added to the circuit, the signal is split up even further and is even harder to hear.

3.2.2 AC Component of the Voice Signal

In the circuit of Figure 3–5, we used an **inductor** to develop the signal that was heard on all the telephones. Its job, in this case, was to separate AC from DC. Alternating current, strictly speaking, changes direction and moves back and forth in the wire, whereas direct current always travels in the same direction in the wire. In all telephone circuits, the actual current acts like a combination of AC and DC. The current keeps flowing in one direction all the time, because the source (a battery, in our examples) has only one polarity of voltage, and "pushes" current in only one possible direction. The amount of the current, however, is not steady. If we imagine an alternating-current source adding its current to that of the battery, sometimes the AC source's current will *increase* the total current in the circuit (when it is going the same direction as the battery "wants" current to go), and sometimes the AC source's current will *decrease* the total (when it is going in the opposite direction the battery "wants" current to go). In the real telephone circuit, it is as though the transmitter were a small AC source adding an AC current to the larger DC current of the battery. Since the actual voice signal is the AC part, circuits that separate the AC part of the current from the DC part are important in the telephone system. There are several ways of doing this.

3.2.3 Separating AC from DC Using an Inductor

We already saw one way—the inductor we used in Figure 3–5 was used because it was able to develop the AC part of the signal without affecting the DC current. What if that inductor were not in the circuit? Figure 3–7 shows the circuit without the coil (the inductor). Each telephone is attached to the positive and negative battery lead and is electrically connected to the ends of the battery. Now, suppose that one trans-

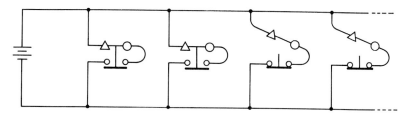

Figure 3-7   Parallel circuit without coil.

mitter is connected and causes variations in the current that flows through the battery. Will this change the battery voltage? The answer is: "not enough to notice." The voltage that runs all the other phones will be a steady, constant voltage, and the receivers will carry a steady current. In other words, none of the other receivers will be able to hear the transmitter that we connected.

To hear the transmitter in a parallel circuit, the changing current from the transmitter has got to make some difference in the voltages that arrive at the other receivers. One way to do this would be to add resistance to the battery. Now, if the current through the battery changes, the battery's voltage will change also—and the receivers will be able to hear what is going on at the transmitter. Unfortunately, the extra resistance we have added to the battery wastes power and cuts down the voltage available to the telephones, even when no one is speaking.

The inductor is a better way to do this, because it's sensitive *only* to the AC component of the current. The inductor looks like a piece of wire to the DC current, and that means it doesn't waste battery power when no one is talking. It also means that the voltage available to the telephones is the full voltage of the battery (or at least a lot more than if we use a resistor—real inductors do have some small amount of resistance). In the circuit of Figure 3-5, it appears that to the AC current, the battery looks like it has lots of resistance (it's called **inductive reactance**, actually), but to the DC current, the battery does not have much resistance, and very little DC power is wasted.

3.2.4 Separating AC from DC Using a Transformer

Another device that can separate the AC and DC components of the current in a telephone circuit is the **transformer**. A transformer is basically an inductor with more than one winding of wire wound around the same core. As with the inductor, a change in the current flowing through the coil makes the magnetic field around the transformer expand or contract. A voltage is *induced* in the coil of the **inductor** (hence the name) by a changing magnetic field. In the transformer, *both* windings are immersed in the same magnetic field. The "powered" winding (with current flowing through it) is called the **primary**, and the "unpowered" winding is called the **secondary**. The magnetic field really doesn't make any distinction between one set of windings and the other. When the field expands and contracts, it induces just as much voltage on each winding of the secondary as on windings of the primary. This is important, because it means that the *variations* in primary current (the AC component, or "information" part of the current) can be transferred from primary to secondary without any electrical contact between the two. The coils or windings of the secondary receive electrical energy from the magnetic field—this is energy that is put there by the primary. What does *not* make it across from the primary to the secondary is the DC component (the steady-state level of voltage or current in the circuit). On each side of the transformer [see Figure 3-8(a)] the DC voltage used in each circuit may be different. Each circuit in our example has a different battery and operates at a different voltage. This doesn't stop the person on the left-hand phone from talking to the person on the right-hand phone. Even though each circuit has a different battery, operating voltage, and isn't electrically connected to the other circuit, it's still possible for the AC part of the signal to cross the distance from one to the other. This is something we couldn't do with the telephone circuit in Figure 3-5. Although we had phones talking to each other, all the phones had to operate at the same voltage. It wouldn't be possible, for instance, to talk into a 48-volt phone and listen at a 24-volt phone in that circuit (Figure 3-5).

On the primary (left) side of Figure 3-8(a), someone is talking and producing a voice wave. On the secondary (right) side, the AC part of the voice wave that transfers through is added to the DC voltage of the battery. The two voltages together reach the handset and, since the AC part is all that can be heard on the receiver, the DC voltage does not matter. Now, there is an important feature of the transformer that we haven't discussed yet. There is no reason why the person using the phone on the secondary (right) side of our picture couldn't talk into his handset. If he does so, the variations in secondary current will transfer through to the primary (left) coil. The right-hand coil now becomes the primary and the left-hand coil becomes the secondary, because the AC voice wave is originating on the right and being received on the left. It really doesn't matter which side you call the primary and which you call the secondary. Whoever is receiving has the voltage of the DC (battery) supply and the AC (voltage induced on the transformer) applied to his phone. Whoever is transmitting is modulating the DC current of his battery (making it vary), with the result that the current has a DC and an AC component. The AC component is caused by the voice, and you can picture the resulting current as an AC wave superimposed on a basic DC level. The AC part "gets across" from one winding of the transformer to the other, and is received, regardless of what DC level it is added to on the receiving side.

Transformer-coupling a 24-volt and a 48-volt system

In our example [Figure 3-8(a)] the two systems couldn't be connected directly. If they were [Figure 3-8(b)], the total voltage in the circuit might be 72 volts. If each phone received half the voltage, the 24-volt phone would get 36 volts, which is 50% too high, and might not be *good* for that phone! So rather than a direct hookup, a natural way of connecting together phones that operate from different voltages is to use a transformer. The transformer "conducts" the signal back and forth between handsets, without conducting the (possibly dangerous) DC levels between the two systems.

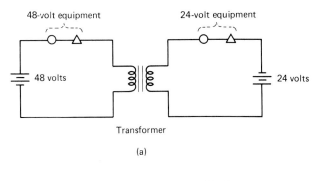

(a)

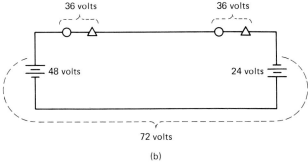

(b)

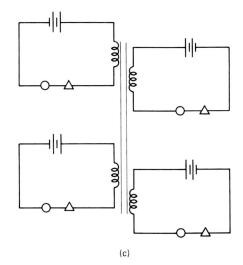

(c)

Figure 3-8 Transformer coupling. (a) Coupling a 24-volt to a 48-volt system using a transformer. (b) Why *not* to couple a 24-volt system *directly* to a 36-volt system. (c) Four independent phones, transformer-coupled.

Another point can be seen in Figure 3–8(c). The transformer with four phones on four windings is not really like a series circuit. If one phone is on-hook, the other three phones can still communicate; the circuit is not broken everywhere, just in the system attached to that one winding.

In Figure 3–9(a) we see a system of several phones on each side of the transformer. The phones on the left winding are part of a 24-volt system, and the phones

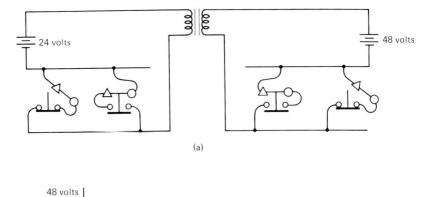

(a)

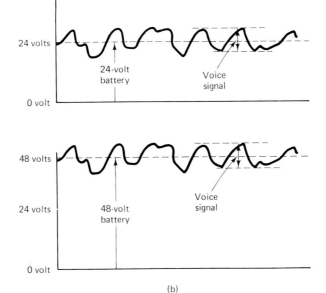

(b)

Figure 3-9   (a) 24-volt (left) and 48- volt (right) phone system. (b) AC and DC levels for 24- and 48- volt systems.

on the right winding are part of a 48-volt system. Each side is exactly like the circuit in Figure 3-5. Where is the inductor? It's the winding of the transformer, of course! The coil we needed earlier to make the parallel hookup work is now also used to **couple** (connect) the 24-volt and 48-volt systems.

Figure 3-9(b) shows the AC and DC levels on the 24- and 48-volt sides of the transformer. The AC part is called the "voice signal" and the DC part is called the "battery." Notice that the voice signal (AC part) is the same size on both sides of the transformer. This is true only if the transformer has the same number of windings (loops) of wire on both the primary and the secondary. Such a transformer is called a **one-to-one transformer**.

3.2.5 Separating AC from DC Using a Capacitor

Another device that can separate the AC (voice) part of the electrical current from the DC (battery) part is a capacitor. In Figure 3–10 we see another way to couple signals from a 24-volt to a 48-volt circuit. On each side of the diagram is a circuit similar to the phone circuit shown in Figure 3–5. The AC part (voice signal) is "coupled" from one side to the other through two coupling capacitors. You can think of them as conductors that pass the AC part of the voltage but block the DC part from getting through. In this diagram the two capacitors allow the voice signal from each side to get through to the other side without the DC voltage in the 48-volt side being able to affect the devices on the 24-volt side.

Notice that you have an inductor in each voice circuit, anyway. If you have to use inductors on each side, why not make them the primary and secondary of a transformer, and eliminate the need for coupling capacitors? As a matter of fact, that's exactly what's done most of the time. That doesn't mean that there's *never* any use for capacitors in telephone circuits—you might even find them used for exactly this purpose—it's just that *transformer coupling* is a lot more common than *capacitor coupling* in telephony. In related fields of electronic communications, audio amplifiers and radio, for instance, *capacitor coupling* is used far more often than *transformer coupling*.

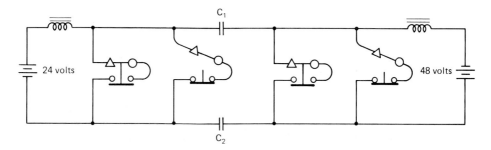

Figure 3–10   Capacitor coupling.

3.2.6 Capacitors

Why does a **capacitor** block DC and pass AC, and what's a capacitor, anyway?

A **capacitor** is an open circuit, which means that it has two sides that aren't connected to each other electrically. That might make you wonder how it does anything at all. Actually, the capacitor is a very specially constructed open circuit, with a large amount of area on the two "ends" that face each other across an open space. Imagine a battery attached to two "wire stubs" with a gap between them [Figure 3–11(a)]. When you try to run current through the open circuit, the voltage of the supply trying to "push" the current appears across the open ends. The positive end of the battery is connected to the positive stub, and the negative end is connected to the negative stub. With nothing but wire between the positive battery terminal and the positive stub, the stub becomes just as positive as the positive battery terminal. The negative stub becomes just as negative as the negative battery terminal, for the same reason. Now, ask yourself: Could the stubs have been positive and negative to begin with, even before there was a battery? The answer is: Of course not! As long as there was nothing pushing a surplus of negative charges into one stub and positive charges into the other stub, they would tend to remain electrically balanced, and uncharged. How does the negative stub end up negative? Some negative charges had to move into it. In fact, enough negative charges had to move into that stub so that further negative charges from the battery would be "pushed back" by the negative charges already there. The charges in the stub would have to "push back" just as hard as the battery was "pushing in." This is why the voltage of the stub is the same as the voltage of the battery. We know that negative charges had to move into the negative stub, until the negative charge of the stub repelled other negative charges just as hard as the battery was pushing on them. We also know that the same thing had to be happening in the positive stub. For a while, as the negative stub was becoming negative and the positive stub was becoming positive, charges had to be flowing into (or out of) both stubs. This doesn't go on for very long, however, in Figure 3–11(a). The stubs aren't very big, and get "filled up" quickly, to a point where they push back as hard as the battery pushes in. In Figure 3–11(b), the stubs have been expanded into broad, metal plates. A great deal more charge can fit into these than into the wire stubs. More current has to flow, for a longer time, into the plates before they charge up enough to push back as hard as the battery pushes forward. This is a capacitor. It is an open circuit that doesn't conduct across the gap, but the wires leading to the plates conduct for a while, until the positive and negative plate are as positive and negative as the battery is. We call this **charging** the capacitor. When the voltage on the plates is the same as the voltage of the battery, the capacitor

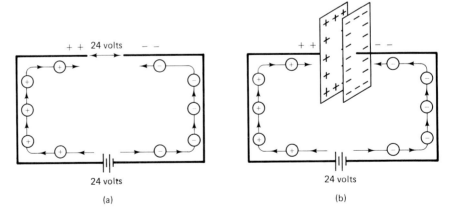

Figure 3–11   Charging a capacitor. (a) Wire-stubs with applied voltage. (b) Large wire-stubs or plates make a larger capacitor.

is charged. That means that its *charge* pushes back on the electrons in the wires as hard as the battery pushes forward.

If the voltage applied to the plates gets smaller (if a smaller battery were connected, for instance), the plates' charge would actually push back *harder* than the battery is pushing in. The charges would actually move backward (back into the battery) until the plates had lost enough charge so that they didn't push harder than the battery anymore. When charges leave the capacitor plates like this, we all it **discharging**.

Now, about the AC and DC. DC is steady, direct polarity, the kind you get from a battery. In a short while, the plates of a capacitor charge up and no more current flows as long as the DC circuit has a steady voltage. Once the capacitor has charged, there is no DC current flow.

If the DC voltage varies, so that the capacitor charges and discharges, it never quite gets a chance to arrive at the voltage that's being applied. A varying voltage is always taking on a new value before the capacitor reaches the level where all the current stops. By "wiggling" the voltage around, a current can be kept "sloshing back and forth" into and out of the capacitor. The current changes direction as, first, the voltage charging the capacitor is bigger than that already on its plates, and then, second, the voltage charging the capacitor is actually *less* than it already has on its plates, and it "spits back" some of the charge. As long as the voltage attached to the capacitor keeps wiggling around, getting bigger and smaller, a current will keep sloshing back and forth in the wires leading to the capacitor plates. This kind of current is called "alternating" because it alternates, or changes, from one direction to another, and back again.

3.3 IMPEDANCE AND ITS IMPORTANCE TO TELEPHONY

3.3.1 Comparing AC and DC Circuits

In Section 3.2 we saw two kinds of devices, coils (inductors) and capacitors (also called "condensers" by old-timers in the electronics field). We saw that these devices work differently with AC than with DC. The inductor developed an AC voltage drop when AC current ran through it, although in a steady DC circuit, the inductor (coil) is just a piece of wire. Similarly, the capacitor (which is just an open circuit to DC current) allowed an AC current when an AC voltage was applied to it. In both cases, the DC appearance of the device (inductor = short circuit) (capacitor = open circuit) is not the same as its AC appearance. In the AC circuit the coil and capacitor both have something called reactance, which works much like resistance in DC circuits.

Suppose that we imagine an "all-changing" current that is constantly switching polarity and value at a high rate of speed. This is a pure AC current. The charges in the wire, we can imagine, rock back and forth and never actually end up moving down the wire more in one direction than in the other. In a circuit like this, a coil will develop a voltage that opposes the changes in the current, and a capacitor will develop charges on its plates that "follow" the voltage of the source driving that current in the wires.

What is an AC voltage and an AC current? If the charges never get all the way around the circuit, is there a current at all?

3.3.2 AC Voltage and Current

The current you get from a wall outlet is exactly this type of a current. If you've ever gotten an electric shock from an AC outlet, you know that this current is *very real*. That current can also light a bulb, but the generators at the electric company never make a single charge get all the way around the circuit between the light bulb and their generator. Somehow, the **power** still reaches the bulb and lights it up. This happens because during the tiny moments when the charges rock in one direction or the other, they **do work** while traveling in either direction. Thus the AC current in the light bulb does a little work when the charges rock to the right, and a little *more* work when those charges rock to the left. All this work heats up the bulb, and it glows. The power distributed to the bulb heats it up and makes it glow just as brightly whether the charges move through it in one direction all the time, or change directions. The important thing is that while they are moving, the charges heat the bulb. The motion itself is more important than the direction of that motion.

Now we ask ourselves what DC current would give the same amount of power (brightness) to the bulb as our AC current? If we operate the same light bulb using a DC battery, and vary the voltage until the bulb is exactly as bright as it is on our AC generator, we have found a way to measure the AC voltage. We simply give it a value in volts that is the same number as the volts of the battery. In the case of the electric company, a 120-volt battery would light the bulb just as brightly as the AC from the outlet.

Now what about current? We can describe what lights up the bulb in terms of current, too. If a 120-watt bulb is lit up by a DC battery, the current will be 1 Ampere when the bulb shines as brightly as it does when attached to the AC wall outlet. Since it takes 1 Ampere of DC current to light the bulb to this brightness, we can call the AC current 1 Ampere, also. This means that our AC current rocking back and forth through the filament of the bulb does the same *work* and lights the bulb to the same *brightness* as 1 Ampere of DC current. Although the current and voltage change direction, and are positive and negative an equal amount of time, the terms "120 volts AC" and "1 Ampere AC" are just as meaningful as the volts and Amperes of DC, because they have the same effect on the light bulb. We call these two figures the **effective** [or root-mean-square (rms)] **voltage** and current of the AC source.

What we see here is an AC definition of voltage and current by describing the work they do lighting up a bulb. Remember that the voltage is the "pressure" that forces charges through the filament of the bulb, and current is the flow of charges that this "pressure" produces. We used the letters *V, I,* and *R* to identify volts,

Amperes, and Ohms. We still use V and I to describe AC volts and AC Amperes, but what about Ohms? Can we still say that

$$V = I \times R, \quad I = \frac{V}{R}, \quad \text{and} \quad R = \frac{V}{I}?$$

It seems logical. After all, the light bulb is just as bright with 120 volt AC and 1 Ampere AC as it was with 120 volts DC and 1 Ampere DC. If we do this, the light bulb is rated at 120 Ohms (because V/I is 120 volts divided by 1 Ampere).

3.3.3 Inductive and Capacitive Reactance

What about the coil (inductor) and the capacitor, whose behavior is different in an AC circuit than it in a DC circuit? The coil, for instance, is just a piece of wire, and although it is wound around a core, it's still very much like other pieces of wire in a DC circuit—that is, it's a *short circuit*. A circuit with very little resistance is a short circuit. It doesn't develop very much voltage when a DC current passes through it. In an AC circuit, however, the same coil develops much more voltage for the AC Amperes that go through it that it would for the same amount of DC Amperes. Has its resistance changed because it's in an AC instead of a DC circuit?

In a way, yes. There are more Ohms. If a larger V is divided by the same I, the number of Ohms we get is a larger figure in an AC circuit. We can't call this resistance, however, because it's not there in a DC circuit, and actually has different values depending on how fast the AC is changing. We call this rate of polarity change the **frequency** of the AC source, and measure it in cycles per second. (A *cycle* is what happens when the voltage goes through both a positive and a negative polarity once.) Like current, frequency has a measuring unit, the **Hertz** (which is the number of cycles per second) and a letter to represent it, the letter F.

The Ohms of the coil depend on the Hertz (frequency) of the AC source. Faster frequencies produce larger Ohms values. As the frequency increases, the Ohms value gets larger. This happens because a fast-changing current induces bigger voltages in the same coil, and as V gets bigger, V/I gets bigger, too. The Ohms we calculate from V/I increase directly with the frequency, and if there is *no* frequency (DC), there should be *no Ohms*. This type of Ohms value is still worked out with Ohm's law (V/I), but is no longer called resistance; it's now called **inductive reactance.**

The capacitor is another device whose AC behavior is different from its DC behavior. In a DC circuit the capacitor is an open circuit, and current cannot get through it from one side to the other. In the AC circuit the capacitor has current as well as voltage. There is something to put into both parts of the V/I formula. We get Ohms from this, but like the coil, they depend on the frequency of the AC source. In this case, the capacitor, which didn't conduct at all in the DC circuit, conducts better and better as the frequency is increased. We can think of DC as AC with a frequency of zero. The capacitor's ability to conduct current at DC (zero frequency) is also zero, but the ability to conduct increases as the frequency goes up. Ohms don't measure the ability to conduct—in fact, they measure the opposite—the ability to *resist* conduction. As frequency increases, the Ohms get smaller. This is actually an indication of better ability to carry current. As the I in V/I gets larger, the whole number gets smaller. At very high frequencies, the capacitor may even look like a short circuit instead of the open circuit it is to a DC current. Ohms like this are called **capacitive reactance** instead of resistance.

3.3.4 Impedance

Since resistors, capacitors, and inductors all have various kinds of Ohms in AC circuits, you might wonder if there's a name for the Ohms in AC that covers everything. There is. The word is **impedance**, and it covers *resistance, capacitive* and *inductive reactance,* and any combination of the three that might be found in a circuit.

Most AC circuits have a little bit of all three things. There is resistance in even the best of inductors, capacitance across the ends of every resistor, and inductance even in the straight wires connecting everything together in a circuit. The word "impedance" is probably the best one to use when describing Ohms in any kind of AC situation.

In telephony, AC current is usually mixed with DC. The Ohms called inductive reactance and capacitive reactance depend on the AC part of the current only. The size of the DC component makes no difference in these values. If we want to figure the AC Ohms (impedance) of anything in a telephone circuit, we must use only the AC volts and AC Amperes, and ignore the DC component.

3.3.5 Distributed Capacitance on Wire Pairs

Capacitance is not only found in capacitors. The capacitance you *want* is usually put where you want it in the form of a capacitor, but there are also unwanted capacitances that occur in telephone circuits. These capacitances are not put into the circuit deliberately, and are not located in one single place like a capacitor. Pairs of wires, for instance, are like the plates of a capacitor. When the wires are close to one another and run long distances, what you have is like a long, skinny capacitor. Even though the wires aren't *intended* to be a capacitor, they have metal conductors with an insulating gap between them. Any pair of wires that are run side by side for a long distance will have more and more capacitance the longer they run. For the voltage on one wire to be different from the voltage on the other wire, charges will have to flow even if the wires are not connected to one another.

Since there is a "distributed capacitance" between long line pairs, signals sent down these wires will act as though they are attached to a capacitor. Figure 3–12(a) shows how a pair of wires acts as a capacitor. There is a little bit of capacitance between every foot of the twisted pair in Figure 3–12(b). We already know that for higher frequencies, the capacitor looks like a small number of Ohms, and for lower frequencies, it looks like a large number of Ohms. Long wire pairs are unable to carry the higher frequencies as well as they carry the lower frequencies. The reason for this is that the high frequencies develop a lot of AC current in the wires (even when there's nothing connected to the other end of the pair). The lower frequencies

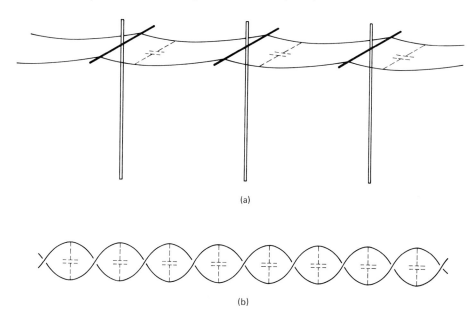

(a)

(b)

Figure 3-12   Distributed capacitance on wire pairs.

develop voltages on the wire pair, but not much current flows if there's nothing attached at the far end of the wire pair (see Figure 3-13). The AC current in the wires uses up power by heating the wires, and also by heating up the generator (which has internal resistances of its own). The energy lost as heat is no longer available when the signal reaches the distant end of the wire pair. This is called **attenuation**, and it gets worse as the wire pairs get longer (because they become larger "capacitors"). Notice that the lower frequency does not run as much current through the wires, so less power is wasted between the generator and the distant end of the pair. At the distant end of the pair, this means that the low-frequency part of a signal will get through better than the high-frequency part. For the same loudness, a high-pitched (high-frequency) voice will be attenuated more than a low-pitched voice. The loss of power as frequency goes up is called **high-frequency roll-off**. It's an undesirable side effect of the fact that the wire pairs have capacitance.

3.3.6 Loading Coils

You will recall that coils work in exactly the opposite way capacitors do. Their Ohms value gets larger as the frequency rises. Could a coil be combined with the capacitance of the wire to cancel its effects?

    Loading coils placed along the wire at intervals (a few thousand meters apart on real telephone lines) will insert Ohms into the circuit which are *inductive reactance.* For instance, loading coils of 44, 66, and 88 millihenries are used at intervals of 1000, 1500, and 2000 meters. The Ohms that caused the high-frequency roll-off were *capacitive reactance.* In this case, the inductive reactance increases as the capaci-

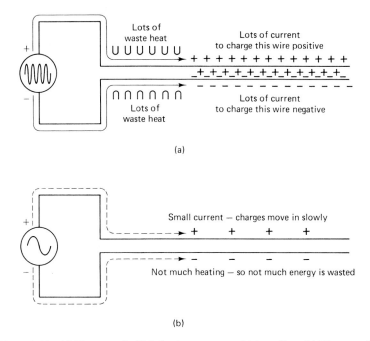

(a)

(b)

Figure 3-13   (a) Diagram of a high-frequency source driving a line. (b) Diagram of a low-frequency source driving a line.

tive reactance decreases, and the lines conduct equally at high and low frequencies (with less high-frequency attenuation). A signal at a high frequency will now get through just as well as a signal at a low frequency. This is shown in Figure 3–14. The loading coils are actual inductors added to the lines; the capacitance is an automatic part of the lines themselves.

We used the word "impedance" to describe the Ohms of a circuit "seen" by the AC component of the current that it carries. In the telephone circuit we want the Ohms to be at the distant end of the pair and not in the lines going there. Neutralizing the effects of capacitance by adding inductance makes the wires' impedance actually get smaller. If this is done correctly, the only losses in transmitting a signal from one end of a pair to the other will be due to resistance in the wire (which cannot be gotten rid of). These losses occur equally at all frequencies, and do not distort the signal being transmitted. (**Distortion** is what happens when the ability of a circuit to pass high frequencies is not equal to its ability to pass low frequencies.)

3.4 POWER AND DECIBELS

3.4.1 Electrical Power

We have used the word "power" casually in describing electrical circuit action. In a technical sense, power has a very definite meaning. **Power** is the rate at which (in

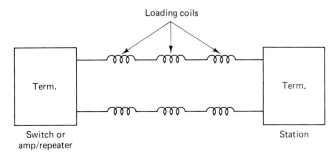

(a)

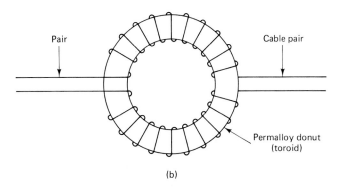

(b)

Figure 3-14   Loading coils.

an electrical circuit) the electricity does *actual work*. Electrical work actually appears in other forms when we see it. Think about types of work you see done by other forces (besides electrical ones). You *work* when you get something *done*. If you push a piano across a room, you've done work in moving the piano. If it didn't weigh anything, and you didn't have to use any *force* to move it, you wouldn't be *working*. If you pushed like crazy and the piano didn't move, you might be tired out, but you wouldn't have done any work either. (If you disagree with this, think about what *evidence* you have that you did anything. There might be a little pool of sweat on the floor, but the piano is just where it was before. There wouldn't be any evidence that your work was effective in moving the piano, because it's still where it was.)

    Work is equal to the **force** you push on the piano with, multiplied by the **distance** you moved it. This is a difficult formula to accept, sometimes, but it's true. The work you do is effective in moving the piano only if the piano moves. Force without motion (of the piano) does *no work*. The work you do flexing and straining your own muscles, on the other hand, generates *heat*. It's not a *visible* form of work, and it's not the kind of work you're trying to do, but it's the only result you get if the piano doesn't move. You end up doing a job on yourself, not the piano.

    Invisible forces can do work, too. Gravity, for instance, can move objects just as effectively as you can by pushing on them (and a lot faster, too!). If I pick up a book and hold it at eye level, then release it to fall on the floor, it doesn't stay in place when I release it; it moves faster and faster toward the floor, and slams into it with a loud thud. The *work* done by gravity is evident when the book slams into the floor. The book has acquired **kinetic energy** (it is moving at a high speed when it reaches the floor). The work the book does when it is slamming into the floor is work *you* did *lifting the book in the first place.* You did the lifting work against gravity, and put the book into position at eye level. The work *you* did gave the book **energy**. **Energy** is the capacity to do work. At eye level the book has quite a capacity to do work (on your toes!) if you let it go (and don't get your feet out of the way!). If you don't get your feet out of the way in time, the book can do *quite a job* on your toes. As long as you don't release the book, the work you did in lifting it is all **potential energy**. **Potential energy** is the form of energy stored in the book *just by virtue of its position.* When the book has been released, its potential energy is converted to kinetic energy (movement). "What has this got to do with electricity?" you ask. Electricity can *work*, just like gravity. Electric fields between charged objects work just like gravitational fields between massive objects. Imagine a world where there is no gravity, but the world is charged negative. To "fall" toward this world, you have to be positively charged (unlike charges attract each other). If we do the same experiment with the book (assuming that it is a *positive* book, like this one!), and let it drop from eye level to the negative "ground," the book will fall under the attraction of electrical forces and act just like it does in a gravity field. The *work* done is the same in its nature—a *force* acts on the book and moves it a certain *distance*—and the book hits the ground with a bang (and possibly a flash). Did the book in our electrical world have potential energy when we held it at eye level? It had **electrical potential**. Like the situation with gravity, we would have to do work to lift the book against the attracting electrical forces, and the higher we lifted the book, the larger is *potential* would become. If we had *two* books, and lifted them, and dropped them at the same time, they would do twice as much work when they hit the ground, provided that they were both equally positive. The electrical work we get charged objects to do can be increased in two ways. Use more charges (two books instead of one) or raise the book to a *higher potential*. Each book has the ability to do work on two accounts: first is the **position** it is in—namely, how far off the ground it is. Second is the amount of **charge** it possesses. We really didn't need to drop two charged books, just charge the first book *twice as positive,* and the *force* of attraction gets larger. What we're really doing in the second case is letting the book drop the same *distance* with a greater *force* on it. (This is something you can't do with gravity.) In the first case, the force wasn't changed, but the distance was doubled. In either case, more work is done, because

$$\text{work} = \text{force} \times \text{distance}$$

and it doesn't matter which one, force or distance, you double; you'll get twice the work in either case.

The potential on each unit of charge in the book depends only on its position.

When we doubled the charge, we doubled the force on the book, and doubled the amount of work it could do if released—but each charge in the book still possessed the same amount of potential energy at the same position. This potential energy per unit of charge has a name—it is called **voltage**. The drop that we let the book fall through (from eye level to the ground, for instance) is now called a **voltage drop**. We know that the amount of work each charge does is the voltage drop times the charge being dropped:

$$\text{work} = \text{voltage drop} \times \text{charge}$$

Now let's suppose we want to keep doing work *steadily*, at a constant *rate*. We could do this by going to the library, and "charging out" some books. Let's suppose we take one book a second and drop it from eye level onto a paddlewheel. The books bear down on the paddles, turn the wheel, and are presumably picked up and spirited away by librarians while we're not looking, so that we'll have a supply of books from the same source, endlessly. We now have a source of **power**. **Power** is the rate at which work is done continuously. There's very little use for electrical work that doesn't keep going continuously. The rate, **work per second**, which we do work as time goes by is called **power**:

$$\text{power} = \frac{\text{work}}{\text{time}}$$

You divide the amount of total work that's been done by the time it has taken to do it, and you get a number called power. In an electrical situation, this power comes from dropping charges continuously through a voltage drop. To know the rate at which you're doing electrical work dropping the books and turning the paddlewheel, you have to know how far up you're dropping them from (the voltage) and how many charges you are dropping per second. The flow of **charges per second** through this system is the **current**. We already know that current is the rate of flow of charges per second, and now we have another way of figuring out the rate of work being done as you turn that paddlewheel with that downpour of charged books:

$$\text{power} = \text{current} \times \text{voltage}$$

which is really measuring how much work each charge does (voltage) times how many charges do that work per second (current). Sometimes, the letter P is used to represent power, just as the letter E is sometimes used to represent voltage, and the letter I represents current. If we use P, I, and E instead of the *words* "power," "current," and "voltage," we get

$$P = I \times E$$

That's as easy as pie to remember!

Power is measured in units called **watts**. A current of one Ampere through a load that causes a one-volt voltage drop will deliver one watt of power to the load. Two volts and half an ampere would work just as well, and, in fact, so would any two numbers that multiply out to one. Since power is voltage times current, sometimes another unit, the **volt-amp**, is used to describe power delivered to a load.

Volts, amperes, and watts are units of electricity measured in the metric system. The British system measures power in **horsepower** (more generally used to describe automotive power than anything else). One horsepower is 746 watts. If you need to recall this number for any reason, remember that it's half of 1492 (certainly, one date we all remember).

3.4.2 Decibels

The amount of power that a person can perceive covers a wide range of **wattages**. For instance, the dimmest light a person can see from a bulb in a darkened room illuminates the eye with a power less than ten-millionths that of bright sunlight. Admitted, it's no *good* to look at the sun, but this illustrates our point: that a very wide range of power can be seen as light. Other types of energy are perceived by our senses in the same way; sound power ranges from the very minimum most people can hear to the threshold of pain. The wattage represented by these two extremes (which is actually watts per square centimeter of eardrum) goes from one thousand-trillionth of a watt (0.000000000000001 watt) to a thousandth of a watt (0.001 watt). These all sound like small numbers, but the larger one (a thousandth of a watt) is what you would hear standing in the same room with a 1000-watt stereo speaker. Louder sounds than this would probably break your eardrums (a result due more to *pressure* than to watts).

When we look at these numbers, the small ones and the large ones, comparatively, are in a range of about a trillion to one. A trillion is not a number people deal with all the time, and is hard to visualize, so a special scale, called the **decibel** or **dB scale**, has been designed to cover the range with numbers a lot smaller than a trillion. We describe loud or soft sounds using decibels to simplify the numbers we handle—and because people's perceptions respond to power in various forms (sound, light) in proportion to the decibels rather than the watts.

For every three decibels, power increases to twice its former value. For instance, if a 10-watt audio output were doubled to 20, you would add 3 to the decibels of sound. (Actually, it's 3.0103 . . . but 3 is close enough.) If you doubled the power again, to 40 watts, you would still only add 3 to the number you had at 20 watts. Each time the power doubles, you add 3 decibels to the power (measured in dB), and if the power drops down to half, you would take away three decibels. Three decibels is just about enough change in acoustic energy for most people to notice. Even though there is twice as much power being delivered to the eardrum, it only *feels* like a little more. If the change in loudness is just enough to notice, the power (watts) has probably just about doubled.

3.4.3 The dBm Scale

What about numbers that don't double, quadruple, or halve in convenient packages of 3 dB? Also, what about the number of dB from which you add or subtract 3; where did it come from in the first place? When 10 watts goes to 20 watts, you add 3 dB of sound power—to what? We have to have a place from which to start measuring. Zero decibels is usually defined for people as the smallest sound (brightness, or whatever) to which the "average" person can respond. In our case we said that it's about a thousand-trillionth of a watt of power received at the eardrum. In electric circuits, decibels are used to describe electrical energy: watts, volts, or amperes of electricity are not directly perceptible to *people* (if they were, we wouldn't need voltmeters). To decide on a starting point for the measurement of electrical dB, we can't start at the threshold of human perception—there isn't any. Instead, we choose to start from any convenient point on the electrical measuring scale. For watts, a **milliwatt** is a convenient point from which to start, and in telecommunications decibels of power are usually measured in reference to one milliwatt. Zero decibels—the starting point—is one milliwatt of power. If the power rises to 16 milliwatts, on the absolute dB scale, the power doubled (to 2), doubled again (to 4), doubled a third time (to 8), and doubled a fourth time (to reach 16). This means that the power doubled four times when it rose from one milliwatt to 16 milliwatts. We should add a 3 to our starting value (zero decibels) four times—which brings us to 12 decibels. When we measure power on this absolute scale, and say that the electrical power of a signal is 12 dB, when we mean 16 milliwatts, we are using the dBm (decibels in milliwatts) scale. If there's any doubt what "dB" refers to, an electrical specification should say "dBm" rather than "dB," to clarify what's being described.

3.4.4 Computing dB Gain or Loss

There's another kind of dB used in electrical jargon, usually in the form "12 dB up" or "3 dB down," which means that the power increased or decreased; it doesn't refer to any specific amount of power. For instance, if someone says "The output of the speaker increased 3 dB when we matched impedances," we can tell that the power doubled, even though we might not know the specific number of watts that was used. Decibels are used in this way to denote improvement or attenuation of signal strength, not the absolute value in watts of the signal. In our example of a 1-milliwatt signal rising 12 dB to reach 16 milliwatts, if it were to rise another 12 dB, the signal would be 256 milliwatts. The power would be 16 times as big (at 256) as its starting value (16) and the 12-dB figure reflects that fact. It isn't where the power started or ended that's important, but how many times bigger it is.

What about something that gets 10 times bigger? All we know about is how to add 3 dB every time the power doubles. To handle a number like 10, you can't just do it by doubling. If you have a dB meter, you just measure it. If you have a calculator, you use this formula:

$$dB = 10 \times \log \text{(multiplier)}$$

where "multiplier" means how many times bigger the power was after you increased it than it was before. What if the power gets smaller? You get a negative number for dB if the multiplier you use is smaller than 1. For instance, suppose that power goes down from 6 milliwatts to 1 milliwatt. The dB loss (negative dB) is

$$dB = 10 \times \log (0.16666)$$

where the number, 0.16666, is one-sixth. The power is one-sixth its former value, and your calculator gives an answer of −7.78 dB. Another way of stating the same thing is to say the power is 7.78 dB down, and that means a power loss. If you don't have a calculator, **log tables** are available in most mathematical handbooks. Figure 3–15 shows an abbreviated version of such a table. To read dB from this chart, find the starting and final wattage, read up to the line, then over to the dB values on the left-hand edge of the chart. As an example, the chart shows how the dB for 6 milliwatts is read. To find the dB up or down from any value of milliwatts, just read the milliwatts on the bottom edge of the chart, find the dBm on the left-hand edge, and find the difference between your starting dBm and ending dBm. The difference between the two dBm numbers is the dB gain or loss of the power increase or decrease.

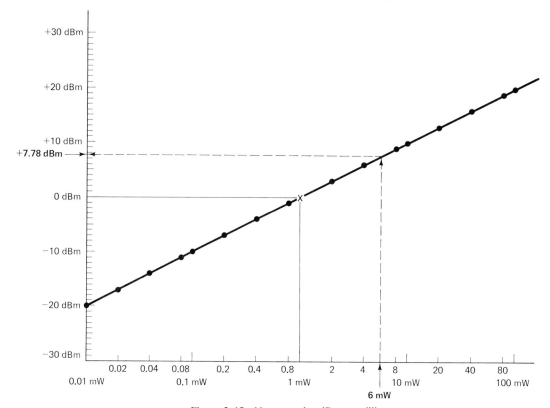

Figure 3-15   Nomograph—dBm to milliwatts.

3.4.5 Computing dB for Voltage and Current

In the telegraph system we described earlier, the losses on the long lines between repeaters (amplifiers) were 50% along a 3-mile length of wire. We started with 24 volts and ended with 12 volts. Does this mean that the power is half as much at the 12-volt "distant end" than it would be if the sounder received all 24 volts? Does voltage loss work the same way as power loss? In other words, is a drop from 24 to 12 volts a 3-dB loss? The answer is *no*.

The power used by the sounder is the voltage multiplied by the current. Twenty-four volts across a 100-ohm sounder coil would produce 0.24 Ampere of current. Twelve volts across the same coil produces only 0.12 ampere—half as much. At the distant end the sounder coil gets half as much voltage, but also only half as much current. The power, in each case, is

$$\text{power (24 volts)} = 24 \text{ volts} \times 0.24 \text{ Ampere}$$

This is 5.76 watts—and

$$\text{power (12 volts)} = 12 \text{ volts} \times 0.12 \text{ Ampere}$$

This is 1.44 watts, which is exactly one-fourth of 5.76 watts. We lost -6 dB of power, not -3 dB! If we had half our original power, we would have a $-3-$dB loss. If we lost half of that again, we would lose another -3 dB. What we would have is half of a half, which is one-fourth. With 1.44 watts instead of 5.76 watts, our dB loss is -3 dB, two times—that is, -6 dB, and doesn't involve any heavy math—you just add up the losses.

Does it work? The *multiplier* we mentioned earlier is 0.25—that is because 5.76 watts times 0.25 is the power we have after the losses—1.44 watts. We use the formula

$$\text{dB} = 10 \log (0.25)$$

and we get -6 dB. (If you have got a good calculator, you actually get -6.0206, but we'll say -6.)

Would it be easier to use a separate formula for voltage? If you ever have to, here's one that works:

$$\text{dB (volts)} = 20 \log (\text{multiplier for volts})$$

In our case, if we put in 24 volts and got out 12 volts, the multiplier is 0.5 (a half). The formula gives

$$\text{dB (volts)} = 20 \log (0.5)$$

After you crunch the numbers, this is the same -6 dB we got before, without working Ohm's law and multiplying Amperes times volts.

This is important only if you are not working in watts in the first place, but it *is* important if you measure losses in volts—or Amperes—the same formula works for Amperes, too!

Whether it is electrical energy, light, or acoustic energy that is being measured, dB measurement is the most "natural" way to identify power changes. Human beings react to these forms of energy with a logarithmic response. The dB scale is a logarithmic scale that matches the way people perceive light, sound, and so on—forms of energy which "feel" like they're going up at the same rate the dB reading goes up.

In circuits with amplification and attenuation, mathematical calculation of volts and watts is cumbersome and complicated. On the dB scale, however, an amplifier merely adds decibels to the signal and circuit losses merely subtract them. Instead of having to multiply, divide, and calculate squares and roots, you can just add and subtract when you're using decibels. It's less work.

As an example, Figure 3–16 shows the journey of a signal on its way from a transmitter to a receiver. It passes down lengths of line where power is lost, is switched through central offices and toll offices that boost signal power (although their main purpose is to switch the signals), and finally arrives at the receiver. To deal with each loss and gain using watts would require multiplication and division for each gain and loss. Instead, all we have to do with decibels is add positive dB for the gains and negative dB for the losses. The total dB is the sum of all gains and losses (which is −12 dB in this case).

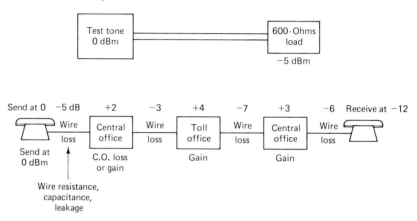

Figure 3-16   Journey of a signal from transmitter to receiver.

SUMMARY

In this chapter we talked about DC and AC technology used in telecommunications, and the basic electrical principles used in telecommunications equipment.

Since this book is for the user, or the technician who's going to service telecom equipment, we're not going to go any deeper into electrical theory. For work with this type of equipment, you're not going to need either deep electrical theory or electronic amplification theory. You're also not going to need higher mathematics, and probably won't mind a bit!

REVIEW QUESTIONS

1. Does a carbon-microphone telephone transmitter convert sound energy directly into electrical energy? Why (not)?
2. How can a varying electrical current be changed into sound?
3. Is a telephone circuit half-duplex or full-duplex?

4. Is sidetone necessary for telephone communication? Is it necessary for telegraph communication?

5. In a simple series circuit, is the current the same at all points in the loop?

6. How is an electromagnet related to an inductor?

7. Does the magnetic field around a relay coil act the same as the field in an inductor?

8. Would it make more sense to connect several phones together in series or in parallel?

9. Can an AC and DC current be combined?

10. Give two common uses for a capacitor in telecommunications.

11. Can a direct current pass through a capacitor? An inductor?

12. Do a one-Ampere DC and a one-Ampere AC current deliver the same amount of power?

13. Would a coil (inductor) oppose the passage of AC current more than DC?

14. What is impedance?

15. For what purpose are loading coils used?

16. Is the decibel a measure of power? Answer this for decibels, both dB and dBm.

17. How much power is dissipated in a load drawing a current of one milliampere at 36 volts?

18. If the power delivered to a load increases by 3 dB, by what factor has the power increased?

19. In telecommunications, what standard unit of power is used most often to describe signals in lines?

20. If a signal is sent at -3 dBm and is received at half the power, what would its measured dBm value be?

4

Digital Telecommunications

In telecommunications systems, digital control of the signal path, and digital encoding of the signal itself, make it necessary for anyone who works with this type of equipment to know something about basic computer concepts.

Digital (computer) methods are being used everywhere to do things no one had ever imagined using a computer for when computers cost $1,000,000 apiece. With computers (at least microprocessors) available for less than $10 a processor, even trivial uses for computing power are taking the place of standard electronics and "watchworks." Telecommunications is no exception. Most people are intimidated by computers—the word "computation" itself suggests mathematics, and most people are scared of math.

Is "computing" necessarily synonymous with "math"? In the early days, when multimillion-dollar computers were all there was, "heavy math" was the only thing anyone would use those machines for that could justify the cost. Today, with microprocessors being used as **control elements**, most of the time the "computer" is not "computing" any computations at all! The majority of the time, a "micro" is switching circuits on and off, that's all. There *are* numbers associated with this kind of operation, but not much *math*.

An example is the way the phone company "catches" all the digits of a number you've just dialed. Each number must be identified, then saved, and as another number is dialed, the ones already saved have to "move over" to make room for the next number. This could be done with a board full of logic circuits: registers, counter circuits, clock pulses to synchronize action, and logic gates to identify special conditions. In fact, all this circuitry could be replaced by a single program running in a single microprocessor. The micro is made, by its program, to *imitate* the board full of logic circuits it replaces. Using a microprocessor and a program to replace single-purpose logic circuits is the major new trend in all telecommunications systems. The telcos used an ESS (electronic switching system) to replace mechanical registers and switching relays with solid-state devices (no moving parts). Early ESS systems were made of *random logic*, circuits designed to solve a problem but with no flexibility to change functions—the whole circuit would have to be rewired if some new characteristic were required. Current designs use microprocessors, which are **general-purpose machines**—they can be programmed to have any characteristics you desire—and these micros can be *reconfigured* to do new tasks by merely changing the program (a list of instructions).

PBXs, like ESS switches, are making major use of **dedicated machines** (microprocessors running a PBX program, in this case). It is this ability to reconfigure rapidly that makes micros an essential part of current communications systems.

4.1 BASIC PRINCIPLES OF DIGITAL INFORMATION

4.1.1 Digital Signal Levels and Logic Circuits

The subject of this chapter is data communication. Before we can talk about data communication, we need to know what "data" is. Then we can see what the difference is between "data communication" and "voice communication."

Data and voice are both ways to communicate information. With voice, we can communicate conversationally, but the information we convey is not always as precise as we would like it to be. Voice is also a poor way to communicate certain types of information, namely, numbers and pictures. To see what we mean by that, try to describe a picture to someone over a telephone. See if they can draw the same picture you are describing. Eventually, with great effort and an outrageous expenditure of time, you may communicate the information needed to make a recognizable likeness of your picture. It's possible you may never do it, if you get frustrated easily.

Data communication transmits information in the form of words, numbers, or pictures equally well. So what is data, and how does it convey numbers and pictures as well as words? Data, simply put, is numbers. How we express these numbers is important, but not as important as what we use these numbers to represent. Let's say, for example, that you wish to send a table of numbers that represents the starting and ending prices of stocks on the New York Stock Exchange at the end of a day's trading. If you do it verbally, it will take forever, and the chance of being misinterpreted by the listener copying the figures is very good. Instead, suppose that you make the numbers electrical, and provide them to a "listener" in **machine-readable** form. The listener, in this case, is a machine that can "read" the numbers you send it much faster than a human can copy spoken words, and without ever getting tired.

Some of the numbers will be used to represent letters in the names of the companies whose stock has been traded. Some of the numbers will be used to represent special symbols, such as the "$" symbol. Some of the numbers we transmit will actually be numbers:

Company	Shares traded	Open	Close	Change
HARTS	9304	22½	24¼	1¾

In this example, the letters "HARTS," the numbers, and the words "Open, Close, and Change" can all be transmitted using a code similar to Morse code.

"Code?", you ask, "I didn't have to use code with voice transmission. Won't that make it more complicated?"

Remember that we said we wanted the data to be machine-readable? First, we will want to use electrical signals to carry the information. This makes it a lot easier for machines to read. Suppose that there are 52 letters in the alphabet using capitals and lowercase (small) letters, and also 10 numerals (0 through 9) and in addition, 20 or 30 punctuation marks that you want to use. Our code had better have at least 90 different patterns, each of which can represent a different symbol (letter, number, or punctuation). If the listener is going to be a "smart" typewriter, we are also going to need code symbols for the spacebar, the carriage return (advance paper and go to start of next line), and even things like ringing the bell. We might as well use 100 different symbols for all these things.

How do we make the code symbols out of numbers? What is the advantage of using code like this? Of what are the numbers constructed electrically?

Answering these questions back to front, let's see what we use to communicate a number electrically. In a long number such as 5,225,344, we have individual numbers within the larger number. We call each individual number a **digit**. From this, we get the name **digital**. Each digit has a value between 0 and 9, although the whole number is over 5 million. Breaking the number down into decimal digits reduces the number of symbols (digits) we must use to communicate *any* size number to just 10 symbols. Imagine what it would be like if we had to use a different symbol for every number from 0 up to 5,225,344. Each number would be only one digit long, but we would have too many different symbols to handle. Besides that, there are numbers beyond 5,225,344—a *lot* of numbers. We would need a symbol for each of those, too. An infinite number of symbols would be required.

The same thing applies to our digital transmission of data by electrical means. We could try to transmit our 100-symbol code using 100 different voltages, one for each symbol. This would not work very well—it wouldn't work at all, in fact—because among other things there would be DC attenuation. You couldn't tell what voltage would arrive at the receiver unless you knew the resistance of the line. That, in turn, depends on the length of the line, and every time you dial a call, you get a line of a different length. Suppose that we transmit two decimal digits, each with 10 voltage levels, instead of one code with 100 levels. In this case, we are doing what we did with the number 5,225,344—we are breaking it up into smaller pieces, with less variety in the pieces. Unfortunately, there is still too much variety. If our signal is attenuated 10%—less than 1 dB—a 9 would become an 8. The solution is to break the decimal figures down still further—using something called the *binary* system—so that each digit is a voltage that's either ON or OFF. It's very hard to confuse a voltage that's *there* with one that *isn't*. Using the binary system, the numbers from 0 to 100 are made by stacking up binary digits, the way we did with decimal digits for the number 5,225,344. In fact, it takes the same number of binary digits to represent 100 as it did decimal digits to represent 5,225,344.

4.1.2 The Binary System

In the decimal system, we use numbers of more than one digit to represent values larger than 9. In the binary system, we do the same thing for numbers larger than 1.

Since there are only two digits in binary, "voltage" and "no voltage," we use 0 (no voltage) and 1 (voltage) to represent all bigger numbers this way:

$$0, 1, 10, 11, 100, 101, 110, 111$$

What you see here is the first eight numbers (0, 1, 2, 3, 4, 5, 6, 7) in binary. We say "in binary *code*" because most people don't read numbers this way, and so, to them, it's as mysterious as a secret code. There is a pattern, however, to these things. In the example above, we have all the possible binary patterns up to three digits. Do you think there might be any others we didn't include?

Perhaps you're right. Did you see the possibilities?

$$0 \quad 1 \quad 00 \quad 01 \quad 10 \quad 11 \quad 000 \quad 001 \quad 010 \quad 011 \quad 100 \quad 101 \quad 110 \quad 111$$

Perhaps you thought of these, and wondered why we said that there were only eight possible patterns.

It turns out that 0, and 00, and 000 are all the same number. So are 1, 01, and 001. The same is true for 10 and 010, or 11 and 011. Why are they the same?

If 11, for instance, is different from 011, where is it different? "In the front," you say. Yes, that's true, the front of 011 has a "0" and there isn't a "0" at the front of "11." This brings up the question: "If there isn't a '0' in front of the '11,' what *is* there?" Answer: "Well, a blank space with nothing in it, I suppose."

That's not a good enough answer. Because if there were a blank space, that would mean that we had three kinds of symbols, 0, 1, and blank spaces. If all we use are "voltage" and "no voltage," there is no third condition for blank space. The blank space becomes the same as a 0 because there's nothing there. Now suppose that the number 11 is written with a zero in the blank space at its front. That's 011, the same as the other number. *Lesson:* If you can't see a number in front of a group of binary digits, it's a 0.

To avoid confusion in the future, any time we deal with binary numbers that have three digits, we'll show all three, even the ones you don't need to see. We'll also do this with four, five, or any other number of digits.

Let's return to the problem of representing 100 different symbols of binary. We said, for some reason, that it would take the same number of digits to represent

100 in binary as it took to represent 5,225,344 in decimal. Here's why: In decimal, each additional place added to a number made the number 10 times bigger. In binary each additional place added to a number makes it two times bigger. A one-place binary number has only two values: 1 and 0. Two-place binary numbers can have four values—twice as many. Using three places, there are eight possible numbers, as we saw before:

000   001   010   011   100   101   110   111

(There are no other three-digit binary patterns, no matter how hard you try to figure others out.)

Each digit placed in a binary number with several places is called a **binary digit**, or **bit**, for short. Here is a list showing how many numbers you can make using binary code with various numbers of bits in each case:

How many bits	How many values
1	2
2	4
3	8
4	16
5	32
6	64
7	128

Each time we add another bit, we double the number of values we can create, because we have all the values used previously, now with a 1 added, and we have all those values again, with 0 added instead of the 1.

Notice that for seven-place binary numbers (7 bits) we can write binary numbers with 128 different values. This will cover all 100 of our symbols—the numbers, alphabet, and punctuation we spoke about before—and will give us additional values to use for control of our "teletypewriter" (the "smart" typewriter we mentioned earlier).

Let's look at one of the symbols, the number 99. If we count binary numbers starting at zero, the number 99, in binary, looks like this:

1   1   0   0   0   1   1

How can we be sure that this number is 99? In fact, how can we read binary numbers at all? The answer lies in the fact that binary, like decimal, is positional. That means that each place in the number has a value; in this case,

1   1   0   0   0   1   1
64 32 16 8   4   2   1

Each place to the left of the units' (the 1) position is twice the value of its neighbor. In binary, there are either bits which you have (the 1, 2, 32, and 64), or bits which you don't have (the 16, 8, and 4). The bits you *have* are shown as 1's, and the ones you do *not* have are 0's. The value of the whole number is the sum of the values of the bits you have, namely:

$$64 + 32 + 2 + 1 = 99 \text{ (decimal)}$$

See how it works?

Now, does it matter what the decimal value of each binary code pattern is? Actually, no. The point is that we should have lots of codes, each different, so that a different code pattern can be used to represent each of the 100 symbols we want our typewriter to type. We could assign number values to symbols in any order.

4.1.3 Alphanumeric Information

There is a standard 7-bit code used for alphanumeric symbols (both letters and numbers). It is so standard that it is called the **American Standard Code for Information Interchange**, or **ASCII code**. (A lot of people call it the ASC-2 code, because the II looks like the Roman numeral for 2, but the II stands for Information Interchange, not 2 of anything.)

A complete list of ASCII is shown in Figure 4-1. In the ASCII code, letters of the alphabet, for instance, start at 65 and go up until there are 26 letters. The actual method is to add 64 + (letter of the alphabet) with A having the value 1, B having the value 2, and so on up to Z, with a value of 26. In this code, Z is coded as 90 (64 + 26). There are actually two alphabets in ASCII. The other one starts at "a" and goes to "z" (lowercase letters). These are coded by adding 96 + (letter of the alphabet), with "a" having a number of 1, "b" having a number of 2, and so on up to "z", with a value of 26 again. This alphabet uses binary 122 (96 + 26) for a "z." The decimal numerals start at 48 and go up until there are 10 numbers. They are coded by adding 48 + (number), with the smallest number being 0, and the largest, 9.

4.1.4 ASCII Data Transmission

To use a telephone circuit to carry ASCII code, we need to ask how the seven bits of each symbol are transmitted. We already know that the presence of a voltage can represent a 1, and the absence of a voltage is a 0. To transmit a symbol where the code for the symbol is made from 7 bits, we can either transmit all the bits at one time, on seven wires, or transmit bits on one wire at seven different times. Transmitting each bit separately, at a different time, is called **serial** data transmission. Using seven wires and transmitting all seven bits at once is called **parallel** transmission. Since telephones are connected through only one circuit, we will use serial transmission to send our characters over the telephone.

The telegraph system we described earlier is another example of data transmission. It is essentially digital in nature, with the "closed key" being a 1, and the open key being a 0, except that *dots* and *dashes* really convey what they are by how *long* they are, not what level the voltage is. Morse code could be viewed as a data transmission code that uses a *trinary* numbering system. The same concepts that apply

Example conversion: Letter "a"
- Type: lowercase (row 3)
- Found in: (column 1)
- "Front bits": 11 (row 3)
- "Back bits": 00001 (column 1)
- ASCII code: 11 00001

Column	"Back" 5 bits	Control codes (row 0)	Numerals/punctuation (row 1)	Uppercase (row 2)	Lowercase (row 3)
0	00000	NULL	b	@	`
1	00001	SOH	!	A	a
2	00010	STX	"	B	b
3	00011	ETX	#	C	c
4	00100	EOT	$	D	d
5	00101	ENQ	%	E	e
6	00110	ACK	&	F	f
7	00111	BEL	'	G	g
8	01000	BS	(	H	h
9	01001	HT	)	I	i
10	01010	LF	*	J	j
11	01011	VT	+	K	k
12	01100	FF	,	L	l
13	01101	CR	-	M	m
14	01110	SO	.	N	n
15	01111	SI	/	O	o
16	10000	DLE	0	P	p
17	10001	DC1	1	Q	q
18	10010	DC2	2	R	r
19	10011	DC3	3	S	s
20	10100	DC4	4	T	t
21	10101	NAK	5	U	u
22	10110	SYN	6	V	v
23	10111	ETB	7	W	w
24	11000	CAN	8	X	x
25	11001	EM	9	Y	y
26	11010	SUB	:	Z	z
27	11011	ESC	;	[	{
28	11100	FS	<	\	|
29	11101	GS	=	]	}
30	11110	RS	>	^	~
31	11111	US	?	_	RUBOUT

"Front" 2 bits:
00	01	10	11

Figure 4-1 ASCII code chart. To convert a symbol to ASCII: 1) Identify its type (numeral, uppercase, lowercase, control code). 2) Find the symbol on the row for that type of character. 3) Read the "Front 2 bits" from the left-hand end of the row it's on. 4) Read the "Back 5 bits" from the top of the column it's in. 5) Put the front 2 bits in front of the back 5 bits to make a 7-digit code.

78

to transmission of data in Morse on a telegraph still apply to ASCII transmitted on telephone wires. Terms such as "simplex," "half-duplex," and "duplex" still have the same meaning. Attenuation and losses still have the same effect on digital data as on telegraph transmissions, and repeaters are required. For digital transmission, repeaters can be much simpler than for voice signals, since they don't have as many levels of voltage to deal with. Signals "repeated" through a digital repeater may thus be *reconstructed* without distortion or noise, rather than merely *amplified* (together with the noise and distortion) as is the case with voice repeaters.

4.2 COMPUTERS AND DIGITAL SYSTEMS

What is a computer, and how does it work? What kind of tasks does a computer do in a digital system? We will look at these questions in the next few paragraphs, and use a simple example to illustrate the answers.

4.2.1 Computing (The Hard Way)

Let's suppose that we want to transmit data using the ASCII code. To simplify things, suppose that the only symbols we are going to transmit are decimal digits 0 through 9. Each code is going to be "digit-plus-48." Calculating the value of the ASCII code for each number, as it is transmitted, is a very elementary task. Suppose that we decide to do it with a pocket calculator. We want to be able to take any number we key in on the keyboard, and have the calculator give us back "key-plus-48." If we have already keyed in a decimal digit from the keyboard, the keystrokes we need to complete this calculation are

$$(+)\quad (4)\quad (8)\quad (=)$$

No matter what digit we keyed in before the ASCII conversion, the next four keystrokes must always be $+$, then 4, then 8, then $=$, in exactly that order, so that we can get the computed ASCII value (in decimal) on the display.

Wouldn't it be nice if we could make some sort of keypresser gizmo that would press those four keys, in exactly that order, after each time we *manually* entered a number by pressing a key with our fingers? That gizmo might look like Figure 4-2. In the figure, a cam drives a set of pushrods that push the keys in $+$, 4, 8, $=$, sequence. If the rods are thin, and do not get in the way of your fingers when you push each key, you have only to key in the number and turn the crank on the cam to get the number converted to ASCII by the calculator. Aside from the rods getting in the way of your fingers, and the fact that this is thirteenth-century technology, it's not a bad way to perform an automatic calculation.

4.2.2 Computing (As It Is Really Done)

Let's use modern technology to solve the same problem. The actual keys on the calculator are switch contacts. Each one can be pushed without any moving parts by attaching a transistor switch to the terminals of each pushbutton, and driving the

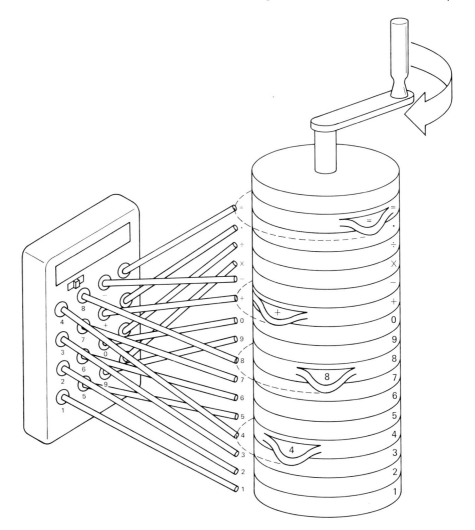

Figure 4-2   Programmable calculator—13th century technology.

transistor with an electrical signal. Each transistor can be assigned a number and driven by a signal from the output of a decoder, as shown in Figure 4-3.

In this case, the decoder is a device that identifies 16 codes and activates one of the 16 keys according to what code it receives. The simplest set of numbers to assign the transistors would be 0000 to 1001 for the keys with the numbers 0 to 9 on them. These are just the binary codes for the numbers on the keys.

The keys with symbols +, −, ×, ÷, ., and = are a different case. There is no "right" code for these keys. The code 0110 is right for the key with a 6 on it, because it's a six in binary code, but what's the right code for a + sign? We assign the symbols any codes we please. There are six 4-bit binary codes we haven't used,

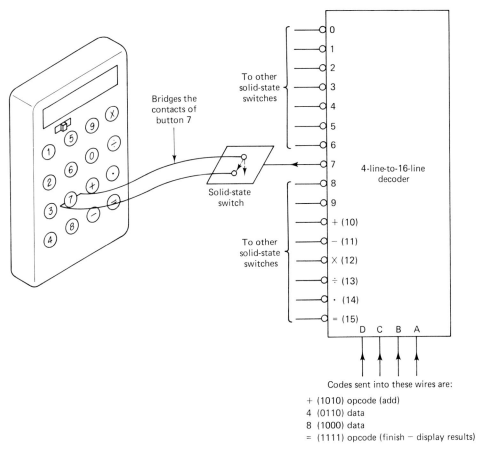

Figure 4-3   Solid-state switches to activate calculator.

and six keys left on the calculator, so we assign each key one of the remaining codes, as shown below:

$$1010 = +   \qquad 1011 = -$$
$$1100 = \times   \qquad 1101 = \div$$
$$1110 = .   \qquad 1111 = =$$

These are called **opcodes** (operation codes), because each one represents an operation, not a number. The opcode 1100, for example, causes the decoder to activate the transistor that "presses" the × key. When the transistor carries current through the same path it takes when the × key is pressed, its the same as pressing the "multiply" key, only without any moving parts.

The binary codes that "press" the keys with numbers on them are called **data codes**. They are numbers plain and simple, and don't represent any other operation except the entry of a number through the keyboard.

The decoder and the transistor switches take the place of the pushrods in our earlier picture. Now we need something to take the place of the cam and the crank. What are we going to use to send the binary code signals to the decoder? It will have to perform the same task as the pegs on the drum of the cam. Each time we want to calculate "number-plus-48," the + , 4, 8, and = keys must be pushed. This means that the numbers 1010, 0100, 1000, and 1111 have to be sent to the decoder, one after another, every time we want to do an ASCII conversion on the number just struck from the keyboard. This requires a **memory** device, a circuit that will "remember" the binary codes and bring them back for us in exactly the same order every time. This is shown in Figure 4-4. The memory device does the job of the cam. Where the pegs sticking out of the cam pushed the pushrods, the memory sends numbers to the decoder, and it pushes the keys electrically. Since the memory device holds the numbers indefinitely, it can "remember" them for us every time we want this calculation done. Now all we need is a "crank" to turn the "cam." We need a way to tell the memory what order to take out the numbers it has stored inside, and when to stop (when it has finished the calculation).

This is accomplished by adding a **counter** to the picture (see Figure 4-5). In our picture, each binary code is stored in the memory in a "pigeonhole" with a number. We can think of the memory like a small post office with post-office boxes, each numbered. To find each boxholder's mail, we just go to the box with the right number. In the memory we are using, the numbers serve to tell us in what order we will "look into each box" to get each binary "command" for the decoder. Our binary commands are stored in boxes numbered 0, 1, 2, and 3. To get the keys + , 4, 8, and = pressed in that order, we send the numbers 0, 1, 2, and 3 to the memory device. The numbers 0, 1, 2, and 3 go to the memory device in binary code. The circuit that does this is called a counter, and this one is called the **program counter**. Its purpose is to "fetch" each binary-code command from its "box" in the memory, and send it to the decoder in the right order.

The type of counter we have counts 0, 1, 2, 3 and is called an **up-counter**. (There are counters that count down, too, like a rocket launch, but that is another story.)

When the numbers go to the memory from the counter, the commands come out of the memory in " + 4 8 = " order, and the decoder, which we now call the **instruction decoder**, makes the calculator do them in that order. In the electrical model we have constructed, the counter serves the same purpose as the crank that turned our cam. It makes the instructions come out in the right order to "push the buttons."

Now, we might ask: "Do the buttons have to be 'pushed' as slowly by the electronics as we would do it by hand?" There is no reason why we couldn't run the counter to count hundreds, thousands, or millions of counts per second, instead of just one or two keystrokes per second. What speed would be too fast for the calculator to keep up? When we know the answer to this, we can attach the counter to a **clock**, which is a pulse generator that makes the counter count from one number to the next. It is the clock that decides how *fast* the counter will count from 0 to 1 to 2 to 3.

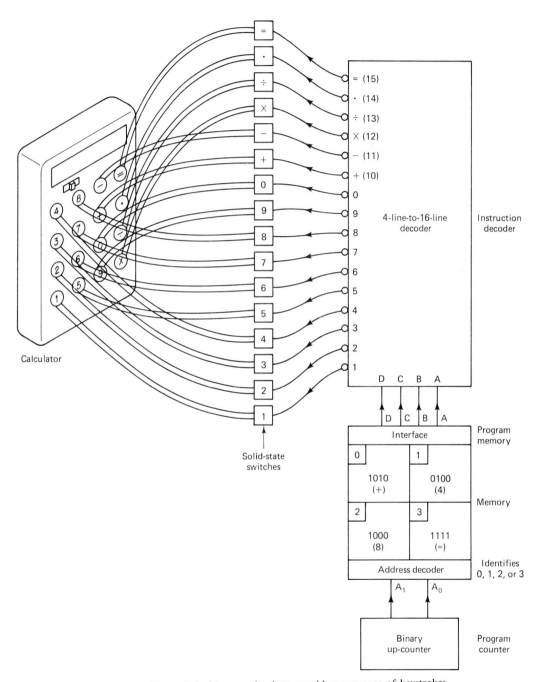

Figure 4-4 Memory circuit to provide a sequence of keystrokes.

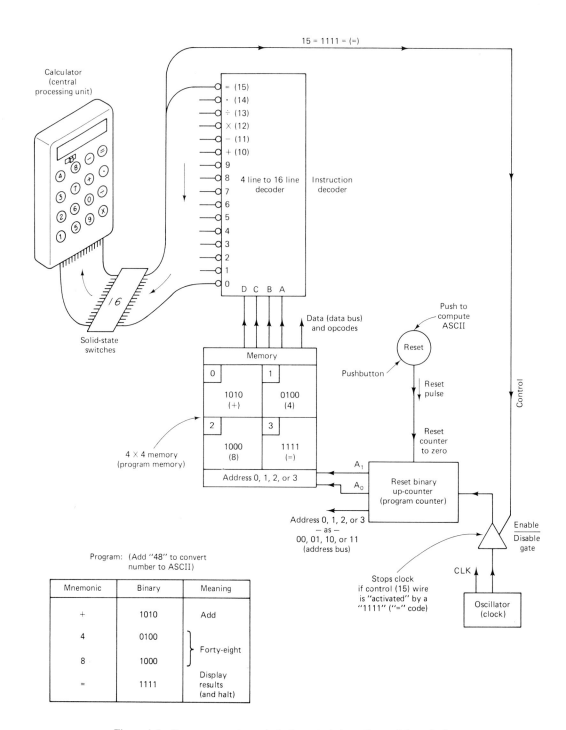

Figure 4-5 Program counter and clock control the actions of the calculator.

84

Finally, we have to decide when to stop. The counter's clock is stopped when the number 1111 appears (15 on the decoder). This is the code for the = key, and " = " is always the last instruction in a set of calculator keystrokes. In Figure 4-5 this is the part of the system that makes our calculator stop.

We know how to stop, but how do we start up again? The counter, as most digital counters, has a control called the **reset**. What that does is sets the number in the counter back to zero. Lots of mechanical counters have this sort of pushbutton, too, like the indicator on a tape recorder. The digital counter uses an electrical signal for this purpose, but otherwise, it works the same way. We start our **program**, which is the list of codes 1010, 0100, 1000, 1111, coming out of the memory by hitting a switch that resets the counter. Once the counter is at zero, the instruction code that enters the decoder is no longer a "stop" code, and the counter can proceed from 0 to 1 to 2 and so on, until a stop code appears on the decoder again.

This is how we **run** the program. We hit the reset switch, and the calculator is off to the races. The numbers 0123 go into the memory and fetch out the codes 1010, 0100, 1000, and 1111 from the memory. These codes activate the decoder and make it "push" the buttons +, 4, 8, and = on the calculator. After that, the number-plus-48 calculation is complete, and the 1111 code at the decoder has shut off the clock, so the program counter does not continue. To people who use this gizmo, it appears that they enter a digit, hit the reset key, and the ASCII conversion appears magically on the LED readout.

What we have now is a **computer**. It has all the parts of a computer, scaled down to a very simple level. The instruction decoder, the program memory, the program counter, and the clock are found in all computers. The pocket calculator whose buttons we pushed with this circuit is called the CPU, or **central processing unit** of a computer.

That's all there is to a computer. Larger computers with larger instruction sets can do more than the six opcodes we used here, but the ideas don't change, they just have larger numbers attached. (If you would like to see more about how these concepts apply to telecommunications switching, skip ahead to Chapter 11, section 11.5.)

4.3 AN EXAMPLE OF DATA COMMUNICATION

4.3.1 Data Communication on the Job

Imagine that you are a service technician and your job is to go from business to business, repairing their telecommunications equipment. You may be dispatched to service emergencies as they come up, so you don't know what your agenda is at the beginning of the day. Your employer doesn't know when or how your services will be needed throughout the day, either. You're dispatched to your first job at 8:30. Now it's 10:00 and you've finished the first service call—how do you let your employer know that you are finished with that call, and how does your employer let you know if there is another job for you to do, and if so, where it is? The obvious thing to do is "check in" in some way.

How are you going to check in? You could drive back to the office and find out, but that could take a lot of gas and a lot of time. Since you're working with telecommunications equipment anyway, why not phone the office? Unfortunately, what you have to say, and what you want to find out, are not really worth tying somebody up at the office for a long time; if we could use data communications techniques, we could get the same information across but use a lot less time doing it than if we used voice communication. We could find out what jobs are "in the queue" (haven't yet been assigned to anybody) and "clear out" the job we've already done by talking digitally with a "fast idiot"—the dispatch computer. The dispatch computer is programmed to take care of you in the following way:

1. You dial-up the dispatch computer's number over the telephone lines. This is done using a **terminal**, which looks like a typewriter and has a place to couple it to a phone line.
2. You give the digital computer a **password** that lets you gain access to the dispatch computer; then you give the dispatch computer your employee number.
3. The computer will "come back" to you with your name to verify that it has identified you correctly.
4. If the dispatch computer has correctly identified you, you will respond to its query; it asks if your current job is completed.
5. You answer "yes," and the computer clears that job and assigns you a new job to do.

Figure 4–6 shows the display you might see on a printout from your terminal at the end of this sequence.

On the printout, you see that your entries are written in all-capital letters in this example. The computer's entries are in mixed capital and lowercase letters. (This was done in the example to make it easy to see what the computer has typed in on your printer and what you have typed on the keyboard—which also appears on the printer.)

The type of computer dialog you see here is typical of dispatching systems, airline ticket reservations by computer teletypewriter, and other data-entry systems where a person at a remote location wants to gain access to a central data base.

4.3.2 What Is a Dedicated Machine?

Inside the **remote terminal**—also called a teletypewriter in this case—you attached to the phone lines is a microprocessor running a **terminal program**. It is taking care of things like ASCII conversion of the keystrokes, converting the data to telephone-compatible signals, and deciding whether to send or receive data at any mo-

```
|O|    DATASYS.345 Version II.  Programmed by R. M.
|--|    logon 10:00:00, 26 Oct. 1987; Enter password please:
|O|    ? SERVICE, DISPATCH
|--|    Enter Employee Number:
|--|    ? 27342
|O|    Hello technician Anderson
|--|    You are presently shown dispatched on job# B-2087
|--|    Do you wish to clear this job:
|--|    ? YES
|O|    Hours worked:
|--|    ? 2
|O|    Trouble found:
|--|    ? HAIRPIN IN FRAMMIS NOZZLE, MODEL 25 TURBOENCABULATOR
|--|
|O|    Next available jobs, in priority order:
|--|    Job# Customer    Nature of Problem
|O|    B-2096  Zion Nuclear    Possible Pepsi syndrome
|--|    B-2104  U.S. Robots    Raster breakup on teletypewriter
|--|    B-2111  Ronco    "$9.95 display" doesn't work
|O|    B-2099  Montgomery W.  Intermittent freemish twizzling
|--|    B-2108  American Can    Trunk 555-8989 dead on PBX
|--|
|O|    Enter job number choice:
|--|    ? B-2099
|O|    Do you need customer location:
|--|    ? YES
|O|    2255 Decimal Place, Scarsdale, Ont.
|--|    END
|O|    Logoff at 10:02:31, 26 Oct. 1987
|O|
```

Figure 4-6 Display printout.

ment. This program decides whether the keyboard or the outside phone line is "printing" on the printer at the moment, and perhaps takes care of things like advancing the paper and moving the carriage to the beginning of a new line. It is "transparent" to you, the user. You don't see this program, interact with it, or even know it's there. The program is just the set of rules by which the teletypewriter operates—the "virtual machine" you see when you operate the keyboard. In a case like this, the program memory is permanent, and you cannot rewrite what it contains. You wouldn't want to change what the memory contains, because you wouldn't want the teletypewriter suddenly to start working according to a different set of rules. A machine that contains a one-program microcomputer such as this is termed a **dedicated machine**. It is dedicated to running one program that does only one thing (act like a teletypewriter) all the time.

4.3.3 What Is a General-Purpose Machine?

The computer we were "talking with" in our dispatch example is a **general-purpose machine**. It is capable of running a variety of programs, and often may be running (multiprogramming and timesharing) many programs at the same time (or at least maintains the *appearance* of running many programs at the same time). If the computer in question has more than one processor (multiprocessing) it can run many programs simultaneously.

In our case, the computer had a great many more programs to run than just the "dispatch computer" program we wanted. Our request to the computer for "SERVICE, DISPATCH" put the "dispatch computer" program at our disposal. Possibly, there may be other technicians calling in at the same time. They can be "time-shared" by this program so that it appears that each has the undivided attention of the dispatch computer. The computer's internal circuitry is so much faster than the transactions you have typed on the keyboard that the computer could be servicing hundreds of other technicians between the keystrokes of your message.

The program in your teletypewriter was inflexible and did not alter its contents. The program in this general-purpose machine must be flexible, because it has to alter its contents every time you (or another technician) "clear" a job as we did in the example, and must update the list of jobs that need service every time there is a call from a customer. People in the office have to update the list when they receive the calls from the customers.

The programmer who developed this program might think of **enhancements** that should be added, and the general-purpose machine should be capable of easily being reprogrammed with these enhancements. For example, if the "Location Request" part of the dialog took place often enough, the programmer might decide to put the location in the original list together with "Customer" and "Problem" every time the list is printed. Being able to do this is a characteristic of general-purpose machines.

4.3.4 How Are Dedicated and General-Purpose Machines Used Together?

In the example we used, you "talked" with the dispatch computer using a dedicated machine, which was a small computer built into the teletypewriter. The dispatch computer was a general-purpose machine, which was able to do a wide variety of things. Even the limited portion of its capabilities that we used while running this dialog was already far more complex than the terminal program in the teletypewriter. What we do not see is the fact that the computer which contains the dispatch computer program also has dedicated machines in its structure. At the distant end of the line, the dispatch computer must also handle ASCII conversion and adaptation of computer signals to telephone-compatible form. The program we are "talking" with is not doing this also at the same time; instead, small dedicated processors attached to the telephone lines are handling these tasks "in the background" while the dispatch computer handles the dialog. It is a little like the way your body is working as you operate the teletypewriter; you aren't thinking about breathing and making your heart contract with each beat while also handling the dialog—those operations go on subconsciously or are handled by local processors while your conscious mind handles the keyboard—in the same way, the dedicated machines in the data communications part of the big computer system handle the subconscious tasks while the main processor does the dispatching.

REVIEW QUESTIONS

1. What are the major reasons for using microprocessors?
2. What is data, and how does it differ from voice?
3. How many different codes can be made using two binary digits? How many using *four* binary digits? How many using two *decimal* digits or four *decimal* digits?
4. Does ASCII code use 7 (or 8) *binary* digits or 7 (or 8) *decimal* digits per character transmitted?
5. Can binary code be used to transmit decimal digits?
6. Are repeaters needed for digital transmissions?
7. Describe the computer-related terms program, memory, and central processing unit (CPU).
8. Name three typical uses for data communications.
9. Would the program of a dedicated machine ever be changed? Why (not)?
10. Is a general-purpose machine used to run more than one kind of program?
11. Would a general-purpose machine be used by more than one person at a time?
12. Why might a program in a general-purpose machine be changed?
13. Why would a general-purpose machine be combined with dedicated machine(s)?
14. Give the 8-bit ASCII code (7 bits, plus *odd parity*) for (a) A; (b) %; (c) 3; (d) g.

5

Private Telecommunications

Systems

5.1 SOME BASIC QUESTIONS

In this chapter we look at the parts of a private telecommunications system, see what they are, and learn what they do. We are going to answer the following questions in this chapter:

What makes up a private telecommunications system?

Who uses a private telecommunications system?

What types of communications are carried on private telecommunications systems?

Who owns and controls private telecommunications systems?

5.1.1 What Makes Up a Private Telecommunications System?

The structure of a private system is much like that of a public system (a BOC—Bell operating company—or one of the other telcos). It is made up of station equipment; this may include voice equipment, typically telephones, and possibly data/video terminals. It is connected by transmission links (lines connecting telephones to switching systems and switching systems to the public network). It is controlled by the switching systems themselves (which are usually located on-premises in private systems). And it may contain optional/peripheral equipment (equipment that enhances the function of the system but is not critical to its performance).

In an organization whose divisions are not all located at the same site, there may be more than one switching system. An example would be a company with offices in downtown Chicago and suburban Chicago, or in Chicago and Milwaukee. The two switching systems can be tied together with tie trunks and calls between the two systems would be handled, from the user's viewpoint, much like calls placed within the system, on-site.

Other equipment that enhances the abilities of the system we will call **peripherals.** These devices could be used for controlling the system, keeping tabs on its activities, or interconnecting the system with other devices, such as paging or dictation tanks. Call-detail recorders, which keep statistics about the number of calls and the duration of those calls, or where those calls were placed, and can also handle billing information, are part of the peripheral equipment that is often found in a private system.

In this section we deal with these four categories:

1. Station equipment
2. Transmission links
3. Switching equipment
4. Peripheral devices

We discuss the different types of equipment in each of these categories, how they work, what they do, and in basic terms, how they are operated. *Note:* Because the PBX (the central switching heart of the private telecom system) requires an overwhelming amount of material to describe its separate features and functions, we will deal with PBX features and functions in a separate chapter (Chapter 6). This will permit us to list in more depth all the abilities of a PBX itself. In this chapter we deal mainly with hardware, defining the overall relationship between the four parts of the system, without getting deeply involved in the details of the more complex parts (the PBX is especially complex).

Until recently, a private telecommunications system was a phone system. Now, any type of electronic communication within an organization and to "the outside" is carried on the telecommunications system. It is a private system because it's carrying the communications for a private organization. The public systems are open to anybody to talk to anybody else who's part of the system. The private system allows its members to communicate with one another without going to the outside system to make the connection. It would be silly, for instance, for someone to send a signal from their telephone to a local office three miles away and back to talk to the person in the next room. It would impractical in a big organization, for every device that carries communication to be wired back to the local telco. The private system is a copy of the public system, in miniature, that provides the same services within an organization that the public system does for its customers. A private system, while it is a scaled-down version of a telco, also has features that are either not present, or not apparent, for the user of the public network. The private system, being smaller than the public system, has more flexibility when we want to change its characteristics. To install data-communications equipment for every user of the public network would cost billions of dollars, and there would not be that much demand for the service. That kind of service isn't warranted in the public network, where most of the customers are consumers without any need to send and receive data. In a private organization, data communication would improve efficiency, cut expenses, and, in a profit-making institution, increase profits. It would be worthwhile, therefore, for the private system to include data communications.

5.1.2 Who Uses a Private Telecommunications System?

Just about every organization uses a private system. Some organizations that use private telecommunications systems are businesses, government offices, hospitals, military bases, colleges, and charitable organizations. If you are an organization, when do you know that you need a private telecom system? When it will save you money, that's when!

For instance, if your organization is spending more on telephone bills than on payroll, there are probably ways to use a private telecom system that will cut your monthly bill.

Improved efficiency is a less tangible aspect of having a private telecom system. Using electronic mail as opposed to carrying the messages through interoffice mail would reduce the turnaround time for a message and its reply from hours to minutes (or even seconds). There would be no need for a staff of people whose job is to convey interoffice messages.

A "Ma and Pa" store or resale shop on the corner is too small a business to need a system like this. A large grocery store would be on the borderline, and a large department store would certainly need a private telecom system. What are the factors that determine who is going to get a system? Cost and efficiency are the primary factors in deciding to use a private telecom system. The Ma and Pa store is a business in one room. Communication between the various elements of the business ("Ma" talks to "Pa") doesn't require any special equipment. In the grocery store, although larger, communications can still be carried on by shouting across the counters or by using a small P.A. speaker system. By the time a store reaches the size of the department store, communications between the various departments is no longer just a matter of shouting over the counter from one department to another. Efficient communication within the premises is going to require a private telecommunications system at least as complicated as an intercom. This is a function of how big, and also how *distributed,* the business is. The Ma and Pa store and its scaled-up brethren are examples of places where efficiency is the primary reason for the installation of a private telecommunicatins system. The same example, however, could be used to look at cost. When Ma talks to Pa, the cost of the communication is nonexistent (as is its value, in many cases). When the cashier sends the stockboy to the produce manager to check the price on a head of lettuce, the time spent carrying the message costs money at the rate of the stockboy's salary. The cost of adding an intercom to the existing telephone might reduce the number of stockboys needed from five to four if there are a lot of price checks done in a day. Communications in the department store would be nearly impossible between departments if there were no private communications system.

5.1.3 What Types of Communications Are Carried on Private Telecommunications Systems?

In an organization that has grown over the years (so that it is now thoroughly overgrown) there may be a number of independent and apparently unrelated communications passing through the same area. The secretaries may have remote terminals for a word processor and electronic mail, and a forest of data lines runs to the processor from all parts of the office. The telephones constitute another forest of wires carrying voice signals from each desk to/from the outside world. In one corner of the office, a security guard sits at a desk studded with video screens watching the entrances covered by video surveillance cameras, while a burglar alarm and fire-detection system runs yet another network of cables from its sensors to the command center of the security network. Paging speakers and intercom might each be independently connected from the speakers to the amplifiers, and so on. In just this simple example, we have six different communications systems living under the same roof but not working together or helping each other in any way. This is a case where everything has been added, one thing here, another there, like a patchwork quilt. The result is a patchwork communications system.

In an ideal organization, these different communications should be consolidated into a single system wherever possible. The voice and data lines would probably be on a common system. The video cables to the closed-circuit TV security cameras would probably have a separate system (a sequencer) at the security desk. Burglar/fire alarms and the security system could be, but usually aren't, part of the telecom system. They are more likely to have their own controller. In most places right now, voice is considered an absolute essential, with data an option. As time goes by, data is also becoming more and more of an absolute essential. We would recommend a telecom system that, at least, combines both voice and data as a prudent preparation for the future. In time, all these separate systems will probably need to be combined into one multipurpose voice/data/video telecommunications network.

5.1.4 Who Owns and Controls Private Telecommunications Systems?

Ideally, the organization that uses the telecom system should own and operate it themselves. This gives them the greatest degree of freedom in deciding how it will be used. We know what ownership is, but what is control? Let's draw an analogy between communications and electrical service. You may buy power from the Edison Company, but your outlets, switches, and appliances don't have to come from the same source. You don't, for example, have to attach an Edison television set to an Edison outlet before they'll let you use Edison electricity. They'll be happy if you pay them for the electricity regardless of what you run it through. By owning your electrical appliances, you can get whatever you want at the best price, as long as it's compatible with the power supplied by Edison. You might get a bargain on a British color TV set, but if it runs on 250 volts and you've got 120 volts, that's no bargain.

In the area of control, you can select any channel you want on the TV set and plug it into power outlets in any room. If you control your own telecom system, you can tailor it to your own needs in a similar fashion. Perhaps you want to change the class of service on a telephone in your private telecom system. The phone has previously been unable to make outside calls (operates just within the organization), but now you want to enable it to direct-dial into the public network. Do you have to call the vendor who sold you the system to make that alteration, or has the system been designed to make that change easy for you to do yourself? If the answer is "We can do it ourselves without calling the vendor," you have control in the same sense that you can change the channel on your TV set without calling the Edison company to do it.

If you own your telecommunications equipment, there are financial and tax advantages, too. You can select the depreciation scheme you wish to employ for tax purposes, and unlike rental equipment, your expenses do not go on forever with equipment you own. You also aren't responsible to anybody for what you do if you want to modify the equipment. Ownership and control are not completely separate concepts. As you can see, you have more control over equipment you own than over rental equipment. You can own your telecommunications equipment, or rent it from someone else who owns it, and to a degree, controls it. Lest it appear that we are beating the drum for ownership of your telecom equipment (because we are), we should mention that there are advantages to rental as well. In particular, equipment that has been depreciated can be sold to an employee or holding corporation, which then leases the equipment back. This *leaseback* arrangement is useful, because rental costs can be deducted from taxes as a business expense.

5.2 A SYSTEM BLOCK DIAGRAM

5.2.1 Composition of a Private Telecommunications System

In Chapter 1 we described the public telephone system in general terms as comprising switches, links, stations, and peripherals. These are also elements of a private system, as shown in Figure 5-1. In describing a private system, we will use the terms "switch," "link," "station," and "peripheral" to refer to categories of system components. In this section we answer these questions about station, link, switching, and peripheral components: What is it (what does it do)? In following chapters we will again see each of these items individually, to answer the question: How does it work?

5.2.2 PBXs—The Switching Component

We will start with the switch component in the private telecommunications system. Most switches in private systems are called either PBXs (private branch exchanges),

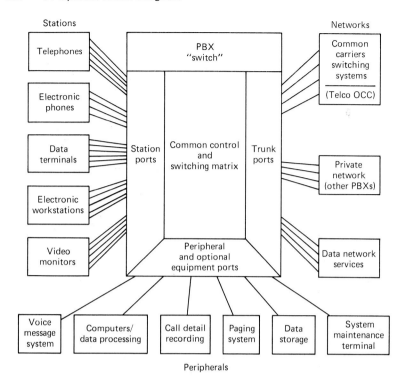

Figure 5-1   Private system.

EPBXs (electronic PBXs), or CBXs (computerized branch exchanges). Generally, the term ''PBX'' is used to identify the switch, whether it is computerized or not. We will use the term ''PBX'' to describe the switching component of a private system.

PBX size

PBXs can be described in terms of either their physical size or their capacity. In general, the capacity of a PBX is how many telephones, trunks, and so on, it can support. You will remember that a trunk is a link to the public network and a line is a link to each individual station (telephone or terminal) in the private system. A small PBX may support 100 telephone lines and 40 trunks or less. A PBX of this size would fit in a large closet. A medium-size PBX, which supports about 400 lines and 100 trunks, would require a 12 × 12 room. The equipment for such a switch would be housed in several cabinets. The largest PBX, which might be used at a university, could have thousands of lines and hundreds of trunks. Such a PBX would require about 1000 square feet of floor space for all the cabinets. Twenty years ago these same switches might have required twice as much physical space. With the miniaturization of electronic components, we should expect to see similar reductions in space requirements in the future.

Port capacity

The port capacity of a PBX is a measure of the number of different devices that can be connected through the PBX. Each piece of equipment connected to the PBX requires a *port*. Each link to the public network, or trunk, also requires a port. Each port requires its own interface circuit designed specifically for the type of equipment. A single-line telephone would require one type of interface, a trunk would require a different type. This interface circuit is generally a printed circuit board that plugs into the PBX.

There are limits to the types of interfaces made to these ports. A PBX with 100 ports, for instance, may use some of those ports for single-line telephones, some for electronic telephones, some for trunks, some for tie lines, some for attendant consoles, some for peripheral devices such as a CDR unit, and some for ports for certain maintenance functions such as a teleprinter for "moves and changes" for a serviceman (see Figure 5-2). The customer or user is generally most concerned with the line and trunk capacity—those are, typically, the capacities that are quoted. The line and trunk capacity can sometimes be traded one for the other. If a system can handle 50 telephone stations—50 lines and 50 trunks—you might be able to "trade"

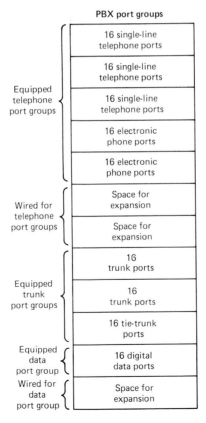

PBX port groups

Equipped telephone port groups
- 16 single-line telephone ports
- 16 single-line telephone ports
- 16 single-line telephone ports
- 16 electronic phone ports
- 16 electronic phone ports

Wired for telephone port groups
- Space for expansion
- Space for expansion

Equipped trunk port groups
- 16 trunk ports
- 16 trunk ports
- 16 tie-trunk ports

Equipped data port group
- 16 digital data ports

Wired for data port group
- Space for expansion

Figure 5-2 PBX block diagram showing ports.

that for 75 lines and 25 trunks. Some capacities are set up so that there is a certain maximum you can't "trade up" beyond; this depends on the manufacturer. One capacity is the number of consoles that can be put on a system. It might be a total of five, but to put five consoles on that system, for each one you add you might have to sacrifice a port of some other type, or even several ports. These are examples of typical trade-offs in system configurations.

In a system that has a maintenance port or a CDR port, which are typically RS–232 ports (for more about this, see Section 10.3.1), there might be one of each. Possibly, in a system that has other terminals used by management in the company to see what's going on in the system, there'll be, perhaps, four other ports to which you connect terminals, and these terminals will give them statistics and data about how the system is being used, or accumulating statistics about how the system is being used that day. There might be a printer tied to one of those ports, or perhaps a terminal.

Bell Labs in Naperville, Illinois, for instance, has as many terminals as it has employees. In a voice/data system like that, there might be as many RS–232 ports as voice ports.

Simultaneous conversations

Another measure of the capacity of a PBX is the number of simultaneous communications that it can handle at once. These could be either voice conversations, or data communications, or a combination of the two. For example, in an all-voice conversation system, although the PBX might have interfaces for 1000 telephones connected to it, the hardware architecture might allow for only 300 simultaneous conversations. With 1000 phones, 300 simultaneous conversations or connections (or paths between stations and trunks through the PBX) isn't too bad. The likelihood of a 301st person going off-hook with their telephone is pretty slim. In the ROLM system, when data is switched on the TDM network, a voice path is broken up into multiple data channels. If the capacity is 300 voice channels in a ROLM, you could have 299 voice channels, and possibly 40 or 50 low-speed data channels in the space occupied by the 300th voice channel. ROLM calls this technique **submultiplexing.** The faster the data is transmitted, the larger the portion of the voice channel which it occupies.

Blocking

One type of **blocking** happens when the system's capacity for simultaneous conversations is reached. All other attempts are blocked. Another form of blocking happens when an attempt is made to place a call or transmission to the public network and all trunks in the public network are in use (see Chapter 11 for more information on blocking).

Expansion capability

Many PBXs can be expanded in size. A PBX with room for ("wired for"), 100 lines and 40 trunks may have circuits installed for (be "equipped for") only 50 lines and 20 trunks. Later, when the business grows, it could be expanded by adding more line or trunk circuits until the cabinet is full, and it can be expanded still further by adding more cabinets, until a maximum expandable capacity is reached. The new cabinets will hold additional line and trunk circuits. This eliminates the need to replace equipment—you just add to it, instead.

Switch location

The room that houses a PBX is often called the **switch room** or the **equipment room.** It is generally centrally located within the organization's premises. The switch room for a large PBX will need space for air conditioning, adequate ventilation, and access around the cabinets. The actual PBX cabinets may take up less than half the available floor space.

PBX function

A PBX, or switch, has the job of connecting and disconnecting the circuits in the private telecommunications system. To do this, it must have some way of telling what source and destination have to be connected. When a number is dialed, some electronic circuitry has the job of making sense out of the pulses dialed-in, and selecting the switching device to route the signal to its destination. Some systems use metallic contacts closed magnetically, called a **relay**. More recent designs replace the moving parts of the relay with a solid-state switching device called a **silicon-controlled rectifier** (SCR). Currently, switching is being done most commonly by "digitizing" the voice signal into data codes, which are reassembled into voice after being switched through logic gates from the source to the destination.

In the **logical switching** variety of circuit, a computer (or more appropriately, a microprocessor) is responsible for "enabling" the various paths through which data will flow. Thus the "operator" who is "connecting" a "transmitter" to a "receiver" is actually a computer circuit. A program gives the computer directions on how to interpret the dial signals and what circuits to enable to make the connection. By changing the program, you can change what will be connected, its class of service, and what features it is permitted to use.

The computer in this system is its **common control.** This means that it is the part of the system that exercises control over which circuits will be connected. This function was formerly performed by relay logic switching control, and later, by random-logic circuits—boards that have a logic circuit that performs only one task. The computer has the advantage of being cheaper and more compact than the earlier relays and random-logic circuit boards and is also more flexible. The program in the system where the computer is the common control element is changeable. It can be altered to change the specifications of the system with much less trouble than re-wiring the random-logic board or relay-logic board.

From the user's standpoint, it won't matter which type of common control is used as long as it works. The only time a user is aware of this common control is when it's a person and the user is asking a human operator to make the connection.

5.3 PBX HARDWARE: WHAT'S INSIDE THE CABINET?

PBX hardware is not necessarily the "nuts and bolts" that you find in a hardware store. The physical components and assemblies that are wired together or plugged into a PBX are its hardware. **Hardware** is a term used commonly around electronic and computer systems to describe various physical components of the system. Most of the hardware of a PBX consists of the printed-circuit cards. Each circuit card has a different function, although many duplicate cards with a certain function may be needed: for example, a single-line interface card—in a system with 20 single-line phones, if each card handles a single phone, the system must have 20 of those cards, which would all be of the same type. Critical circuits such as the computer may be duplicated. If one of the units fails, control can be switched to the other.

Printed-circuit boards

The PBX cabinet is constructed inside according to industrial standard "shelf layout" architecture. This means that there will be a series of shelves with PCBs stacked on them like books on a bookshelf, except that these "books" are held in place by being plugged into card-edge connector sockets located on the back of the shelf (shown in Figure 5–3). The **cards** (PCBs) are slid into the sockets along slots in the shelves and fit into the connector at the back of the shelf. The reason this arrangement is used is easy installation and removal of cards. Enough slots are provided so that a considerable amount of expansion (adding more PCBs than the original design called for) is possible.

Some items that would be found on these cards:

1. **Interface circuits.** These are the cards that contain ports, which permit equipment at different stations to be connected together. Each card may have a single port, or a number of ports may be found on one card.

2. **Common control computer.** This may be two cards, up to an entire shelf of cards for redundant systems. These are the cards that contain the CPU (the central processor) and the memory, which includes ROMs (read-only memories) that "boot" the system. A **boot** is a set of operations that a computer system needs to get going. The "bootstrap" program usually uses a peripheral such as disk to load a more complex program into the RAM. The computer then runs the RAM program to perform all the control functions of the PBX.

3. **TDM/SDM busing circuitry.** These cards permit communication using serial and parallel digital data. TDM is serial data, sent at different times on a single path. SDM are parallel data, sent on multiple paths at the same time. Either the single path for TDM, or the multiwire path for SDM are called *buses*.

4. **Tone generator.** This card produces dial tones, ring-back tones, busy tones, and the holding, flashing, and other tones or beeps that convey information to the user.

5. **Register circuits.** These devices hold numbers dialed in from the stations. Tone

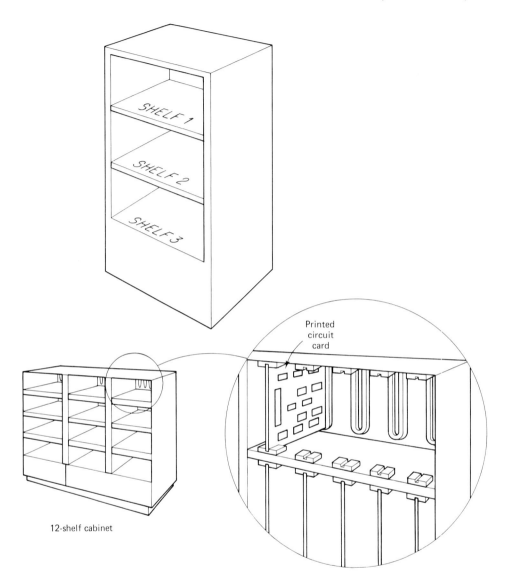

Figure 5-3 Cards in a shelf.

registers take the tones from pushbutton telephone keypads and decode them into digital numbers which are "stacked up" in the register. The numbers are generally stored in the register in the order in which they were dialed.

6. **Power supply.** A power supply does not supply anything, really. What it does, instead, is *convert* one form of electrical energy into another. Generally, the power supply in any electrical device converts AC line voltage to a lower DC voltage. In the PBX, there are a variety of DC voltages converted from AC

to meet the needs of the cards in the cabinet. If the PBX runs from a DC source (say, 48 volts), the power supply now no longer needs to convert AC to DC, but must still provide multiple voltages from a single supply. These voltages (5 volts, 12 volts, 15 volts) are all lower than 48 volts, and are divided down using circuits called *regulators.*

7. **Fuse panel.** This device contains banks of fuses to protect individual supplies, groups of cards, or even individual cards. This panel may also contain circuit breakers and/or lamps to indicate blown fuses or tripped circuit breakers.

8. **Diagnostic indicator lamps.** These display the status of action in circuits found on the cards. These may also indicate malfunctions of circuits within a card, or actions that affect the whole system.

9. **Floppy drives or other magnetic storage media used to record data.** The data may be programs for the common control computer, phone call statistics recorded by a call detail recorder, or even voice messages.

10. **Batteries for memory backup.** These are needed only in emergencies, but emergencies happen. If a power outage occurs, these keep the memory from "forgetting" the numbers it held when the AC disappeared.

5.4 TRANSMISSION LINKS

5.4.1 How Links Are Named

Many types of transmission links are used in a private telecommunications system. These links provide a path for transmission between the station and the PBX. Transmission paths are also needed between the PBX and the public network. There are a number of ways the transmission links are named, and the same type of link may be given different names in different circumstances.

One way to identify links is by the types of equipment they connect together. For example, a **station link,** or station line, connects a station to the switching system. An **off-premise line** connects a station that is not located on the premises of the business or organization to the PBX on-premise.

Another way of identifying a link is by the type of transmission it carries. For example, a link could be a voice or a data link. Usually, when talking of a voice link, you are talking about an *analog* link, whereas a data link usually uses *digital* transmission. Although an analog link can be used to carry digital information, using a *modem,* or a digital link can be used to carry voice, with the help of a *codec,* the name "voice link" implies analog transmission, and "data link" implies digital transmission.

A third way to define a type of transmission link is by the way it functions, either electrically or mechanically. For example, telephones are usually connected to a **loop-start line.** This is a two-wire twisted-pair conductor, and "loop-start" means that when a connection is made to complete the circuit to which the line is attached, then, at the other end of the line, an interface in the switch will turn on. This circuit will signal its "recognition" by returning a dial tone to the telephone. The connection is completed by picking up the telephone (off-hook), and this complete circuit is called a loop. **Ground start trunks** and **wink start trunks** describe links that are used to activate interface equipment in different ways than by attaching a loop to the line.

Thus the name of a transmission link may contain one, or all three, of the preceding definitions. We will describe these concepts in detail in Chapter 10.

5.4.2 Station Lines

The first type of link we'll discuss is the station line, used to connect stations to the PBX. If the station is a single-line telephone, it's nearly always a two-wire, twisted-pair, loop-start line. If the station is an electronic-type telephone, the link is usually a two-pair, twisted-pair line, with one pair used to carry the voice—analog—signal and the other pair used to carry the data from the electronic set to the PBX. The data pair would carry information such as "turn on the ringer tone in the telephone," or from the telephone back to the PBX, it would carry information such as "what line is being selected" if it is a multiline phone. The data pair would also carry information about what function or feature has been activated—the data could tell the PBX that the conversation on that particular phone should be put "on hold" and another dial tone should be supplied so that a second call can be placed. If the electronic station set requires a power supply, a third pair of wires is used to bring power to the station. Data terminals that connect to a PBX are similar to electronic station sets. No pair for voice exists, since there is no voice in a terminal's transmission, but a pair for data, and possibly for power, is connected to the data terminal. The data isn't connected directly to the terminal, but generally to some sort of interface. Typically, the interface will have an RS–232 plug to the terminal.

Station lines are distributed throughout the premises to each individual station. This is shown in Figure 5–4. Starting at the PBX, a group of stations are connected to the PBX with a standard 25-pair, 50-pin *Amphenol connector.* A group of stations will be connected to the PBX using a pair for each station, for single lines, or the two pair needed for electronic stations or terminals. A group of 10 stations might use 10 or 20 pairs of the 25-pair cable. The other pairs remain unused. The Amphenol connector is then coupled to a type of distribution frame located in the same room as the PBX. That 25-pair cable may then be interconnected on that frame to another 25-pair cable that runs hundreds of feet away to another part of the building, or the individual pairs for a particular station might be connected from that 25-pair terminal to the individual wire that runs to the station. Possibly, four 25-pair cables from the PBX to the main distribution frame might be connected to a 100-pair cable that runs to a remote area of the building, and there it might be broken down to 25-pair cables. Those 25-pair cables might run to another section of the building, and individual pairs might be split off and connected to individual stations.

Another type of station link could use **coaxial cables.** Although very few manufacturers use coaxial cable in telecommunications, it is widely used in strictly data networks.

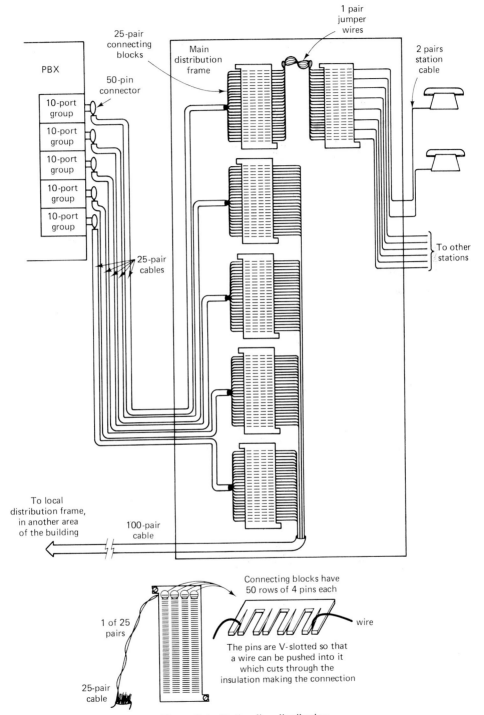

Figure 5-4 Station line distribution.

There are many advantages in using *twisted-pair;* it has become standard in the telephone industry, it's cheaper, more pairs can be fitted into a smaller cable than with coaxial, and twisted-pair is a lot easier to work with. In the future, perhaps, fiber-optic links going to individual stations may replace twisted-pair, but this has not yet been commercially announced. Figure 5–5 shows a coaxial cable (a), twisted-pair (b), two-pair (c), and 25-pair twisted-pair cable (d).

An on-premise extension or off-premise station would also use a twisted pair, although a special interface in the PBX may be required to compensate for the long distance and high resistance of the loop.

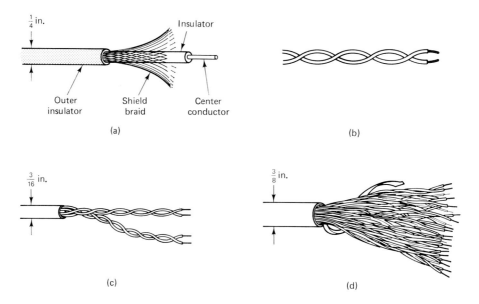

Figure 5–5   (a) Coaxial cable. (b) Twisted-pair cable. (c) Two-pair twisted-pair cable. (d) Twenty-five pair twisted-pair cable.

5.4.3 Links between the PBX and the Public Network

Central office trunks

The most common link between a PBX and the public network is called a **central office trunk.** A trunk is a shared link, and it can be accessed by any station served by the PBX. Central office trunks are voice links, and are usually two-wire (twisted pair), either loop or ground start. Trunks come in many different varieties. People at the telco might call these PBX links lines. People at the PBX would call the central office links trunks. (It's all a matter of viewpoint.) We talked about these in Chapter 1. A central office trunk is a link to the local telco that has an individual telephone number.

WATS trunks

Another type of trunk is the **WATS line.** Physically, the WATS line is the same type of circuit as the other types of trunks, the difference being how calls over that trunk are billed. WATS stands for wide-area telephone service and is a service of AT&T. When they are used for outgoing calling, they are defined in **bands.** Band 1 would allow the caller to call anywhere within the neighboring few states, band 2 would go out farther, band 3 farther still, band 4 even farther, band 5 would be across the continent, and band 6 might serve Alaska, Hawaii, and Puerto Rico.

The users of a WATS line are billed for the time they use the line, not for how far they call. If they call the area that their band of WATS service handles, they are just billed a certain amount for the time that line is used. Another type of WATS line is used for incoming calls **(inWATS).** Usually, this is called *800 service.* The person who receives the call is billed, in this case, for the amount of time the line is used. InWATS calls are billed to the called person; someone dialing an inWATS number is not billed. This is often advertised as "call toll-free," which, of course, is not exactly free—someone has to pay for it.

FX trunks

The foreign exchange—usually called an **FX trunk**—is very similar to a local central office trunk, except that it is to a distant city or area. A PBX located in Chicago might have a foreign exchange trunk going to New York. Let's say that an organization had an office in New York with a local number; then they closed the office and moved to Chicago. The people in New York could still call that local number and would be charged for a local call. The line would actually be going to Chicago, but the New Yorkers wouldn't know the difference. For outgoing calls, Chicago callers to New York would use the FX trunk, and they would be charged for local New York call, although they are actually making a long-distance call. Of course, it will cost the user in Chicago quite a lot for the FX trunk. Usually, you would have to have a lot of traffic going to New York to make that kind of service worthwhile. The charges and billing would be similar to what we discussed in Chapter 1, where we suggested that the Harts Co. might be better off using a WATS line rather than the specialized common carrier. If the amount of traffic and the charges were comparable to having their own dedicated line, Harts could consider having an FX trunk to Milwaukee.

Trunks can be ordered from common carriers as either inbound or outbound, or two-way. InWATS (or 800 or toll-free numbers) and outWATS are exceptions. Right now, you cannot get both-way WATS lines.

All of these types of trunks are connected by the local telco and billed to the customer depending on the time they are used. There are two methods for billing: one is on a time basis—the amount of time a trunk is used; the other is by a **message charge.** Depending on where the call is placed, so many **message units** are charged. A short-distance call might be one-message-unit charge, which would be a certain number of cents, and another longer-distance call might be a couple of message units, which would cost a multiple of the basic message-unit charge.

Tie trunks

Tie trunks, sometimes called **tie lines,** are used to connect a PBX to another PBX. More generally, tie trunks connect one switching system to another. Tie trunks allow communication between one PBX and another. Stations served by PBX A can call over the tie trunk to PBX B to any station connected to it. Tie trunks differ from central office trunks in that there is a flat charge for a tie trunk, usually on a monthly basis. It doesn't matter how long you're using that tie trunk, or how many calls you make.

Tie lines are either two-wire or four-wire links. In a four-wire link, transmission takes place on a separate pair from the "receive" pair. We will describe this in further detail in a subsequent chapter.

Another function of the tie trunk is to provide supervision and signaling between the two PBXs. This is done on another pair. This signaling pair is called an **E&M signaling pair.** Thus a tie trunk can use two-wire E&M signaling or four-wire E&M signaling, which would require either two pairs or three pairs. The type of signals that go over the E&M leads or E&M pair are used to instruct the distant PBX to send a dial tone and wait for digits, which would be the number of the extension you're calling at the distant PBX. When the distant PBX receives the request, it would send back a signal over the E&M leads to the originating PBX that it has received a signal requesting dial tone. The distant PBX then waits for digits to come over the tie line. Those digits would be the number of the called station at the PBX. Either rotary or Touch-Tone® dialing could be used. Rotary pulses could be sent over the E&M leads, and DTMF (Touch-Tone®) would go out over the calling PBX's transmit pair, and of course, received at the distant end on the distant PBX's receive-pair (assuming, of course, that this is a four-wire trunk).

There are a number of different types of E&M signaling arrangements, and ways they function. We cover this in Chapter 10.

5.4.4 Digital Transmission Links

The tendency in today's PBX designs is for the PBX to switch digital as well as analog transmissions. In fact, analog will probably be changed to digital before it is switched, in modern designs. Since the trend in today's PBXs is to switch data as well as voice, we need to discuss digital transmission links. If, in fact, a digital link is used, the digital station equipment (terminals) in the system are connected to the PBX without using modems, and the analog station equipment (telephones) will have codec equipment to "digitize" the voice signal into digital code. Between the PBX and the public network, it is also possible to have a digital link. Systems are now available which have **digital ports,** each of which will switch up to 56K bits per second (bps) of synchronous data. Between the PBX and the public network it is possible for data to be transmitted using a **T1 span,** which is capable of 1.544 million bps. Data are multiplexed onto this type of link, allowing it to carry many channels of information. The tie trunks we just talked about may be replaced by a T1 span.

T1 spans

A T1 span would be used between two PBXs or between one PBX and a central office. This T1 would perform the task of a multitude of analog trunks. Up to 24 voice channels could be carried on one T1 span, whereas a single tie trunk carries only a single channel. Instead of 24 voice channels, the T1 span could carry a great many more channels of data from data terminals. The reason for this is that the voice information (that makes up sounds and words) contains a great deal of redundant information, while digital information is a lot "leaner" and can carry more information with fewer bits. Often, many digitized voice and data channels are being transmitted on the same T1 span at the same time. Once the voice in an analog channel has been digitized, it can be multiplexed in with other data from other types of channels without the T1 span being any the wiser. As long as the ultimate information being multiplexed together onto the T1 span is digital, its origin makes no difference. Of course, the multiplexed information must be demultiplexed when it is received, and the voice information must be converted back into analog form, so it *is* important that the equipment at the transmitting and receiving ends be able to distinguish a voice channel from a data channel. While it is transmitted, however, everything is digital. T1 spans can be obtained from the common carrier. We suspect that even the cable video networks may offer T1 spans, or something similar, as an option in the future.

Usually, a PBX is connected to its station equipment with an analog link, originally set up for telephones. When a terminal is used, a modem is employed to connect the terminal to the analog links. Now PBXs are available with digital ports to accept a digital transmission between the station and the PBX. In that case, if the station is a telephone, we would use a **codec** (coder-decoder), mentioned earlier, to convert the analog signal into digital code. Through the miracles of multiplexing, that same digital link that's going to that telephone could handle all the terminals and telephones in a small office. For example, suppose that the telephone has an RS–232 plug built into the back, where you could plug your terminal in. It bypasses the codec in the telephone, attaching the terminal through a multiplexer into the digital link. You could be talking on the telephone and using the terminal at the same time, if you're good at doing two things at once. The digital link—perhaps the 56K-bps port mentioned earlier—can handle two things (or even dozens!) at once without any trouble. With the 56K-bps transmission, you could find one channel of voice at 48K per channel, and several data terminals at 300 baud, or an RS–232 terminal and an RS–232 printer, all attached to one digital station's line.

Between the PBX and its stations, the digital link might be 56K-bps digital, and between the PBX and a T1 span coming in from outside, the digital link might be 1.544 Mbps. The ports where the trunks are ordinarily connected to a PBX may be designed as analog input/output ports. If this is the case, a device that converts the T1 span's digital into 24 channels of analog might connect it to 24 analog ports on the PBX, only to be digitized again into several 56K-bps channels inside the PBX, so that the voice information can be switched to various stations in the system. At the stations, the voice information in digital form would again be converted to analog in order to be heard. If this seems a tad wasteful, it's only because it *is*. For this reason, many manufacturers are designing PBXs to have T1-compatible ports.

5.4.5 High-Speed Transmission Media

As well as wire links, other means of communication—other **media**—could be used to carry the signal. Microwave and infrared are wireless links that permit information to be sent on a "tight beam" from one place to another. They have the advantage of not requiring crews to string wire across the countryside, or poles to support it, but have the disadvantage of having the signal interrupted by the occasional bird or airplane. Of course, yet another interface converter is needed to render the electromagnetic signal into electrical form. Both microwave and infrared can be passed from source to destination using waveguides—in the case of the infrared signal, the waveguide is called a **light pipe** or **fiber-optic link.** In these cases, the waveguide may either be strung like wire, or placed underground. The same mechanical procedures are needed to pull waveguide or fiber optics, compared to stringing other kinds of wire, except that the increased bandwidths available in the case of microwave and fiber optics make the waveguide lighter than the electrical conductors needed for an equal number of channels at lower bandwidths. Fiber optics, being silicon dioxide, a glass-like substance, are also immune to most forms of chemical corrosion.

The transmission on the microwave or infrared beam could either modulate the light or microwave with many different channels having different frequencies, as is done to separate the channels of radio or TV, or the microwave might, for instance, be modulated using T1 format to carry time-division-multiplexed channels, and be demodulated directly into the PBX's T1 interface.

The high frequencies employed in generating the microwave signal or infrared signal are so many times larger than audio (voice) frequencies that a great many channels of voice can be carried on one of these links. For instance, suppose that a microwave link exists that can carry frequencies from 10 million kilocycles to 20 million kilocycles. In a typical voice channel, less than 3 kilocycles of frequency are used to carry all the information in a spoken message. Using the microwave link just mentioned, we have "room" in the frequency **bandwidth** of from 10 to 20 million kilocycles, for over 3 million voice channels at 3 kilocycles apiece. Nobody's going to use that 3 million channels on one waveguide. First, there will be "dead space" between channels, so that they won't overlap by accident. That will reduce the available number of channels. Second, the demodulator or demultiplexer required to extract those 3 million channels from the multiplexed signal would be beyond anything a telco would be likely to bother with. There are not likely to be any two places in the world with 3 million calls going between them at any one time. Infrared has even higher frequencies, and thus capacity to handle even more channels. Infrared frequencies are in the neighborhood of 100 billion kilocycles or more. That could be divided up into 30 billion voice channels. This capacity is unlikely ever to be fully used on any real fiber-optic link. Infrared "broadcast," that is, transmitted without a fiber-optic light pipe, is unlikely to be used much because of distance limitations and interference.

5.5 STATIONS

A **station** is the place where a communication has either originated, or terminated, or both. It could be a telephone, a computer terminal, an answering machine, a dictation tank, or a picturephone station. There are a number of different types of station equipment on the market. Some are manufactured by PBX manufacturers to be compatible only with their equipment. Others are more-or-less standardized and can be used with any PBX or telco's system. Until recently, the only type of station equipment associated with the PBX was the telephone. There are a number of different types of telephones, some electronic, as we mentioned—some are digital and some are analog—some have electronic "smarts" built in, while others are relatively "dumb"—there are phones with rotary dials and phones with DTMF (Touch-Tone®) dials—there are phones that digitize voice and send it out on digital links, while most phones transmit and receive analog signals on analog links—some phones even have memories and displays.   *Home phones all analog?*

5.5.1 Station Interfacing

A station has to have some way of signaling to the switching system to whom it wants to connect. On the telephone, this is done with a dial; on a terminal, this can be done through the terminal's keyboard. The output of most stations is usually compatible with either a twisted pair or two twisted pairs of wire connected to the switching system. The switching system would then have an interface which can reformat the type of transmission or actually change that transmission to something compatible with the switching matrix of the system. In a digital switching system—which is used with a standard analog telephone—the interface needs to change the analog signal into a digital signal. This is done with a coder-decoder (codec), which is a card or circuit in the switch. If the station is a terminal, the interface needs to change the digital format from the terminal format to something that's compatible with the switching matrix. This could be done with a data terminal interface at the terminal. This is the most likely place, since most terminals are RS–232, which is a standard that doesn't lend itself to transmission over any great distance on a standard cable, and requires more than one pair of wires. A data terminal interface at the terminal requires that only one or two twisted pairs are required to get to or from that terminal.

5.5.2 Electronic Workstations

Electronic equipment manufacturers today are developing what they call **workstations,** or **executive workstations,** that are a combination of telephone, data terminal, and personal computer. These usually have either a telephone feature button section and then a regular terminal-type keyboard, or they use a terminal-type keyboard with various mode and function buttons to select use as a telephone dialer, terminal, or what-have-you. They have large enough memory to be used to store personal phone directories and access messages to the station if the PBX has some sort of electronic mail system. They might have a built-in speaker phone and a directory of frequently called numbers for simplicity of calling. They might store customer information for a sales-type person: addresses and other types of information. They can also be used to access databases of other computers or send messages to terminals that are connected to this system.

Other features might include multifunction calculator, clock-calendar, an appointment calendar, and anything else with which you might think of to program a small computer for the convenience of a businessperson. The workstation usually has a display that shows a few lines, or a page, of text. It may use a video display, light-emitting readouts or LCD characters similar to the numbers on a digital watch. Since these workstations are computers in miniature, they have the capability of memorizing and dialing station or phone numbers, but in the data-communications world it is also necessary to enter passwords and code numbers to gain entry into most dial-up systems. This "litany" of routine log-on procedures can also be memorized in the memory of the workstation. The workstation can now carry out the routine part of the log-on procedure for you, and save you the work of manually typing in station numbers, routine codes, and passwords. The use of passwords and codes in gaining access to a computer system from your workstation is one side of a two-sided coin. You would probably have confidential or business-related information in your database stored in the workstation—information to which you would not want someone else to have casual access. This means that your workstation would also have a layer of "security" between the operator and the data, usually in the form of code words and passwords not generally available to others familiar with the operation of your workstation or terminal device. The routine functions of the workstation, such as the clock or calculator modes, would, of course, be available to any user without need for entry of codes or passwords.

5.5.3 "Smart" Telephones

Other new releases for "smart" telephones include programmable capabilities, and a small display that shows one line, or a few lines, of text. This telephone would also include a speakerphone and a number of buttons for multiple-line appearances, something like a big keyphone. These buttons could be used for special functions: calling frequently called numbers, transferring calls, conferencing, putting calls on hold, saving numbers, repeating numbers, and multiple-feature functions. These electronic phones could also have another plug in which you could plug a standard terminal to make them become more like the executive workstation. Typically, these things are made compatible with the largest of the computer companies' terminals, such as DEC and IBM standard terminals, so you could have an electronic phone with a lot of smarts, and it would have a plug in the back to allow you to plug in a terminal or a printer device. With a terminal and certain "smarts" built into the electronic phone, it becomes difficult to distinguish an electronic phone with workstation-type features from a workstation.

5.5.4 Home Office

One conclusion that the development of these telecom workstations leads to is that you will be able to use such equipment to work out of your home as easily as out of an office. With a digital link over the public network, you can work out of your home and still have access to everything that you had access to at the business. Anybody with a personal computer and a modem has the capability of doing this right now. It is merely necessary to have the right software to load into the personal computer, and a lot of patience, since most modems are designed to run at much slower data rates than specialized workstation equipment is capable of. (You can't normally use your phone for voice and data at the same time.) If you want to use the specialized high-speed terminal or workstation equipment in your home, you can obtain a higher-speed data-conditioned line from the common carrier. This line would have the capacity of running the terminal and printer at speeds about 60 times as fast as the standard modem-and-voice-line combination in use with personal computers. Depending on the capabilities of the system to which you have connected, you may be able to run the terminal, printer, and talk on the phone all at the same time.

In his book *The Third Wave,* Toffler describes the "electronic cottage," where no one goes to work anymore (because gasoline costs $112.95 a gallon?); they just stay home and take care of all their business, including shopping and everything else, by telecom hookup.

5.6 OPTIONS AND PERIPHERALS

The reason we are placing selected items into the following section is that the items we have chosen are not yet found on all PBXs. In some cases, they are stand-alone devices that connect to a PBX, rather than cards found inside the PBX. We can't promise that the options of today won't be standard equipment tomorrow. One manufacturer's option is another's standard feature.

5.6.1 Modem Pools

A **modem** is a device that converts digital signals from voltages that are merely ON or OFF into tones that represent these two states of a digital system in analog form. Transmission of digital information using a modem is called **modulation,** and involves conversion of the digital signal to analog form. Receiving digital information via a modem involves conversion of the tones back into digital ON and OFF states, and is called **demodulation.** From the words "modulation" and "demodulation," we get the name "modem."

In ordinary voice communication, there usually are more stations (telephones) than there are **trunks** to the public network. This is okay because only a fraction of the stations are making "outside calls" at one time. There is a **pool** of trunks from which the PBX may select any one for you to "dial out" on. Similarly, the digital stations (terminals) have to "dial out" from time to time. There are a number of modems in the modem pool from which the PBX may select one to let you "dial out" on your terminal. The modem pool in a digital system of terminals stands in an analogous situation to the trunks used to link voice stations to the public network.

When you have a PBX with the capacity to switch data, digital signals can be routed between terminals through the system, without modems. If the PBX doesn't have the capacity to switch data, there will have to be a modem at each terminal. If your PBX *does* have the capacity to switch data internally, but you don't have a digital line to the outside, you need a modem pool (or at least one modem) to send digital information over the public network.

5.6.2 Battery Backup

In case of a power outage, some telecommunications systems have battery backup to provide power for the duration of the blackout. You might wonder why the telecommunications is considered so important as to be kept going when the rest of the electrical appliances are shut down. In an emergency, the ability to communicate is more than just a convenience—it may be a matter of life and death. What must be kept going? For how long? At what expense? We have to decide whether we need the telecommunications system in a power failure, and for how long we can afford to keep it going. The longer we want to run off batteries, the more batteries we will need and the more it will cost. Some parts of the system that are running under normal conditions are a "luxury" in a blackout. Others may be considered "necessities." "Music on hold" is a luxury; you could get along without it easily. **Bypass phones** to the public network (powered from the central office, which *always* has battery backup) are a necessity. What caused the blackout? Is someone hurt? Is emergency help needed? Is there a fire? These concerns require a link to the public network at all times, even if power is lost.

Some equipment will not run directly off DC (which is all we can get from batteries). In this case we need **inverters,** which make the DC into AC, for running standard 110-volt AC gizmos. As much as possible, we want to get equipment that runs off DC if we plan to have a battery backup system. Whatever we can't get that will run off DC must get power from the inverters.

In a hospital environment, where virtually nothing can afford to be shut down, we need emergency generators. These supply power to all parts of the hospital after there has been a blackout, but battery backup would still be required. It takes time to start the generators, and in digital systems especially, data would be lost unless something maintained power between the beginning of the blackout and resumption of full power under the generators.

5.6.3 Paging Unit

This is probably one of the most used options on a telecom system. The paging unit is connected to a port of the PBX. Anybody using the system could access that port and "make a page" by dialing either an access code or an extension number. This port could be designed specifically for a paging system, or a trunk or an extension port may be used. Often, a building is broken up into **page areas,** several areas that could be paged independently, each having a separate number. An **all call**—which would page into all areas—is used for such things as an emergency announcement, or one for everybody at large, or to try to locate a wandering employee when nobody knows what department to look in.

A hospital, for instance, wouldn't have a regular page going into the surgical areas, but would have emergency pages going into those areas, and there might be a separate page for the doctors' lounge, which would be tried first before bothering the whole hospital environment trying to find a doctor.

Most often the paging amplifier is a separate unit made by an amplifier company. The only thing the manufacturer of the telecom system supplies is a port, and often that port is no different from any other extension or trunk port, as we mentioned. We've seen manufacturers of key systems supply a **page card** a circuit board (card) that has an amplifier built into it, and it will drive perhaps five or ten speakers with a total output of perhaps five or ten watts. Again, some PBXs may have this feature, but not all do.

The **meet me** page system is a system in which someone dials into the page system and makes an announcement. The person being paged picks up a phone, dials into the system, and is connected to the person making the page. This feature makes it unnecessary for you to get the extension number or name of the person paging you. All that is necessary is that you know you are being paged and can dial in to whoever needs you.

5.6.4 Recording Units

In Chapter 6 we discuss automatic call distribution groups and call detail recorders. These, together with voice-message systems, may utilize tape recorders or disk units, which record information on magnetic materials, called **magnetic media.**

There are several types of devices that are used to tape-record information for the optional part of a private telecommunications system. First, there are **cassette recorders;** the inexpensive kind you would use for music, or to record programs from an inexpensive home computer can be used for some options in a private telecommunications system. More elaborate **cartridge recorders** use tape with greater capacity but similar in nature to a cassette. The cartridge is generally larger than a cassette, but is still a sealed unit that can be popped into a slot for recording or playback.

Second, there are **reel-to-reel tape recorders,** which have open reels of tape. While personal use is usually restricted to tape $\frac{1}{4}$-inch wide, and only one- or two-channel

(monaural or stereo) information is recorded for hi-fi purposes, commercial recorders
that use reel-to-reel tape can have many more channels, used for recording either
analog or digital information. Nine-track (nine-channel) recorders are commonly used
for recording digital information. Older digital systems often used seven-track (seven-
channel) recording tape. In both cases, the tape is about a $\frac{1}{2}$-inch wide. Some types
of cartridge recorders use multitrack recording, too, although cassettes and minicas-
settes usually record on one (monaural) or two (stereo) tracks.

Disk recording media use either **floppy disks** or **Winchester disks** (also called
hard disks) to record magnetic signals. These signals are generally in the form of
digital pulses. The recorder for ACD (automatic call distribution) statistics may be
a "floppy" and the statistics for call detail recording might go onto a tape recorder.
Voice-message recordings are generally digitally encoded, and recorded on a floppy
disk as digital code. There is no reason why voice messages could not be recorded
in analog form on an ordinary tape recorder, but digital recording on disk makes
swifter access (finding the message) possible than with analog tape. With digitized
voice-on-disk, it also becomes possible to store reference codes identifying who sent
the message and the intended destination—the person or group to whom the message
should be delivered—together with the message itself. Of course, an ordinary telephone answering machine is a simpleminded voice-message system of sorts, and it
does record the messages on tape with ordinary analog techniques. In a telecommunications system, the recorders would be attached to one or several of the PBX's interface **ports,** where the device affixed to the port would be regarded in the system block
diagram as a peripheral device, but treated by the PBX as just another extension
or trunk.

5.6.5 Dictation Tanks

A **dictation tank** unit is a glorified recorder that you can dial into to leave dictation
for the secretarial pool to type. The DTMF dial may be used to control the tape to
rewind, playback, or edit the tape, much like an ordinary dictation recorder. Some
units can take a stack of 10 cassettes and record on them one after another. Others
work from a turret or carousel/slide-tray type of arrangement. Either standard cassettes or larger tape cartridges may be used. After the tapes are filled up, the stenog-
raphers pick up the tapes and type them. There may be more than one of these recorder
units in the dictation tank, or a unit with more than one port, capable of recording
several voice channels simultaneously.

5.6.6 Maintenance Port (Maintenance Terminal)

Another piece of equipment that is often found on a PBX is called a **maintenance
port** or a **maintenance terminal.** It allows the telecommunications manager to change
parts of the programming of the system. Some of the abilities provided include renum-
bering of extensions, changing the system forwarding conditions of a telephone, and
changing the class of service of a telephone.

It is possible for the telecommunications manager, using this option, to give different features or abilities to a telephone or possibly to a terminal. Although the features available to a terminal or telephone user are usually defined by its class of service, the telecommunications manager can use this option to change the contents of a table of features (in the memory of the control computer) that is assigned to a particular class of service. To illustrate how this works, consider class of service 1. Ordinarily, the only thing phones with this class of service can do is call other extensions within the system. Class of service 4 might have this feature, and "conference," "transfer," "save and repeat," and also be able to make local calls. From this control terminal or maintenance port, the manager might be able to change that group of abilities in a group of phones or a particular class of service, or even create a new class of service with functions and features for a new group.

On the maintenance side of this option, most systems have some type of **self-diagnostic software** built into the operating system of the control computer. When troubles show up, this option would turn on an alarm, possibly on the console or operator's board, and through this maintenance terminal or port, you can question and run tests on the system to localize the troubles. The alarm message might tell you which card has failed or which extension or terminal has failed a test.

Another type of maintenance port or management port could be used for accessing traffic information or statistics that are kept on the usage of the system. This option could be available in addition to a call detail recorder, or it could be a separate function on its own. On the *ROLM* system, for example, these maintenance ports use an RS–232 interface, and the maintenance port a service technician uses can access to just about anything the system can do; a port that's used by the telecommunications manager would be limited to the listing abilities of system traffic statistics. It is possible that the manager might be able to make "moves and changes" through such a port. This would mean that the manager could reassign extension numbers and change the class of service to which the extension's user is entitled. The manager's terminal might have this capability either through a special **hardwired** connection, or the manager might access these special powers through a special set of passwords entered from the terminal or attendant console.

5.6.7 Teleconferencing

Teleconferencing is a feature that is growing more popular with business users all the time. In business and education, this combination voice/video service provides the ability to see and hear the calling parties. A committee composed of members from different cities can meet without the expense of flying all the members to a common conference room. In sophisticated systems, each conference table at the various sites in the conference is equipped with several microphones and is in sight of several cameras. The loudest voice signal from a microphone switches on the camera aimed at that microphone. This gives the effect of a whole camera crew dollying the camera around from speaker to speaker so we can see who's talking without actually requiring the personnel to run such a system.

5.6.8 Security Systems

It is possible to tie the main guard station in an organization into the telecommunications system so that voice and video functions (surveillance cameras) are monitored by the same PBX. The functions of the common control computer in the PBX can be used to switch cameras to the guard's video monitor, and to route calls to the guard or from the guard to outside authorities.

5.6.9 Call Cost Accounting

Call cost accounting is a system for keeping track of how much money is being spent on communications. This could be done on an organization-wide basis, or, more often, separate accounting can be done for each department and billed to that department's telecommunications budget.

5.6.10 Environmental Monitoring and Control

Sensors and actuators can be attached to ports of the PBX or optional equipment attached to those ports. The sensors would be used to input such conditions as temperature, time, status of fire, and intruder alarms into the system, and the control computer would activate such actuators as heating and air-conditioning systems, alarms, automatic telephone dialers, and the like, to respond to the inputs from the sensors.

Early systems that did this kind of monitoring and control were developed by Honeywell Corp. and others and were stand-alone control systems. Now this type of monitoring and control, including printed report generation on data terminals and printers, is sometimes included as an option in the telecommunications system.

5.6.11 Other Applications: Some Examples

The private telecommunications system may be outfitted to carry out tasks Alexander never thought of. Once the lines are in place, signals may be sent along them to control all sorts of things from remote locations throughout the workplace by using your telephone or terminal as a controller. Following are some examples:

1. Someone pushes the doorbell button and is linked into the telecom system automatically. Once you speak to him, and determine that you want to let him in, you dial the number of the door lock, and the door is unlocked.
2. In a hotel, cleaning personnel dial a code on the house phone after a hotel room is cleaned and put in order, to indicate that the room is ready for occupation.
3. Electrical appliances can be turned on and off remotely, such as turning lights on, coffee machines, and so on. This is done primarily by controlling connection of power to electrical outlets.
4. Voice-synthesis units could be switched onto the paging system automatically

to carry emergency "alarm" messages, such as which stairwells are safe to use in case of a fire, or what routes of escape are best to take from a particular section of a building.

REVIEW QUESTIONS

1. What are the four parts that make up a private telecommunications system?
2. What organizations use private telecommunications systems?
3. Do all private telecommunication systems handle voice *and* data?
4. How would video probably be transmitted in a private telecommunications system?
5. Name two advantages and two disadvantages for organizations that own and maintain their own telecommunications system.
6. What would be the minimum port capacity of a PBX with 100 stations?
7. What requires a port in a PBX?
8. Describe two ways in which a call could be blocked in a PBX.
9. What is the difference between wired-for and equipped-for capacity on a PBX?
10. What is used for the common-control element of PBXs currently being designed?
11. What does hardware refer to in a PBX?
12. What is the purpose of a tone generator?
13. What types of memory could be used in a PBX?
14. What are three advantages of using computer software logic instead of hardwired logic in a PBX?
15. What are three attributes by which a transmission link may be identified? (*Example:* WATS, ground-start, trunk.)
16. Give examples of five types of trunks and what they are used for.
17. Is a T1 span an analog or a digital communications link?
18. What are the advantages of high-speed transmission links in multiplex transmission?
19. Can a personal computer with a modem be considered an electronic workstation?
20. Why is battery backup needed on some systems?
21. Name three applications where recording units are used.
22. Give some examples of peripheral devices.
23. Why would you want to have a maintenance port in a PBX? What would it be used for?
24. Name three functions that might be found on cards in a PBX cabinet.

6

PBX Features and Functions

The primary purpose of a PBX is to connect stations to other stations and to other networks, either public or private. There are, however, many other functions performed by a PBX. These functions were developed to control or reduce costs, and to improve efficiency in a business or organization. Many of these special functions are optional. The special functions and features may require changes in the hardware (adding cards or circuits to the PBX) or software (changing the program in the computer to do what the special functions require).

PBX manufacturers have a special name for each function. These names are not standardized. Some manufacturers may use the same name for similar functions, but they may not operate in exactly the same way.

The functions that are developed for the people who use the telecommunications system are called **user features** or **station features**. Functions that improve the efficiency of the telecommunications system are called **system features**. An example of a system feature is the ability of the PBX to select the least-cost route for a phone call going out to the public network, perhaps with a WATS line, as opposed to a direct-distance-dial line.

The most commonly used station and system features are often included as standard equipment by the manufacturer. The station feature "call transfer" gives the user the ability to transfer the call to another telephone within the system. Transfer has become a standard station feature. A system feature that keeps records of each call that is made is available on all PBXs, and is always an option. In this chapter we look at the most useful station and system features.

Anything that a PBX does that goes beyond simple call and receive is a feature. When we speak of **station features**, we're talking about something that someone who

118

uses the system does. When we talk about a **system feature**, we're talking about something that's going on everywhere in the system, and it's likely to be "transparent" to the user; users who are taking advantage of a system feature may not be aware of it.

Since PBXs are used by a variety of different types of organizations, which need different features depending on the nature of their business, manufacturers have put together "feature packages" that answer these needs for certain businesses. For instance, a "hotel package" or a "university package" would contain features tailored to those types of organizations. This would make it easier for the manufacturer to define a system, because for certain common types of organizations, you could order a package and get all the features needed in your business.

Most feature packages are software—computer programs you would buy and load into the system. It's possible to buy either a feature package with a group of programs to implement multiple features, or programs could be purchased for each feature individually. An example of this would be automatic call distribution (ACD), a system feature that is available for most systems, and purchased separately.

In the first section of this chapter we look at station features (user features) that someone at a phone, terminal, or console might use.

6.1 STATION FEATURES

When you want to access a station feature, you must have some way to tell the system what that feature is. When you are using a featurephone or electronic phone that's relatively simple, there's usually a separate button—with a label for that feature— that you press. When the button is pressed, the phone then signals to the system, identifying that feature. These signals may either be a series of tones, or digital data. If you are using a standard single-line telephone, you must dial an access code for that feature. Each feature would have a different access code. For example, to put a phone in the "do-not-disturb" mode, you either press a separate button labeled "do not disturb," or with a single-line phone, you would just dial an access code such as #5 for the "do not disturb" feature.

Some features need to be activated while you are engaged in a conversation, or at least connected to a line that's ringing a distant phone or another phone in the system. In this case, the PBX must first be told that the conversation in progress will be interrupted, and that a feature will be accessed. This is done by using a "flash" button, or doing a "hookswitch flash," before dialing the access code (with a single-line phone). This isn't necessary on electronic and featurephones, because these phones are "smart enough" to do this for you, if necessary. With electronic phones, since the signaling is on a separate digital path, it's not necessary to suspend the conversation to call up a feature.

When you "flash," the conversation or line you were on is put on hold and a new dial tone is returned. When the new dial tone is on the line, it's telling you that the system is ready to receive the access code. This is just like the ordinary dial tone telling you that the system is ready for you to dial a number when you first pick up the phone. The reason this is necessary is that the part of the PBX which registers the numbers of the access code you're dialing is connected to your phone only when you first pick it up or when you have a dial tone. After the connection is established, that part of the PBX "drops off" and is "free" for some other phone to use. With electronic phones, this register is connected to the line at all times. When using the hookswitch to flash, the timing is sometimes critical. Many users have trouble getting a "feeling" for flashing. If you hold the hookswitch down too long, the system sees that as a "hang-up" and disconnects the call. If you don't hold the hookswitch down long enough, the system will ignore it, and the only thing that will happen is that there will be a click, and you will be connected back to your conversation. On phones that have a flash button, this timing is electronically controlled so that it doesn't matter how long you hold the button down. It will activate a circuit that controls the interruption of the line for a preset period (the correct length of time for a flash). Some PBXs do not require the hookswitch flash, but do require a flash button. In one type of system, flash buttons usually open and close the circuit briefly. In another type of system, the flash button momentarily puts a "ground" on one side of the line so that it doesn't disconnect the call in progress but the PBX "sees" the ground on one side of the line as a signal to put the call in progress on "hold" and wait for the access code to be dialed. With this kind of flash button, the chance of losing the call by holding it down too long is eliminated. With the "old-fashioned" rotary-dial phones, flashing may not be necessary. It is, in some systems, but the rotary phone does open the line for a brief period of time, and this tells the PBX to await an access code. In the case of a featurephone, when the feature button is pressed, if a flash is necessary, the phone will send that flash signal and then send the tone signals to the PBX in the right sequence. Featurephones that are "universally" adaptible to many systems are programmable so that the feature button can be programmed to send the right access code for any system. Electronic phones are typically made for a specific PBX, by that PBX manufacturer.

Besides the hold and transfer features (which are the most common station features available on PBXs), you can expect a PBX to do some, or all, of the things described below.

6.1.1 Conference Calls

You are able to talk to more than one other person at one time. You begin by dialing one number, and when the person answers, you flash, if it's necessary, depending on the type of phone you have, and then you dial another number. When the second person answers, you flash again (if necessary, depending on the type of phone), and you key-in the access code for a conference or hit the conference button on a featurephone. You can continue this way, adding people as you go.

A common option is the ability to drop off the last person dialed. This is useful if, for instance, the last one who answers your call is an answering machine which says that your party is not available. You would certainly want to drop it from the queue without being "hung up" (ending the entire conference call—forcing you to make the earlier calls all over again). This seems worthwhile in terms of time, money, and not least, embarrassment.

6.1.2 Call Forwarding

You can make calls to one extension ring (and be answered) at another designated extension. Most systems have two types of forwarding. In the first, often called **system forwarding**, a program in the system routes calls in a predetermined pattern that is not changeable by the user. In system-forwarding, which is hard-programmed into the PBX system, if the phone rings for a predetermined time, the call would be routed to another extension. If the phone is busy, the call might be routed to the same extension, or to a completely different destination, depending on the program. The two different conditions (called ''ring—no answer'' and ''busy'') would give rise to two different courses of action by the PBX, or perhaps the same course of action. Another condition for system forwarding is the origin of the call—whether internal or external—if someone within the building is calling, internal system forwarding might not be programmed to take effect. The calling party might just receive a busy signal. For an external call, the calling party might be forwarded to a secretary.

The second type, called **station forwarding**, allows the user to change the forwarding scheme programmed into the system. Users can program their phones by keying in another extension, and can change or cancel it whenever they please. In the morning, you might forward calls to a secretary, and in the evening, to a message center, for instance.

6.1.3 Camp-on Queuing

When you reach a busy line, the system can ''test'' the line and automatically place a call to it when it becomes available. Your phone provides a signal to you that the call is going through. If you want the system to call you back whenever the line is free, you have to ''tell'' it so. When you call the party and the line is busy, you hit ''camp-on'' and hang up. Then, when the line isn't busy, the system will call you back and place the call for you.

6.1.4 Automatic Camp-on

Suppose that you are making a call within the system and get a busy signal. You listen to the ''busy'' tone, and after a programmed length of time, the system puts a beep on the phone you are calling, signaling that you are trying to get through. You just hang on, and eventually, the person you are calling hangs up on the other call and your call goes through.

In the last two cases, camp-on queuing and automatic camp-on, we describe what happens when you are calling other people within the system and their phones are busy. Both of these features are also available when you are calling outside the system and all outside trunks are busy. You place a call and get a ''fast-busy'' signal (also called a ''reorder'' or ''all trunks busy'' signal). You could hit the same feature button, ''camp-on,'' or key in the code on a single-line phone for camp-on, and when the line becomes available, the system would send the number you dialed. Thus ''camp-on'' could apply either ''busy—internal extensions'' or ''busy—external trunks.''

When applied to a busy trunk, this features is also known as "trunk queuing." You cue for a trunk to become available, either on-hook (listening to the busy signal) or off-hook (hanging up and letting the system call you back).

6.1.5 Distinctive Ringing

You receive different patterns of ring signals (for example, two shorts and a long) to identify calls from different sources (inside, outside, your boss, etc.)

6.1.6 Speed Dialing

You can store often-dialed numbers in the system's memory. You can then dial them in an abbreviated form. Three versions exist. One is a "station" form, where each station can store, say, 10 different phone numbers in that station's own little directory. A "system" version also exists, in which the directory contains, perhaps, all the clients of a company. Somewhere in between is a form containing several "group" directories, where the most commonly called numbers from the engineering department might be held in one directory, sales in another, and so on. There are actually three arrangements, then: for station, group, and the entire system.

6.1.7 "Do Not Disturb" Mode

You can make your phone unavailable for any but the most important calls. (looks like a "busy" but can be identified by the caller as a "special" busy signal). An alternative method would simply not ring your phone, although the caller would hear a ring. It would appear that you are not answering your phone.

   This can work in conjunction with forwarding. If you put your phone in the "do not disturb" mode, then instead of a busy signal, your caller could be forwarded to your secretary.[1]

6.1.8 Call Waiting Signals

A light goes on at your phone, or you hear a beep on your phone while you are talking, to let you know that someone else is trying to reach you.

6.1.9 Automatic Reminder

This feature is especially useful in a hotel/motel package. You program in when you want the phone to ring you back and wake you up or remind you of a meeting. Essentially, you are using the phone as an alarm clock.

---

[1]It is common to allow a boss/secretary talk path, even in the "do not disturb" mode.

6.1.10 Park (or Call-Park)

This is similar to transferring a call to another phone, but without ringing that phone—instead, the call is placed on "hold" on that phone. To use it, you would key-in the feature for "park" and the extension you want the call to "park" on. For example, suppose that you get a call in your office, but the caller needs something you must go to your shop to do. You "park" the caller on the shop phone and pick up the extension in the shop when you get there, just as if you had put the call on "hold" on that phone.

Another example: You get a call for somebody; you park it on their phone, then you access the paging system to tell them they have a call holding on their phone. Then they would go to their phone and take off "hold" the call that you had parked there.

6.1.11 Call Pickup

This feature is useful in an office where there are many phones. If another phone in the room rings and no one is there to pick it up, instead of someone else having to run across the room to pick it up, the person can access the feature "call pickup," then dial the extension number that's ringing and be connected to the caller who was ringing the other phone.

In an abbreviated form of "call pickup"—called "group pickup"—if you access the feature, it will transfer to your phone any caller ringing a phone in your immediate area. There can be a number of groups in one system. The phones in the sales department may be one pickup group, the phones in engineering another pickup group. That eliminates the chance that a totally unqualified person might be answering a call for another group by accident.

6.1.12 Executive Override

This permits a user privileged to use it the ability to "walk in" on anyone's conversation and "take over." You would key-in the "executive override" feature on a busy extension. A beep or tone would be put onto the busy extension to let them know that you are about to "walk in on" their conversation. In a few seconds you would be joined into the conversation. This is obviously something that would be used only in cases of extreme urgency—such as a security officer breaking into an executive's conversation to report a fire—or extreme ego, such as an executive who's sure that nothing anybody else has to say could be half as important as what he or she wants to say, right now!

The tone is included to warn the busy phone's users that their call is about to be broken in on, to avoid misinterpreting this as an eavesdropping technique.

6.1.13 "Meet Me" Conference

An announcement is made on the paging system of a conference on a particular extension. Whoever dials that number is joined into a "conference call" conversation.

   An example: You receive a rumor that the president is about to announce the nationalization of your company in a televised press conference. You page everybody in a policy-making position to tune in the president's press conference on their TV sets, then dial the "meet-me conference" number to discuss the company's course of action as the press conference proceeds.

6.1.14 Repeat Last Number Dialed

When you hit the feature button for last-number dialed, it will just redial it for you. This is useful where you want to redial a person whose number is busy. "Camp-on" only works only within the PBX system; it doesn't extend to the outside telco numbers. The next best thing to "camp-on" for outside numbers is just to keep redialing from time to time, until the called number stops being busy.

6.1.15 Call Transfer

There are two ways to transfer a call. A *blind transfer* happens when the call is sent to the other extension to ring on the phone you're transferring it to, and you just hang up. Another type of transfer is called a **consultation transfer**. You access the feature "transfer," then dial the number of the phone to which you want the call to transfer to; then you have the opportunity to wait for the extension to answer so that you can announce the call. After you announce the call and hang up, the transfer goes through.

   You wouldn't necessarily have to key in different codes for blind or consultation transfers; the same transfer code or button could be used for either one—it would just be whether you waited to consult, or not.

6.1.16 Auxiliary Equipment Access

This could be called "paging access" or "dictation access." It allows the user to dial into the paging system, dictation tank, or other system.

6.1.17 Night Answer

This is a feature that allows a person to take incoming calls when the operator has gone. You use a special code or button for "night answer," or you might access the "pickup code" feature (say, "pickup 0"). When the system is in "night answer" mode, incoming calls might ring in over the P.A. system so that anyone in the building after hours could pick up the call by using the night answer feature.

6.1.18 Intercom Group

This feature puts a group of phones together and gives each phone a separate intercom number. The members of the group can dial among themselves just by dialing the intercom number. Within the group, members can reach one another with fewer digits to dial. One possible use for a "com group" would be to permit members a higher priority of access to one another than people outside the group have. A member of the group could, for instance, ring in over a "do not disturb" that would exclude calls from extensions outside the group. All phones with 300 extensions could, for example, be in the same com group. If so, extension 343 could call extension 351 by simply dialing "51."

If members of an intercom group have privileged access to each other, the converse may be true. It is possible that members of one intercom group can not call members of another intercom group. This is called "com blocking" or "intercom group blocking." One com group—let's call them the "honchos"—can make sure that another com group—let's call them the "peons"—cannot bother them. It is as though there were two separate PBXs.

6.1.19 Call Intercept

When someone calls a busy number, it might be intercepted by the message center, or it might be intercepted by the operator. This is similar to call forwarding. Another use for "intercept": If someone dials a number that doesn't exist, the system can send the call to an operator or connect the caller to a voice-message recording "the number you have dialed does not exist. . . ." This feature can also be used to redirect calls to busy or nonexistent extensions from outside when the system has the "direct inward dialing" feature.

6.1.20 Two-Line Ability on a Single-Line Phone

Normally, a single-line phone can handle only one call at a time. In a PBX system with this feature, one call can be placed on hold while another call is made; then the user can toggle back and forth between the two extensions, alternately talking to one and the other. This is normally a feature that would require a two-line phone with buttons, but it can be done on a single-line phone with the mediation of the PBX.

6.1.21 Private Call Feature

This is a kind of countermeasure to somebody coming into a call using executive override. By keying in the code for "private call," the user makes the system keep someone from using "executive override," or "camping on" and putting a "camp-on tone" onto the line. The operator cannot monitor the call—this feature basically excludes everybody from that conversation. It's not only an important feature for ensuring privacy of a conversation, it is also essential for protecting a modem-linked data transmission from being "blown" by the beeps or clicks of someone breaking into a data transmission, thinking that it's a conversation. If, for instance, you are using a modem with an acoustic coupler to transmit data, a "camp-on" tone will interrupt the transmission of data.

6.1.22 Dial Tones, Ring, and Other Signals

There are several different dial tones that you might hear while using a phone in a PBX system. One is the dial tone you hear when making an outside call. If your system requires an access code to make an external call, you would hear the internal dial tone before dialing that access code, then you would hear an external dial tone after dialing the access code. Another dial tone would appear if you picked up your phone and there's someone on hold. For instance, when there is someone "parked" on your line, you would get a "call holding" tone, and could connect to the "parked" call by dialing a special code to connect.

With distinctive ringing tones, you would receive one type of ringing for an outside call, and another type of ringing for an inside call. Perhaps a third type of ringing would occur for an intercom call, and a fourth type of ringing would tell you that it's a call-back from a "camp-on." When you're "camped" on somebody's extension and your phone rings you back telling you it's free, it might give you a distinctive tone (or type of ringing) for that purpose.

Other tones you might hear while using a phone in a PBX system, as mentioned before, are:

1. A "barge-in" tone, telling you that someone's about to come into the conversation.
2. A "camp-on" tone or beep, telling you that someone's waiting for you to finish your current call.
3. An "all trunks busy" or "fast busy" when you try to dial outside and all the outside lines are busy.
4. An "error" tone, telling you that you have dialed a nonexistent number, or a mistake in your dialing, telling you that you have done something wrong.

In addition, of course, there are the usual tones found on the outside system for dialing or busy signals.

6.1.23 Voice Message Feature

This is a relatively new feature, one that has just been added to many private business systems. When you don't answer your phone, someone can leave a voice message. When you return, you can look in your "voice mailbox" and see what messages there are. As an enhancement of this feature, the "voice message" can be sent to a group—say everybody in a department—or when you gets a "voice message" in your mailbox, you can forward the message to someone else or distribute the "memo" to a group of people. Another handy ability associated with this feature permits an outside salesperson to call in on an outside line and be connected to the voice message center in order to collect messages and leave messages for other people in the company. The salesperson could do this at any time of the day or night, since an operator wouldn't have to be on duty to handle the messages.

The ''voice message'' (VM) feature basically solves the same problem for a PBX system that a ''phone answering machine'' solves for the person who can't always be available to take calls. In some sophisticated VM systems, the sender can find out if previous messages left in the system were picked up by the person for which they were intended. This gives senders some assurance that their messages were delivered.

Since this feature is quite new, we don't know what abilities will become standard on a VM system. One ability on the ROLM system permits a new user to get spoken instructions from the system, and allows the neophyte to call up a ''help file'' of spoken directions explaining its use. The whole idea is to make the VM, and the PBX in general, as user-friendly as possible. Some of these systems, such as ROLMs, are integrated with the PBX, and others, such as Octels, are stand-alone.

6.1.24 Data Communications

When data terminals are connected to a PBX to switch data from one terminal to another, many features that apply to voice switching can also apply to data switching. For instance, when a called data terminal if busy, the camp-on feature can be used, much as the camp-on feature from the telephone.

You can place a call from a terminal in much the same way that a call is dialed with a phone. The terminal is turned on, then you could type in the number of the terminal or computer system that you want to call and the call would go through. The call could be placed to an internal terminal or to a host computer, or it could be placed to a computer on the outside. Other features might be used for data as well like speed dialing. Different types of com groups might be defined and programmed for terminals. If someone is trying to access data from a particular computer, he or she would have to have a terminal with access to the right group.

Computers often have multiple *dial-up lines*. These are like trunks for data access. The computer in the accounting department might have five ports (or dial-up lines) through which the computer could be reached. These ports would probably have a group number. You could dial the computer's group number and be connected to any one of the ports.

With the telephone, different tones are given to the users, telling them the condition of the line. With a data terminal, digital information can be fed back from the system to the terminal, informing it in much greater detail than is possible with tones on a voice line. For instance, messages could be printed on the terminal telling you that the number you are trying to reach is busy—and not only that it is busy, but that three other people have already camped-on and are waiting. You could then decide to camp-on yourself or to try the call later.

When you are using a telephone, you get a dial tone to tell you to go ahead and dial. On a terminal, you get a printed message that tells you to dial, or key-in, the number of the terminal you're calling.

6.1.25 Ring-Down Station

When a telephone is set up as a "ring-down" station, a call is automatically placed when you lift the handset. No number need be dialed. This feature is used in "emergency phones." The Red Phone from the White House to the Kremlin is a good example of a ring-down station. A more prosaic example is an emergency phone in an elevator that would automatically ring the building maintenance or security offices if picked up.

6.2 ATTENDANT CONSOLE FEATURES

A call to a PBX is usually answered by an operator who "extends" the call to the extension being called. The operator uses a **console,** where as in "the old days" the operator would have used a switchboard.

The attendant console has many features which are similar to the station user's features. In addition to those, there are some features that are designed specifically to make the job of the console operator more efficient and easier.

Some of the features found at a console which are the same as those found at any other station are:

1. Executive override
2. Call transferring
3. Camping-on
4. Paging ability
5. Call-waiting signals

The attendant console operator has the job not only of answering calls and extending or transferring calls to the people within the organization, but also needs to control some of the functions of the system. These control functions may be as follows: Control of station forwarding—the console operator may be able to set up another extension to forward or to set up a phone on a "do not disturb mode" or to set up a phone so that it will not be able to make outside calls. A console operator may also be able to take a group of outgoing trunks out of service, to make them unavailable to the users of the system.

6.2.1 Console Operation and Indicators

If you operate a late-model console, you will see indicators that give you information about the trunk you are answering and the status of the lines to which you might want to transfer the call. A typical console might look like the one shown in Figure 6–1.

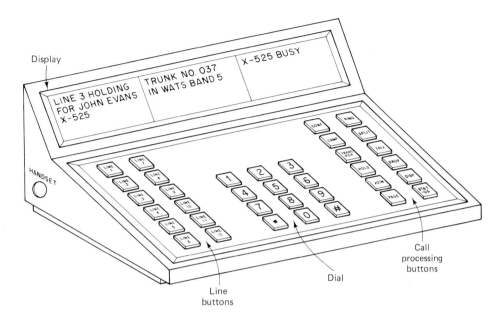

Figure 6-1 Typical operator's console.

The controls and indicators on Figure 6-1 do the following things:

1. **Line buttons.** These buttons indicate an incoming call by a flashing lamp. The operator connects to the caller by pushing the line button. A bell or tone also is heard when a call comes in.
2. **Status displays.** These indicators show the operator who is calling by displaying the caller extension number. If the call is coming in from the outside (from the public network), the display may show the trunk number, such as 555-1123.

There are also indicators in this group that show when outside trunks are busy or when a group of outside trunks is busy. There will be indicators that will show if an extension to which you are trying to transfer a call is busy, or in the "do not disturb" mode, or has been forwarded.

The status of the hardware in the system might also be displayed on the attendant's console. This would include alarm indicators, such as "blown fuse" lamps, or indicators for malfunctioning phones. If one of these alarm indicators appeared, you would call for service of the system.

3. **Hold.** This button is used to hold a call. The line button on which the call came in will wink until the operator connects to it again.
4. **Transfer.** The operator connects an incoming call to an extension by using this button, following by dialing the extension number of the called party.
5. **Camp-on.** When the operator encounters a busy extension and cannot com-

plete the connection, the incoming call can be set up to "camp" on the line, waiting for the current caller to get finished. The caller is now next in line to ring the extension after the current call is completed.

6. **Conference.** A conference call permits multiple-person conversations by connecting more than two phones at once. Each additional extension dialed after pushing the conference button will be added into the same conversation.

7. **Join.** When paging somebody, the operator can use this button to join the party with whoever was calling for that person.

8. **Serial call.** When a person calls in and is transferred in this mode, at the end of the conversation, the person would be returned to the console operator to be transferred to another extension.

9. **Dial and handset.** These are used the same way as with any ordinary telephone.

6.3 SYSTEM FEATURES

System features are the characteristics of the PBX that make it work efficiently. They are often unnoticed by users, or may be taken for granted. Some of the things that the system does for you are things that previously were done by an operator. For example, the operator at an attendant console may answer incoming calls.

An operator is not always necessary to answer incoming calls. A business may dedicate an incoming trunk to a particular extension, thus making operator handling of the call unnecessary. This is possible when a caller dials the particular phone number of the dedicated trunk. This is similar to public network switching, which no longer requires operators to connect calls. There are other methods for extending a call directly to an extension without requiring the services of an operator. Two of these methods are **direct inward dialing** (DID) and **direct inward system access** (DISA).

6.3.1 Direct Inward Dialing

This system feature, which originated with Bell's Centrex system, gives outside callers the ability to call directly to an extension number within the system. To do this, the outside caller would dial an office code and a four-digit number, which would be the same as the person's in-house phone. This feature requires a special incoming trunk that can be supplied from the local telco.

The telco provides information from its central office to the PBX that extends the call to an appropriate phone. The disadvantage of this is that the extensions are actually separate telephone numbers to the telco and are charged for on a monthly basis in addition to trunk charges. There must be space for all the telephones in the telco's numbering scheme. An entire exchange, such as all numbers starting with 753, might be dedicated to just one company or organization, such as a university. In that case, the telco's numbering scheme would reserve numbers 753-0000 to 753-9999 for the University's DID trunks.

6.3.2 Direct Inward System Access

Unlike DID, direct inward system access (DISA; sometimes called "remote access") does not require special trunks from the telco. A specific trunk number would be assigned as a DISA trunk in the programs of the PBX. If you call in on a DISA trunk, the dialed PBX returns an internal dial tone and waits for you to dial the extension number. You can dial any extension number you want once you have the internal dial tone. This permits callers not to only call people inside the company, but also to call out on the company's WATS line (or whatever) through the connection. To prevent abuse of this privilege by unauthorized callers, DISA trunks may require the caller to dial an authorization code. After the authorization code is dialed in, the caller is given an internal dial tone and may place a call.

6.3.3 Distribution Groups

A distribution group is a group of extensions that all share a common function (such as the order takers of a catalog sales department, or airline reservationists). The distribution group would have an extension number, possibly, to direct callers to any one of the persons in that group. One application of a distribution group would be to calls that come in to an engineering department. The operator could transfer to the pilot extension number of the group, then the call would be distributed to the first available extension or to one of the idle extensions in the group. The routing of calls is time-shared so that no one agent gets more calls than the others within the group. No one has priority over anyone else, and calls are apportioned so that everyone gets an equal amount of traffic. This may be decided by trying to give everyone calls that total to an equal amount of time, or by trying to give everyone an equal total number of calls. We might name these two systems "time-weighted" or "call-weighted" systems. This is the type of system to use where everybody in the group does the same job and all have equal priority at getting calls. A distribution group is the main building block of automatic call distribution.

6.3.4 Automatic Call Distribution

ACD is used in situations where there are a number of agents with similar functions. It distributes calls to the agents and keeps statistics on what happened to each call. Airline reservationists, travel agents, claims adjusters, and customer service departments all have a number of people with similar functions. Their main purpose is to answer incoming calls and handle customer requests, take orders, and so on. The ACD system may be offered either as a separate piece of equipment, or built into the PBX. Typically, most PBXs offer the ACD feature as an option.

Incoming calls are routed through the ACD to the first available agent. The outside caller calls a main number, and that main number is directly connected to one of the agents' extensions within the group. The ACD distributes the calls equally among all the agents. No agent gets more calls than another. The ACD also delivers voice messages, such as: "All agents are busy; we will handle your call as soon as one becomes available."

In this way, calls can be stacked up for agents so that, as soon as an agent's phone becomes available, the earliest call in the stack is immediately transferred to that agent's phone. An incoming caller in the stack might be connected to a music source for the duration, with periodic voice message (recorded) assurances "No, you haven't been forgotten," and then a return to the music source again.

Some systems have a second group of agents that are backup to the primary group of agents, or for calls that have waited too long. Many ACD systems keep data on each agent's performance, number of calls taken, average time spent on each call, and so on, as well as overall system performance statistics, such as average number of calls per hour, how many agents were available to take calls at any given moment, and how many calls overflowed to the "backup group." These statistics can be compiled for predetermined periods—on the hour, or every eight hours, or accumulated for a 24-hour period. Most of this is under the control of the manager of the ACD group. The simplest ACD systems may just distribute calls, and the more complex ones will provide the detailed statistics just mentioned.

Most of these ACD features are determined by stored programs in the PBX. The recording devices that the caller is connected to (when listening to music or being given recorded "please wait" messages) are external to the PBX, whereas the call-processing (switching) functions are all internal. Indicators may be given to the ACD administrator or manager to indicate calls waiting, or the administrator might have a data terminal on which to call up statistics for the group of agents. A business might have many ACD groups, with many managers, each with a terminal. There are probably as many different ACD group configurations as there are organizations with need of this feature.

6.3.5 Hunt Groups

In the section on distribution groups, we said that all the calls would be distributed to the extensions with equal priority. Sometimes, we want to distribute calls to extensions in a very definite "pecking order" that does *not* give equal priority to all the extensions. The calls, in this case, go to a **hunt group.** A hunt group is a group of extensions that are accessed by a pilot[2] or an extension number that directs it to the first idle (accessible) extension that appears in that group. A call to the group will hunt sequentially through all the extensions listed in that group. The first extension that it finds idle is where the call will ring. A good application of a hunt group is a group of secretaries capable of answering an executive's telephone. When the executive's phone is busy, or when he or she is not there and no one is available to answer the phone, the call-forwarding sends the call to a hunt group. The first person in the hung group would be the executive's personal secretary. If the secretary is not available, the next extension in the hunt group might be another secretary who would

[2]Not always necessary on hunt group.

be familiar with that executive's work. The third choice might be a general secretary or a message center, or perhaps one of the other people in the department that share the executive's responsibilities.

Ordinarily, when an executive forwards his or her phone to a secretary, a call coming in doesn't ring the executive's phone, but rings the secretary's, and if the secretary is not there, the caller is out of luck. With the hunt group, if that secretary is not there, the call will continue to hunt until it has exhausted all the possibilities in the group. Thus there is a much better chance of getting through to someone who can be of help to the caller.

When you call the main business number of an organization, the central office (telco) has a hunt group of trunks going to the organization. Their switch does what the PBX does with a hunt group within the organization. It selects the next available trunk to that organization if it finds a busy trunk on the first choice. Central offices sometimes call a hunt group a "rotary group."

6.3.6 Pickup Groups

An abbreviated form of call pickup, called group pickup, was described earlier in the chapter. Since we're describing various types of groups here, we thought it would be a good idea to repeat the description of a pickup group (and also that of a com group in the next section).

If you key-in the code for group pickup, the PBX will connect to your phone any phone ringing in the group. There can be a number of pickup groups in one system. The phones in the sales department may be one pickup group, the phones in engineering, another pickup group. The idea of enabling anyone in the group to pick up an unattended call from another group member's phone is that the call will then be answered by someone who does similar work and can thus deal with the call as a competent agent.

6.3.7 Intercom Groups

This feature puts a group of phones together and gives each phone a separate intercom number. The members of the group can dial among themselves just by dialing the intercom. Within the group, members can reach one another with fewer digits to dial. One possible use for a com group would be to permit members a higher priority of access to one another than that available to people outside the group. A member of the group could, for instance, ring-in over a "do not disturb" that would exclude calls from extensions outside the group. All phones with 300 extensions could, for example, be in the same com group. If so, extension 343 could call extension 351 simply by dialing "51."

If members of an intercom group have privileged access to each other, the converse may be true. It is possible that members of one intercom group can't call members of another intercom group. This is called **com blocking** or **intercom group blocking.**

6.3.8 Call Detail Recording

Call detail recording (CDR) or *station message detail recording* (SMDR) used by the telecommunications manager or the business's accounting department to keep tabs on the expenses incurred in the use of the phone system. It can be used for both data and voice. A supervisory report is generated on each call made from any station in the system. Its purpose is to make each user accountable for use of the station.

There are a number of CDR-SMDR systems available. The simplest is a listing device that is connected to a printing terminal. Each time a call is made, the details about that call (not including the conversation) are printed out on paper at the end of the call. A typical printout may look like this:

Extension number	User	Dep't.	Time call was made	Number called	Duration of call	Trunk cost
2357	G. Smith	E–35	14:05:37	919–555–2222	3:34	WATS $.86

It shows the extension number and the person making the call, the department, the time the call was placed, the number that was called, the length of the call, the trunk on which the call went out, and the cost of the call. If the system has the ability to charge calls to a specific account, when the user places the call they might enter an account number so that when the call is listed on the printer, an account number to charge that cost to would also be printed.

Another type of call detail recording is a little more sophisticated than the list device; that is, the details of the calls are accumulated and formatted and can be accessed by a data terminal. The telecommunications manager would typically have this data terminal. He or she would be able to list out details by a person's extension number, get all the calls that person placed, how long they talked, the cost of the call, and so on. Details could be listed from a specific group of extensions. Perhaps all the extensions of a particular department would be arranged in a group. The total cost of all calls for that group could be listed. Many types of statistics are available on an advanced CDR system.

The selections and format of the call detail printout can be set by the telecommunications manager. These statistics could be used to allocate a fair share of the telecommunications cost back to each department that incurs the costs. Many companies have separate departmental budgets; call detail recording is almost the only way to bill each departmental budget according to its real use of phone resources, sort of having separate trunks and separate telephone numbers for each department (thus giving each department its own phone bill). The CDR could also be used as a tool for studying calling patterns of workers to determine where more trunks might be necessary. It could be used to control telephone abuse. Users are less likely to call their friends in Hawaii if they know that the company manager will have a printout of the number they called and of how long they talked.

Another type of CDR device is used in a hotel-motel system, where the individual extension of each room accumulates the records of the calls that were placed. When the guest checks out, those calls can be added to the person's bill. Some CDR systems are built so that they interface directly with the billing systems in hotels. These billing systems (computers, usually) have standard communications interface ports, and the CDR can be designed to interface with those same standard ports.

One type of CDR recording device is a magnetic tape or floppy disk recorder. Details are recorded on a floppy disk, and later, are played back into a computer using a program that reads in the details and formats them. That program can also be arranged so that the details for a single extension can be singled out and/or the details for single departments, details for long-distance calls, or details for local calls— all types of statistics can be derived from the details. A program that can do this kind of sorting out and searching for, or organizing, specific details is called a **database management system** (DBMS) program.

Again, the most common use for call detail recorders is for a company to use the CDR to charge back each division's cost and to cut down on phone abuse.

6.3.9 Voice-Message System

This is a relatively new feature. When you don't answer your phone, someone can leave a voice message. When you return, you can look in your "voice mailbox" and see what messages there are. As an enhancement of this feature, the voice message can be sent to a group—say everybody in a department—or when you get a voice message in your mailbox, you can forward the message to someone else or distribute the "memo" to a group of people. Another handy ability associated with this feature permits an outside salesperson to call in on an outside line and be connected to the voice message center in order to collect messages and to leave messages for other people in the company. The salesperson could do this at any time of the day or night, since an operator wouldn't have to be on duty to handle the messages.

The voice message (VM) feature basically solves the same problem for everybody using a PBX system that an individual phone answering machine solves for one person who can't always be available to take calls. In some sophisticated VM systems, such as the ROLM PhoneMail system, the sender can find out if previous messages left in the system were picked up by the person for whom they were intended. This gives senders some assurance that their messages were delivered. We expect that VM will shortly be available on virtually every PBX on the market.

The VM unit might be an add-on to a PBX, or built within the system. Typically, the voice messages are stored on digital disk storage using a condensed digital representation of the analog voice signal.

Messages can be distributed much like the phone distribution groups with a voice-message distribution group. All the members of a certain department might be members of this voice distribution group, so that when a message is put into the group, it is accessible to all the members of that department. When they call in for their messages, members of the group might get a message that was distributed to them either on a group basis or on an individual basis. The sender of the message can also get a memo or message back that the message was received, as mentioned before.

As with other parts of PBX systems, the VM system has a capacity rated in how many messages, of what length, it can keep track of. A representative VM system might have the capacity of recording 5000 60-second messages. Since the number of messages entering and leaving the system may be more than one at a time, a number of ports will be needed to handle incoming and outgoing messages. This number will also be part of the system's capacity.

Since these systems are relatively new items, we do not know what abilities will become standard on a VM system. One ability on the ROLM system permits a new user to get spoken instructions from the system, and allows the neophyte to access a "help file" of spoken directions explaining its use. The whole idea is to make the VM, and the PBX in general, as user-friendly as possible.

6.3.10 Automatic Route Selection

When WATS, MCI, local or leased trunks are available, selection of which type of trunk is used (according to which is cheapest) takes place automatically. MCI, for instance, provides attractive long-distance rates. An automatic route selection (ARS) system (also known as least-cost call routing) would place a long-distance call over an MCI trunk if one is idle, before it would be placed on the more-expensive WATS trunk. If no WATS trunk or MCI trunk is available, the call might be sent over a more expensive common carrier's trunk.

Depending on the class of service (see the following section) of the extension making the call, that extension might be restricted from making long-distance calls altogether, might be permitted to use any available route, or might be forced to "camp on" until a least-cost route is made available. All these abilities are programmable in computerized systems.

6.3.11 Class-of-Service Assignment

Stations are assigned various classes of service, which will determine what features they are allowed to use or whether they are permitted to use certain privileges (such as long-distance calling). A feature such as speed dialing might be available to an extension with a higher class of service but not to one with a lower class. As mentioned above, the class of an extension limits the abilities of the user of the ARS system. With a very low class of service, some extensions may not even be permitted to make any outside calls.

The abilities permitted to the user of any system feature are limited by the class-of-service designation of the extension.

6.3.12 Traffic Data Statistics

This is similar to the CDR, but records the statistics related to traffic on various trunks. For example, this feature could tell the telecommunications manager how many calls were placed over a WATS line, how often calls were not placed because all the trunks were tied up, or how often calls overflowed from a less expensive to a more expensive type of trunk because the lower-cost trunks were busy. These traffic statistics could help the manager determine, for trunk utilization, if adding or removing certain kinds of trunks and service would improve system efficiency.

Traffic statistics could also be kept for common equipment within the PBX, such as the DTMF registers. Before a call can be dialed, a register must be connected to the phone. If all registers are busy at many times during the day, the manager can see that more DTMF registers will be needed in the system.

6.3.13 Network Abilities

This is a feature that allows one PBX to be connected to other PBXs by tie trunks. When the users in one PBX system call the users in another, they do not have to dial over the public network, but are able to access that other PBX over tie lines. In many systems it would appear that calls placed to another PBX (perhaps in another city or another part of the state) would appear no different to the user than a call to another extension within the building. Some systems carry the caller's class of service and phone number, together with who is being called in the other building. Between divisions of the same company, perhaps having buildings in different states (such as New York and New Jersey), the use of this arrangement would cut down on costs immensely compared to calling done over long-distance trunks.

These network arrangements have names such as "satellite systems," "network systems," "tandem systems," or "centralized attendance service." All the attendants (operators) of a certain company could be at one location, connected to the various PBXs at each division. Calls to any of the divisions would go through the centralized group of operators, and then be extended to the appropriate PBX via the tie trunks.

6.3.14 Digital Trunk Interfacing

Most trunks available are single-channel analog trunks. Multiplexed trunks from the central office can be connected to some PBXs using **T1 interfacing.** T1 is a wideband serial data channel that will be described in more detail in a subsequent chapter. Both voice and data can be transmitted on a T1 channel. A single T1 span could carry a number of voice communications and data terminal transmissions on the same link. Such trunks can be connected directly to the PBX if it supports digital communication. In a PBX with this ability, both hardware and software are necessary to support this feature.

SUMMARY

From time to time, in referring to a feature, we have used the word "capacity." What determines how much capacity the system has for certain features? For example, we mentioned that the capacity of a voice-message system is related to the overall memory size of the VM system disk drives. As a note on all features and their capacities, the capacities of such features as speed dialing, group speed, or system speed directories are allocated a certain portion of the PBX's memory, so they're limited by its memory size. The size of the various distribution groups, pick groups, ACD groups, com groups, and hunt groups are similarly limited (the number of stations that can be put into each group) by the amount of memory allocated to each of them. Some of these are configurable—the memory assigned to each purpose can be traded-off between different features. In other cases, the system memory allocations are preset. A system may originally be configured, for instance, so that it is possible to program 10 pick groups, each with 20 members. After all the group spaces are filled, the system has reached its programmable capacity. Some systems can be programmed so that as long as there is memory space left, you can go on adding to com groups, pick groups, and so on, as needed.

Another description of the capacity of a system is the type of dialing it supports and the type of terminal or port that it supports. A system can support single-line phones, electronic phones, multiline phones or keyphones, data terminals, or perhaps even certain types of sensors could be supported by the in-house telecommunications system. The type of station that is supported could also vary, as stations could be either rotary, tone dialing (DTMF), ASCII data, or EBCDIC data.

As with a DID trunk, when the system is accessed by a DID call, the system might be accessed by a phone with either DTMF or rotary pulses. The system has to be able to accept either type of input. With DISA, typically, DTMF is the only type of signaling that it will accept.

REVIEW QUESTIONS

1. What is a station feature? What is a system feature?
2. Name two station features that would be most useful to save time for a business user.
3. Name two station features that would be of most convenience on a home phone.
4. What station feature would you use that could make it easier for people to reach you when you move around to many locations during the day?
5. What station feature would you use if you don't want to take any calls?
6. Why are access codes needed for activating some station features? Why don't *all* station features require access codes?
7. What is flashing, and why is it needed?
8. Is the Red Phone from the Kremlin to the White House likely to be a ring-down phone?
9. What is the difference between camp-on queuing and automatic camp-on?
10. What is the difference between internal queuing and external queuing?
11. How does the attendant console differ from other stations?
12. Name two system features that allow an outside caller to call an extension inside a private system without requiring the assistance of an operator.

13. What system feature would probably be used to calls for a pool of clerical workers taking calls for airline reservations?

14. If a company wanted to charge its employees individually for outside calls, what system feature would they use to do this?

15. What advantages would a traveling salesperson gain by having a voice-message center at his or her home office?

16. What system feature can be used to minimize the cost of common carrier service?

17. What is the significance of overall memory capacity to the number of station and system features available in a PBX?

18. Besides the number of ports, what does the term system capacity refer to?

7

Station Equipment

Station equipment is the part of the telecommunications network where information enters or leaves the system. The most obvious example of a piece of station equipment is the telephone. Usually, one station is connected to another through the telecommunications network, and the two stations interact (communicate back and forth), but this is not always the case. A telephone answering machine is considered a piece of station equipment. People call in to it and leave a message; later someone accesses the answering machine and listens to the messages. The type of communication that's entered into a station doesn't even necessarily have to be voice, as it is with a telephone. A teletype, CRT terminal, or video camera/monitor arrangement (picturephone) are all considered stations. None of these data/video communications devices are new inventions, but their use in conventional business telecommunications networks has become significant only recently. What business has adopted today to aid in office automation may find its way into household "telephone" devices in a few years. We might not recognize the home telecommunications station as a telephone, anymore—it would have a video or CRT type of display and a handset for voice communication, but the DTMF keypad might be replaced by a full ASCII alphanumeric keyboard, making it resemble a typewriter more than a dial. You might call up the local library and have them download volumes of printed information into your terminal, including, perhaps, pictures, color pictures, or even moving pictures.

Services like this are presently available on a limited basis for schools and universities in the PLATO network, and through videotex to home users, and KEYFAX to leaseholders on cable-TV networks in some areas. We anticipate future growth in this segment of telecommunications to provide everybody with access to stations having these capabilities.

In this chapter we look at the various types of station equipment available now and anticipated in the future, and how they function. We concentrate first on the types of station equipment that are most widespread today, and then go on to basic ideas and concepts that are related, and which are becoming a part of telecommunications systems. We see how telephones, modems, and other devices found in station equipment work. Together with this, we look at the basic principles of how CRT and television operation, and how various devices in the station convert one form of energy to another, such as the microphone, which converts voice signals into electrical signals.

Standardization

Since there are many manufacturers of station equipment, and many pieces of station equipment are similar, certain devices have become standardized, such as the telephone and data terminal. Standardized devices are generally interchangeable between one manufacturer's system and another. The types of station equipment that connect to the public network, because of the standardization of the public network, must also be standardized, or have some kind of interface between the station and the public network to allow them to be compatible. Until recently, when the public network could handle only voice communications, any digital communication was first changed into an audio signal through the use of a *modem*. That interface device (the modem) was necessary to connect any digital station (terminal) to the public network, and modems were designed to conform to standards so that they could be intelligible to one another, as well as being able to work within the design limits of the public network.

Both in the United States and abroad, organizations have been set up to oversee the standardization of telecommunications equipment. It is the task of these organizations to make sure that instruments conform to existing standards, and that proliferation of new types of equipment unable to communicate with one another is avoided. CCITT and the EIA are two such organizations. The EIA (Electronics Industries Association) is a body formed by American private enterprise, rather than a government bureau of standards. Major corporations in the electronics field have cooperated to form the EIA, which is responsible for such things as seeing that the manufacturers of resistors all use the same color code in marking a resistor. The CCITT is an international forum (Consultive Committee International for Telephone and Telegraph) which is part of the International Telecommunications Union. It recommends standards for international communications systems, including data.

Victor, mature

Of course, new technology is developed by competing companies, using techniques that may be incompatible with one another. These devices are not "mature" technology, and have not had a chance to become standardized. In these areas, the tendency in free-world countries is for competing technologies to be allowed to fight it out in the marketplace, until the "victor" becomes the de facto standard. The standard is then assigned a name by the standardizing organizations, and other companies are permitted or licensed to produce equipment that conforms to this standard. At this point, we say that the technology has become "mature."

In this chapter, and throughout the book, when we are dealing with technology that has become standardized, we can go into those technologies in greater detail than we can for new ones. Clearly, where all examples of a particular device are standardized, we can say with some assurance what they all have in common. When technology is still in the stage of prestandardization, it would be waste of time talking about individual vendors and their particular methods. Instead, we will try in these cases to describe the general type of problem the technology attempts to solve, rather than going into the current methods being used to solve them.

7.1 THE SINGLE-LINE TELEPHONE

Rotary telephones are slowly being phased out by DTMF—Touch-Tone®—dialing, but will still be with us for many years. Understanding this older (but simpler) technology is a good place to begin understanding how any other station equipment works. In Chapter 3 we introduced some basic concepts about the telephone, particularly how the receiver and transmitter work. It might be good to refer back to that chapter at this point. Figure 7–1 shows a block diagram of a single-line rotary telephone and all its components. The purpose of the transmitter and receiver was made clear in Chapter 3. The ringer, of course, lets you know someone is trying to call you. The hookswitch is an electrical make-and-break connection that, in effect, turns the phone on and off. The dial is a way of sending out coded information over the phone. This information goes to the switch, which determines from these codes what number is being dialed, or in a PBX environment, what feature is being accessed. The network is a circuit board mounted inside the phone that has places to make the connection from the transmitter to the receiver, and to connect the ringer terminals, outside line, and dial. Part of the network is a hybrid transformer used to help develop the voice signal, and some coupling capacitors, which allow the ring-signal component to be carried to the ringer.

7.1.1 Parts of the Telephone

The transmitter and receiver convert the energy from sound waves into an alternating current that's transmitted out over the phone line. The transmitter and receiver (the handset) are not connected to the line until you pick up the phone, which connects the phone to the line through the hookswitch. The ringer is always connected to the line, even when the phone is hung up. Although there is no path for direct current to flow through the ringer, when an AC signal of 100 volts at 20 cycles appears on the line, it is passed through the coupling capacitors to the ringer. (*Reminder:* Capacitors block DC, but pass AC.)

When the phone is actually picked up (when it "goes off-hook"), a DC current begins to flow. That current is sensed by the switching system and is interpreted by the switch as a signal to stop ringing. The DC current that flows through the line

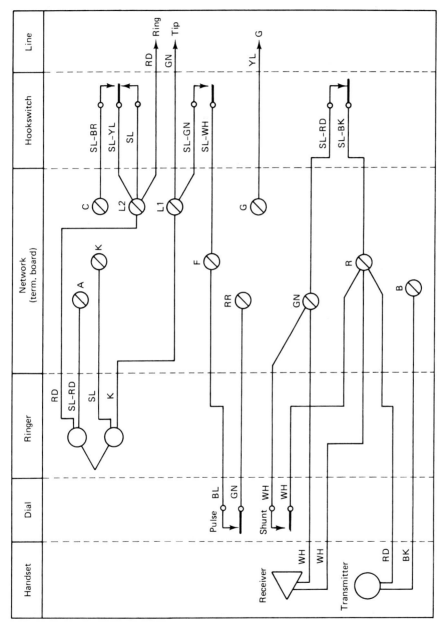

Figure 7-1 Single line rotary telephone set. Wire colors coded: RD = red, WH = white, BR = brown, GN = Green, BK = black, YL = yellow, SL = slate, BL = blue.

when a telephone goes off-hook is also going through a contact on the rotary dial that is normally closed. When the dial is turned and released (the moving part is called the "finger wheel"), the line current is interrupted. As the dial returns to its original position, the current path is connected and disconnected at a rate of about 10 pulses per second, and is on and off for about 50% of each pulse. These pulses in the line current carry the signaling information to the switch, which interprets the numbers dialed. If you dial a "5," for instance, a string of five pulses in line current are sent out from the dial. A counter at the switch counts the incoming dial pulses. This **dial-pulse register** will identify a "5" by counting five pulses. Another contact on the rotary dial disconnects the receiver from the line, so that the dial pulses, which are about 50 volts, are not heard (50-volt pulses are much louder than ordinary voice signals on a receiver). The dial mechanism is a spring-loaded device, designed so that it returns to a constant rate of speed. There is a small mechanical governor on the dial that does not let the spring rotate the dial too quicky or too slowly.

The **ringer** is a coil (or actually, two coils wound around the same core that pull the clapper back and forth between two bells. Some phones have only one bell, but the principle is similar. At one time, the ringing signal to a phone could be different frequencies so that the bells could be set up at different resonant frequencies on a party line. This arrangement allowed only one bell to ring at a time, although all the phones received bell voltage. This is seldom used today.

The **hookswitch** has more than just one pair of contacts. Some contacts in the hookswitch are spared, and normally not used. They might be used to control some device connected to the telephone—possibly a recorder or an "in use" lamp on a panel that says which phone's in use. Other contacts in the hookswitch might be needed for the connection of a speakerphone—a hands-free unit that we'll discuss later. The main function of the hookswitch is to connect the handset to the line when it's picked up.

The main function of the **network** is as a terminal board to connect different parts of the telephone together. It also has a "hybrid transformer" and "equalization circuits" to compensate for the line resistance the DC current must pass through in long open-wire loops. When the telephone is relatively close to the switch, the line resistance would be low and the line current would be relatively high. As a result, a voice signal from this phone would be much louder than one from a phone an "average" distance from the switch. A longer line going to a more-distant telephone would have more line resistance and less line current. Its voice signal would be weaker than a phone located at an average distance. Since the phone company can't arrange for all their phones to be the same distance from the switch, they add an equalization circuit to the line to make all the analog voice signals relatively equal, within the range of different line lengths coming into the switch. Another function of the hybrid and network circuits is to return a small part of your analog voice signal to the speaker (earpiece) part of your handset so that you can hear yourself talking. The returned sound energy to your receiver is called **sidetone.** The advantages of being able to hear a small part of the voice signal are:

1. You know your phone is working—it has a "live" sound, even when nobody else answers. This gives you some assurance that your phone is actually attached to a line and some current is flowing.
2. If the "right" amount of energy is returned to your receiver, you can tell if you're speaking too loudly or too quietly.
3. In the case of a phone with Touch-Tone® dialing, sidetone also feeds back a portion of the DTMF tones from the dial and lets you know that the dial is working properly.

7.1.2 The Rotary Dial

As we mentioned, the rotary dial pulses the line as the dial is released. As the number "1" is dialed, there is one pulse, when the number "0" is dialed, there are 10 pulses. Figure 7-2 shows the signal produced on the line when the number "312" is being dialed on a phone with a rotary dial. The standard line voltage applied to a telephone line is 48 volts, so these pulses are nearly 48-volt pulses. Current depends on the length of the line and the internal resistance of the telephone set. The current of each pulse is more important than the voltage—the switching network is current dependent. The line current is interrupted in three short pulses, then again for one pulse, then again for two pulses.

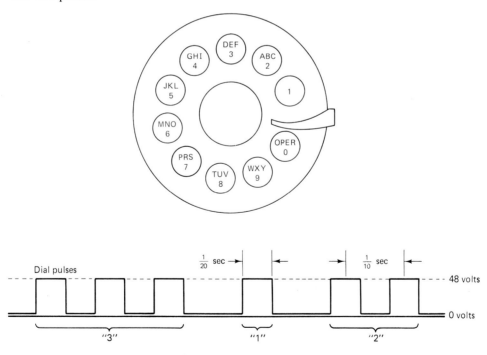

Figure 7-2 Rotary dial.

7.1.3 The DTMF Dial (Touch-Tone®)

With the DTMF dial (dual-tone, multifrequency), instead of pulses, two distinct frequencies are sent out over the line. These frequencies are shown in Figure 7–3. This Figure shows how the frequencies are combined when the number "312" is dialed on a DTMF dial. Eight different frequencies are used in this system. Each digit or key produces a combination of two of these frequencies. Four of the eight frequencies are attached to the four rows of the keyboard, and the remaining four are attached to the columns of the keyboard. With four rows and four columns, a DTMF dial would have 16 keys. No phones have 16 buttons; they all have 12 (the numbers 0 through 9, plus the "*" and "#"). There are, as you can see in the figures, four rows and three columns on a DTMF keypad. The frequency for a fourth column is already there on the dial. To have four more keys (they would be called A, B, C, and D, perhaps), all that would be needed are the four additional buttons and contacts connecting the four rows to the additional column.

Tones are generated in several ways. Some DTMF dials have only one contact per button, and others have a number of contacts per button, as well as some common contacts that are closed if any button is pushed. One pair of these common contacts cuts down the level of sound at your receiver while you are dialing so you don't have to hear the loud tones generated as each button is pushed. Another set of contacts connects power from the DC line to the DTMF oscillator circuits. Yet another set of contacts may be connected to auxiliary equipment, such as a speakerphone, but would allow the dial to function without lifting the handset.

Standard DTMF dial architecture

Standard telephone keypads have two contacts per button. When you push button 1, for instance, one contact turns on the 697-Hertz oscillator, and the other one turns on the 1209-Hertz oscillator. All the buttons on the top row in the diagram, that is, buttons 1, 2, and 3 on your telephone keypad, are all connected to a **row wire** that runs horizontally beneath them to the 697-Hertz oscillator. On the second row of buttons, that is, 4, 5, and 6, the keys all have one contact that connects to a row wire that goes to the 770-Hertz oscillator. Columns of keys that are aligned up and down, such as the column that contains 1, 4, 7, and *, have one of their contacts attached to a **column wire** that runs vertically beneath those pushbuttons. The 1209-Hertz oscillator is switched on whenever any of the keys (1, 4, 7, or *) in that column is pushed. In this design, every key is connected to a row wire and a column wire, and activates two oscillators when it is pressed. In the diagram, the fourth "phantom" column is shown as "dotted keys." A standard keyboard does not have those keys, so use of the 1633-Hertz oscillator is optional.

Manufacturers usually pack a wiring diagram and data sheet in with each telephone. These show the various connections on the network, and are labeled on the network itself so that you can make cross-references between the diagram and the network on the real phone.

We should point out here that most single-line phones are the same type of phone

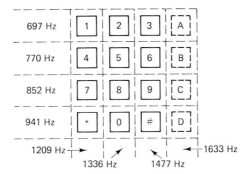

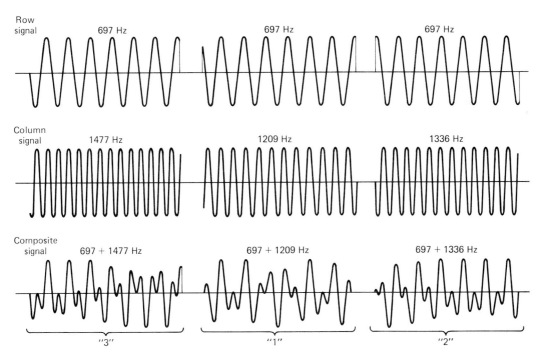

Figure 7-3 DTMF or Touch-Tone® dial.

as that used in a business or a residence. Their styles may vary, but they will work on the same type of line, and with the telco's central office switch or with a PBX in a private system. Sometimes phones used with a PBX require slight modification, but most non-PBX phones can be used with a PBX, and require no modification at all.

7.2 MULTIFUNCTION PHONES

Unlike single-line phones, featurephones, electronic phones, and multiline phones have "bells and whistles" that make them different from the phone found in your residence. You could not use most of these features if you connected one of these phones to the public network. In some cases it wouldn't even be possible for you to connect one of these phones to the socket into which your household phone plugs. Even so, at the heart of any of these "advanced" phones is the basic instrument described in Section 7.1. In this section we describe the features that have been added to that basic instrument, and what each feature does.

7.2.1 Multiline Phones (Keyphones)

We begin looking at multiline phones with the original multiline phones, 1A2-type keyphones. Mechanically, everything is switched and controlled with pushbuttons and contacts. Since this technology has been around quite a while, it has become standardized over the years. We discussed multiline phones in Chapter 1 (as keyphones), and now that you know how to use a multiline phone, the only thing that remains is to tell you how they work. Basically, the workings of a multiline phone are quite simple. The common parts, like the handset, ringer, and network, are almost identical to a single-line phone. The difference is in the group of mechanically coupled pushbuttons, called a "gang" of pushbuttons, that connect the network to one of the many lines that come into the phone. These lines (wire pairs) are each run to the phone individually; all the pushbuttons do is select which line you want to talk on.

On the simplest multiline phone, there would be a row of buttons like the channel selector in an automobile radio to "pick" a pair of wires, and depending on which pair of wires you picked, that would determine the line you selected.

More common is the multiline phone with a gang of pushbuttons, each backed up with a lamp that lights when you're using a particular line. To operate, this kind of phone requires three wire pairs for each pushbutton.

One pair of wires is needed to bring power to each lamp. When you pick up the extension, its light goes on on your phone and on all other phones with that line appearing on their button sets. If a call comes in, all phones having a button for that line will flash that button. The light on that button will stay on continuously when anyone answers that line.

The second pair is a voice pair that carries the analog voice signal for a conversation. You may recall that these two wires are called the "tip" and "ring" wires from the fact that, at one time, one was connected to the tip of a phone plug and the other to the ring around the shaft of the plug. (The name "ring" has nothing to do with the signal that rings the phone.) The voice pair, of course, is the same kind of voice pair used by a single-line phone. Rotary pulses or DTMF signals would go out over this line exactly as they do on a single-line phone.

The third pair is used to show the status of that phone. It communicates with a control system called the **key-service unit** (KSU), which is needed to handle things like ringing all the multiline phones when a call is coming in, and putting calls on "hold" from a multiline phone. The key-service unit is located in a cabinet or closet in the office where the phones are used. The ultimate source of each line, which would be either the central office's switching system or local PBX switching system, wouldn't interconnect with the telephone directly but instead, would interact with its' associated **line card** in the KSU. The third pair of wires (the **control pair**) gives the line card a status of the line selected. This pair is connected to a "make" contact in the keyphone that closes when the line button is pressed. Each line button has contacts for tip, ring , and control. This control signal tells the line card of the KSU that the line is off hook . In turn the line card will connect power to the lamp lead to light that line button of the keyphone. This control signaling is usually called **A-lead control.** Figure 7-4 shows a simple keyphone diagram.

Another difference between the multiline and the single-line phone is that the single-line phone ringer was connected (AC coupled) across the tip and ring connections of the line, and in the multiline phone, the ring signal is brought in on a separate pair. This pair is called the **ringer line** or **ring pair.** Although every button has a lamp lead, a voice pair, and an A-lead control line, there is only one ring pair regardless of the number of lines going to the phone. A 10-line phone, for instance, wouldn't have 10 individual ringers. One pair goes from the KSU to the phone, and rings it whenever a call comes in for any of the lines for which it has buttons. If line 20 is ringing, for instance, a relay at the KSU can direct the ring signal to any or all phones that have line 20 on one of their buttons.

What happens to the three pairs that are serving each line in a multiline phone when a call is made? The pair for control, the *A lead,* as it's called, is connected to ground when the phone goes off-hook, through the hookswitch contact. Also, when the phone is hung up, the A lead opens up, telling the KSU to drop the line. Another purpose of the A lead is to signal the line card in the KSU to put the call on hold. If you press the HOLD button on the keyphone, the voice pair (tip and ring) remains connected to the handset, but the ground is disconnected from the A-lead wire on the pushbutton key.

The KSU interprets this as a HOLD, and puts a resistance across the voice line to keep carrying loop current. When you release the HOLD button, the line button is also released (pops up), and this disconnects the caller on HOLD from your phone. The resistance connected at the KSU line card keeps the loop current flowing so that the caller is not disconnected from the KSU.

When you put a caller on HOLD, the KSU also connects the lamp for that particular circuit to a "wink-lamp" generator instead of a "steady-lamp" generator. When you picked up that extension, you'll recall, the lamp for that line button lit up. Now the light is winking on the button for that line holding the call. When a call is made to a line, the lamp flashes (a flash is different from a wink) on that line. These three different lamp generators provide information to you from the KSU about the status of every line for which your keyphone has a button.

Mechanical keyphones with a pair of wires for each line are being rapidly replaced by electronic phones and with electronic key-service units that approach the complexity of a PBX. Up until recently, even today, while multiline phones and KSUs are run "behind" the PBX, not all multiline phones are controlled by the PBX. We'll probably see less and less of that in the future.

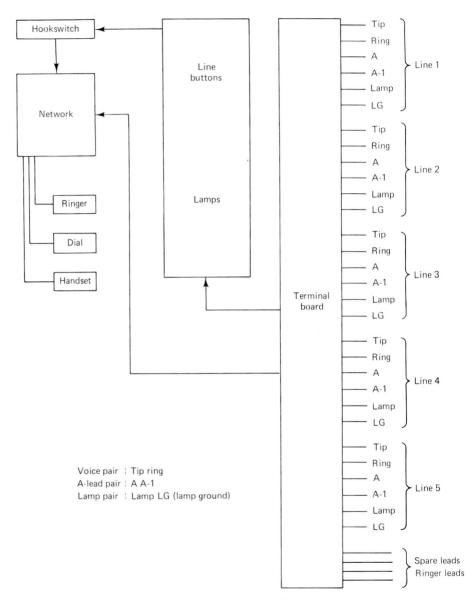

Figure 7-4   Keyphone block diagram.

7.2.2 Featurephones

A featurephone is a cross between a standard single-line phone and an electronic phone. It uses the normal dialing techniques, either rotary pulses or DTMF, to get information to the switching unit. These pulses or tones are instructions to the switching equipment no different from what would be dialed on a standard phone, except that the sequences, the instructions, can represent a whole sequence of keystrokes or dial rotations. Pressing a single key or a short sequence of keys causes the phone to "play back" sequences of pulses or tones stored in an internal memory device. The sequence could be used to access a special feature, or just to dial an often-called number. Also, if a signal other than tones is required by the PBX, such as a flash, the featurephone can also initiate that and then send the stream of pulses. These codes can be programmed into the featurephone in RAM memory by the user, or can be stored at the time of manufacture in ROM memory, which cannot be changed by the user.

Featurephones, in general, don't require a separate power supply. They run off the line current just as any single-line phone would. When auxiliary power was needed, one solution was to put nickel-cadmium batteries into the phone which were charged by the line current slowly while the phone was sitting idle, and then, when the phone was in use, a drain was put on the batteries (see Figure 7-5). The batteries were then recharged when the phone went on-hook. With modern, low-current, LSI electronics, this type of battery storage is no longer needed.

Some featurephones have been manufactured so that they display time (a clock), or the buttons on the phone could double as a calculator, with the feature buttons handling math functions such as " + " and " − ." Some featurephones with AM-FM radios and alarm clocks are marketed by Sears and Penney's.

There are cordless featurephones that need a radio receiver/transmitter mounted at a line outlet. The transmission to the phone is good for 300 to 1000 feet, depending on the model. These cordless units could also have features such as speed dialing and "save and repeat the last number dialed." For featurephones, the "smarts" needed to handle these features are built into the phone and not gotten from the local telco or PBX system.

What is the difference between a featurephone, as just described, and an electronic phone such as the ones we are going to describe in Section 7.2.3? It's not easy

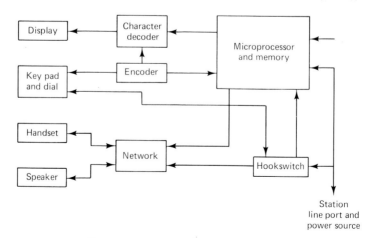

Figure 7-5 Featurephone block diagram.

to draw a line that separates the two, since the features—especially the speed-dialing and last-number-called features, which are the result of electronics built into the featurephone, are often available from the PBX, and can be invoked by pushing a button on an electronic phone. The primary difference is that a featurephone can attach to almost any private or public system that uses universal rotary pulse or DTMF dialing—with PBXs or the public telco switch—while an electronic phone would communicate with the switching system, using codes or signaling that is not usually compatible with the public network's signaling standards, and probably not compatible with signaling standards on other manufacturer's PBXs.

It seems that the market always tends to combine featurephones with decorator-style phones. Often, this appeals to people; if their phone is going to have unusual or special abilities, they often like it to look unusual and special. Putting a radio inside a telephone is one example. It's not uncommon to find a phone in a cigar box, and I'm sure that someday you'll see a phone that looks like a banana. At the 1983 Consumer Electronics Show (CES) in Chicago, we saw a banana phone, and others that were even more exotic or weird. We should also mention that there are a number of new manufacturers of phone equipment, many overseas, marketing and distributing in this country. Prices have come down radically on this type of equipment. The newer manufacturers have done a lot to reduce the price, often by the reducing the quality compared with the price of the—up until recently—standard telephones, which are built with high quality so that they last for years. The newer, cheaper phone is made such that if you were to drop it, it would break—but this lower quality has reduced the price considerably. In this year's CES, about half the exhibits in the McCormick Place Exhibition Center were cordless phones and featurephones or designer phones. The primary market for the phones we saw exhibited appeared to be consumers—the "home phone" or retail market—rather than businesses. Featurephones such as these would be compatible with business systems, but it would be hard to imagine the banana phone fitting into any business except the Chiquita Banana Co. Possibly, another reason why featurephones are not often seen in business systems is that many PBX manufacturers market their own electronic phone (one that works with their system only). Also, because the features in a private system are controlled by the PBX, and they can be mechanically dialed from a standard phone, having that extra ability to push just one separate button, to activate a feature rather than two, does not seem to make dollar-sense for most users. The time it takes a featurephone to output a code to access a feature of a PBX would take about as long as for a user to dial on a standard phone. But when it's done by an electronic phone with a digital code to activate a system feature, the response is almost instantaneous. Using a featurephone to memorize a complete long-distance telephone number, and using that featurephone to dial the number out through a PBX, would take twice as long as with an electronic phone. The reason for this is that the featurephone would have to output tones of normal duration. Even though you hit only a few keys, the featurephone is going to pulse-out all the tones you *would* have dialed. To auto-dial a long-distance number, the phone would have to output standard tones and intervals over the line to the PBX, and most PBXs would buffer those tones and check them out against a memorized list of numbers that the user is allowed to dial. If the PBX has least-cost call routing, the "smarts" in the PBX will have to determine where that long-distance call is going to, and then route that over the most economical trunk, say MCI or WATS or whatever. Then, after making the connection to that trunk, the PBX would have to output that sequence of DTMF or rotary tones all over again. You can see how a featurephone with auto-dialing could take twice as long as dialing out on an ordinary trunk without a PBX.

If we have a PBX to attach our phone to, all those extra smarts in the PBX should make the call faster, not slower. This is clearly *not* the case with a featurephone in a PBX system, and the alternative is to use electronic phones, which duplicate these features but use internal forms of signaling, which are much faster. We discussed speed dialing as an example. If the memory of the speed number were in the PBX itself in a digital format that the phone could use, you wouldn't have that long sequence of outpulsing which the featurephone uses. It is also worth mentioning that retail packages for home telephones now have a number of features that are controlled by the telco's switching system, and these work the way our PBX works in the example above. They require no home featurephone, and they give you abilities such as being able to forward your calls and have a special speed-memory for often-called numbers at home, using the telco switch's capabilities.

7.2.3 Electronic Phones

As we indicated in Section 7.2.2, it isn't easy to draw a line between featurephones and electronic phones. The biggest difference is the location of the "smarts" that do the "nifty feature stuff." In the featurephone, the smarts are largely self-contained. For whatever special functions the phone is supposed to perform, the circuitry must fit inside the phone. The electronic phone, on the other hand, signals the switch (the PBX) to activate special functions using digital signaling rather than rotary or DTMF. The restriction on special functions is now "What will fit inside the PBX?" instead of "What will fit inside the phone?" As we mentioned earlier, different vendors sell systems with electronic phones—and the electronic phones "do their thing" in completely different ways from one vendor to another. This means that it would be a difficult—probably an impossible—task to describe what the "right" way is for an electronic phone to do its task.

The basics are: The electronic phone has to be able to signal the PBX or switch—in a fast digital-signaling format, rather than DTMF or rotary pulses—to turn on different functions. Of course, the electronic phone still has to be a station, too. It must be able to carry out normal voice/data communications, as with any station. In addition, the smarts in the digital part of the electronic phone probably require the use of a microprocessor, complete with a program to control all the actions of the signaling function, as well as turn lights on the phone on and off, and activate and deactivate any special PBX features at the control of buttons on the phone (which must be scanned by the microprocessor). With its memory, programs, and microprocessor, the digital portion of the electronic phone is a small computer, and accordingly, needs the power required to run a small computer. There are various ways in which these needs can be handled by electronic-phone design:

1. The phone connects into the system through three different links. An audio pair connects the phone to ordinary voice lines, a digital pair goes to the PBX for control and signaling, and a power pair runs the microprocessor and the electronics of the phone.

2. The phone uses only one pair. It is a digital pair, and carries all the signaling and control information, while a codec digitizes voice transmission, which travels over the same pair as the digital signaling until it arrives at another codec, which reconstitutes the digital signal back into voice, and the whole thing could be line-powered, since digital signaling can be done with circuits that consume less power than analog transmission circuits use. This power can be derived entirely from the voltage of the DC line to which the phone is attached.

3. A design similar to that in (2) might again use a digital link for signaling and voice communication but have separate power for the electronics (plugs in locally, instead of deriving power for the electronics from the line voltage of the link).

How does an electronic phone work? By our definition, an electronic phone has some digital type of signaling link to the system, and also has a separate link for voice (which might be an analog link, or another link for digitized voice, or that digitized voice might be multiplexed on the same link that carries the signaling information on the state of the telephone). The switching system and the electronic telephone need a receiver and transmitter for this digital information. Typically, when the phone is not in use, or is idle, some type of "idle-channel" data is being sent to the phone, and either retransmitted back, or else the phone itself generates an idle-channel signal to the switching system. In digital systems the signals between the main system and an "outboard" device which indicate "ready" or "busy" status are called **handshaking** or handshake signals. The absence of a handshake from the telephone when it is "interrogated" by the PBX would indicate that something is wrong—perhaps, that the telephone is unplugged. This handshaking gives the switch some way of looking out at its telephones to see if they are, at least, plugged in and functioning. Modern switching systems have testing routines that would indicate a failure of an electronic phone. Usually, these take the form of programs stored in the PBX's control computer. Such programs are called **diagnostic routines.**

The idle-channel data is always being received and transmitted through the link. Something has to change when the phone is being used or information is being requested. If the phone goes off-hook, some signal would go over that data link, telling the switching system to connect a dial tone to that audio link. When you hear the dial tone, that's your ordinary signal to start dialing a number. The numbers are a digital-code representation of the number you are calling. They go out over the digital link, and the switching system interprets these codes, then "echoes" them back to the display, showing you what numbers you dialed. When a call is placed to an electronic phone, a signal is sent over the digital link telling the phone to turn on its ringing generator. (As opposed to ringing a bell, electronic phones have some sort of tone generator.)

The digital information that is exchanged between the electronic phone and its PBX can be transmitted over standard twisted-pair telephone wiring. All wiring is connected with quick push-on connectors, also standard to the telephone industry. This is advantageous, since the same technology that is used to connect standard single-line phones can also be used to connect electronic phones. The existing wiring can be used if a phone is changed-out from a single-line to an electronic, providing, of course, that there are enough pairs. (Remember that although single-line phones need only one pair, some electronic phones use three pairs: for voice, power, and data.)

A clock is a common feature on electronic phones. A signal is sent over the link, requesting the current time, and the PBX returns that information over the line to the electronic phone, where it is buffered and displayed for a few seconds (whatever time is needed to read it).

Let's examine the workings of an imaginary "typical" electronic phone. The "Frobozz 500" has three pairs of wires: a voice pair, a digital pair, and a power pair [see Figure 7–6(a)]. A 40-character LCD display shows one line of alphanumeric information above the keypad. The familiar 12-key dial resembles a DTMF keypad, but isn't. Instead of producing tones, these keys produce 6-bit serial BCD code transmissions when struck. These dial codes travel on the digital pair of wires. Two additional columns of four buttons to the right of the keypad transmit the digital codes that activate special station features in the PBX. Typical of these features, we see buttons for *hold, call transfer, conference, call pickup,* and *call forwarding.* Serial digital codes sent by these buttons are interpreted by the PBX and used to activate the station features desired.

When the phone is on-hook, the microprocessor in the smart part of the phone is running continuously. Every time the PBX sends the phone a code that interrogates it regarding its status, it responds with a "ready" code that says the phone is not in use, but *is* available [see Figure 7–6(b)]. If the phone is taken off-hook, its status changes. The microprocessor senses the change in the hookswitch, then its program switches to a different code when interrogated. Upon receiving this new code, the PBX places a dial tone on the voice line. Now that the dial tone is on the voice line, suppose you dial 252. The codes for 2, 5, and 2 are transmitted in the following way: The microprocessor sense switch closure in the keypad in the same way that it sensed the activation of the hookswitch, and according to the location of the switch that was closed, it carries out a program routine that transmits digital pulses in a pattern that is used only for that key. The PBX has, stored in its common control computer, a table of digital patterns that match the ones in the phone. It identifies each number as it comes in, "echoes" the digit back to your phone, and the microprocessor in your phone displays the digit on the LCD display. When a complete line number has been dialed, the PBX rings the telephone you dialed. "Echoing" the dialed digits back to your phone's display through the PBX is a secure way of knowing whether the numbers received at the PBX are the ones you thought you dialed. If you, or the PBX, messed up, you will know by what appears on your LCD display. Mean-

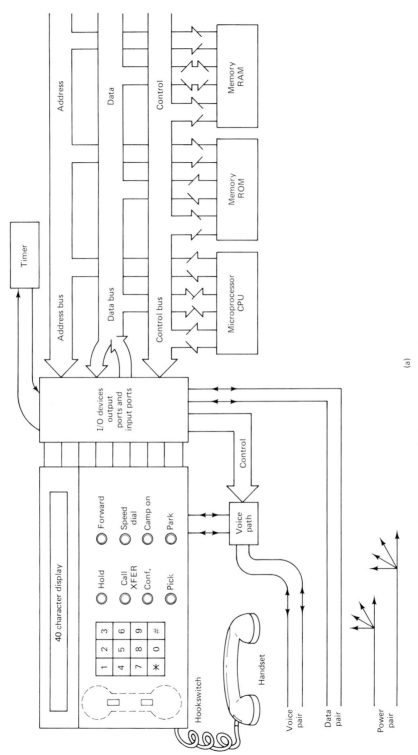

Figure 7-6 Block diagram of electronic phone hardware (a).

(a)

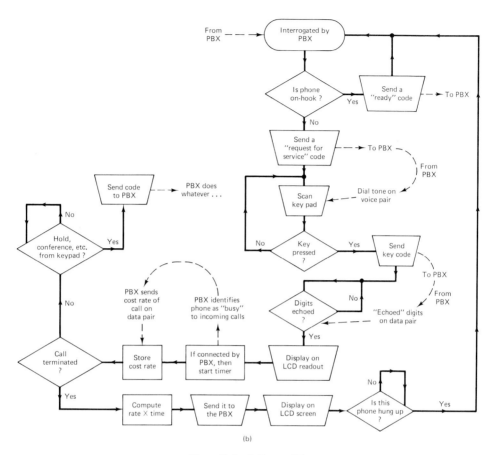

Figure 7-6 Software (b).

while, the phone handshakes with the PBX, which now "knows" that the phone is busy. The PBX also continues to monitor the data link for any action, such as HOLD, that might take place during a conversation. It also knows if your phone was disconnected, not just hung-up, and when you hang the phone up, to remove a busy signal from any other people trying to call you, permitting them to ring your phone. Ordinarily, this would be done by sensing analog loop current line, but in the electronic-phone system, direct transmission of information via the digital link is used to keep the PBX updated on the status of the phone. Our PBX might sense the loop current on the voice line anyway, to "back up" the digital information on whether the phone is off-hook or not, and also, so that if the voice pair is disconnected, the connection would be terminated regardless of the status of the digital path.

The microprocessor in the electronic phone is able to take part of the work load off the PBX's common control computer. For example, a "smart" phone can keep track of the time duration and cost of a call, as well as buffering and displaying information that's returned from the system on the display panel. Let's suppose that we want the system to tell you what a phone call on an electronic phone is costing you, as time goes by. You dial the call—and the PBX, which contains a table that estimates the per-minute or per-second cost of a call on any trunk—sends your phone a figure that says how much the per-second cost of your call is. A memory device in the electronic phone "buffers"—that is, it holds the number transmitted from the PBX—and a program in the microprocessor consults its built-in clock and computes the cost of the call based on the elapsed time and the per-second cost of the call that was downloaded to it from the PBX. From time to time, the microprocessor has to display the current cost of the call on the display. The microprocessor would display this number to you, updating it every second, or every penny, nickel, or dime of the way, until the call is completed. The PBX would not be burdened with this updating task, and the program

$$\text{cost} = \text{rate} \times \text{time}$$

could be handled easily, even by a very simple microprocessor system in the phone. The phone does this work for the PBX, and leaves it free to do more meaningful tasks. If the PBX wants to "know" the total cost of the call at its termination—and in a system like this, the PBX will almost surely keep statistics on this sort of thing—the electronic phone can upload those data to the PBX through its digital link.

Another way to handle the same problem is to have a "not-so-smart" phone handling only data display, handshaking, and digital encoding of feature buttons. In the example just mentioned, the common control computer of the PBX would be handling all the computation of cost, and the electronic phone would download that information from the PBX through the digital link every second or so. The PBX would be doing all the work, and the electronic phone would merely act as the "slave" of the PBX.

This is a completely different philosophy. Everything is done centrally, and no peripheral processing is done on outside, distributed processors. This way, no accidents could befall the cost-of-this-call data before it is transferred to the PBX, because the data is already in the PBX itself as it is computed.

Which method is better to use? At this time, the question of central processing versus distributed processing is not resolved. Security-conscious system designers prefer to have the information all contained at one central location at all times, not distributed throughout the network. The cost, in terms of added overhead for the common control computer in the PBX, is worth it to these designers in terms of having the data in a secure location. Efficiency-conscious system designers prefer having the "smarts" of the system handled by distributed processing rather than central processing. Since the station has to contain a microprocessor, memory, and operating-system program anyway, they reason, why not let the microprocessors in stations throughout the system take care of the overhead so that the common control computer can handle more users more efficiently (and perhaps at a lower cost)? This is one of those trade-off situations where you may find either viewpoint reflected in a real system (sometimes a little of both).

7.3 OPTIONAL EQUIPMENT CONNECTED TO TELEPHONES

7.3.1 The Amplified Handset

The **amplified handset** is a common optional device connected to telephones. It's usually a one- or two-stage transistor amplifier that's powered by the line current of the phone itself (there is no separate power supply for it). Some amplifiers have to be connected to the network of the phone. Various leads are moved around and the phone is rewired to accommodate this amplifier. There's a volume control that you use to control the loudness. The standard control is built right into the handset. A finger-wheel volume control is built into the "grip" of the handset that you can control while you're talking. The amplifier built into the handset usually provides enough amplification and requires no modification to the telephone. In modular phones, especially, the amplified handset can be installed very easily. The standard handset is simply unplugged and the amplified handset plugged in its place.

A device that's often used with an amplified handset is a **background-noise-reducing transmitter**. This is designed to be used in noisy areas where the background noise overwhelms the conversation. It replaces both the transmitter and the cap of the handset. This handset contains a type of directional microphone that's designed to minimize noises from any direction other than directly in front of the handset. Usually, there are no modifications needed to the wiring of the telephone—the standard transmitter and cap are unscrewed, and the background-noise-reducing transmitter and microphone are screwed into the handset. Using the amplifier and noise-reducing transmitter in a factory area, press room, computer printer room, or other noisy area works out pretty well. In a noisy area, amplifying the "receive signal" often isn't enough, since there is a type of feedback that picks up the surrounding noise and amplifies it. This is deliberately done in the telephone—you may recall that it's called **sidetone**. The user of the amplified handset with noise-reducing transmitter can turn up the volume without increasing the sidetone noise in the area at the same time.

The receiver of an amplified handset can vary depending on the process of manufacture. As the quality of parts and manufacture vary, this causes variations in how efficient the receiver is in converting the electrical signal back into the acoustic energy (sound). Some receivers are louder than others, and some have different tonal qualities also. Typically, from manufacturer to manufacturer, there is also some degree of difference. The same applies to transmitters with regard to efficiency. Some pick up more effectively, and produce a greater variation in current for the same sound level.

7.3.2 Speakerphones

The **speakerphone** is another common piece of equipment that is tied together with the telephone. It is especially useful to users with disabilities, and in certain business applications. As we have mentioned previously, the speakerphone allows someone to talk without using a handset. It eliminates the handset and contains a speaker that amplifies the distant party's voice so that anyone in the room can hear the conversation (see Figure 7-7). There is also a microphone, similar to a transmitter, that picks up anybody's voice in the room. One problem that has to be solved with speakerphones is the problem of acoustic feedback. If you've ever played with a microphone and amplifier, you know that the microphone can pick up the music in the area, amplify it, and through feedback, turn it into a shrill screech or oscillations and completely block out any recognizable sound. The same thing is possible with an arrangement like the speakerphone, so design precautions have been built into the speakerphone to prevent that from happening.

Some of the first speakerphones were built so that the volume level, the placement of the speaker, and the placement of the microphone and its adjustment would keep the unit from going into spontaneous oscillation (the feedback condition). This makes it difficult to adjust the speaker volume past a certain point, so this wasn't a very good method. Later models have been made to switch between transmit and receive. Since only one person is talking and the other person is listening at any given moment, this usually works out pretty well. Since the microphone is disabled when the speaker is going, and vice versa, feedback from speaker to microphone becomes impossible in this situation. The speakerphone designed this way has to be able to switch quickly from "transmit mode" to "receive mode" when you stop talking and start listening. The electronic switching mechanism of the speakerphone actually looks at the signal level coming to it from the transmitter and the levels coming in to it from the line. Depending on which level is higher, the switch mechanism decides to shut off the microphone and turn on the speaker, or vice versa. Listening to someone using a speakerphone, for instance, if you're talking to someone who has a speakerphone at the distant end, you might notice that the first syllable of every word they speak is either quiet or cut off. If the first syllable is cut off, that happens because the speakerphone isn't switching quickly enough. Another thing that you might notice is an "echo" or "barrel sound" because of the reverberations and acoustics of the room in which the speakerphone is used. As you might expect, more expensive speakerphones switch more quickly and have better sound quality than do the less-expensive units.

Some electronic telephones and featurephones have the speakerphone feature built right into the phone. Stand-alone units have to be wired up to the telephone.

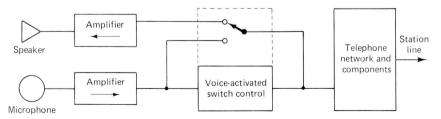

Figure 7-7 Simplified block diagram of a speakerphone.

This may require some changes inside the telephone to adapt it to speakerphone use. Often, this isn't necessary unless you want to be able to use the dial on the telephone. In this case there would be modifications to the telephone—also, the dial would have to be of a type that is compatible with a speakerphone. Since, in effect, the dial of an ordinary telephone is disconnected from the line when the phone is hung up, there must be a way that the speakerphone can turn on the dial in the telephone when the speakerphone is turned on, without picking up the handset. These modifications require the telephone to be opened up so that its leads can be moved around to new points on the network, and leads from the speakerphone can be attached to the network or spliced into the leads that are removed from the network. This is called **on-hook dialing**, with the speakerphone activated. If it's not necessary that you be able to dial out while using the speakerphone, the connections to a speakerphone are very simple. You just pick up the handset on the phone, place the call, and turn on the speakerphone. Once this is done, the speakerphone is "holding the line"—the line current is now going through the speakerphone (the loop current)—then the handset can be hung up and the current through the speakerphone continues to hold up the line connection. After the conversation, the speakerphone can be turned off, and the loop current will stop, effectively "hanging up" the line.

The actual connections of a speakerphone to a telephone are usually packed with each speakerphone. Depending on the manufacturer, these instructions can be very complete or a little sketchy. Also, another variable is the type of phone to which you are connecting it. The connections would have to show how it would go on different types of phones. Getting into the special phones that are unique to a particular vendor and not very widespread or common makes for another problem. If this is the case, and the vendor of the special phone feels that there might be a demand for connecting a speakerphone to it, especially a different manufacturer's speakerphone, then the vendor or manufacturer of a unique telephone might supply instructions that would show how to connect somebody else's speakerphone to their phone. This sounds like a lot of trouble and it usually is.

7.3.3 Automatic Speed Dialer

Another item that's often connected to a telephone is an **automatic speed dialer**, or **auto-dialer**. The automatic speed dialer has a memory device in which you can store a number of phone numbers, and dial them in abbreviated form. This could be a separate unit connected to the phone, or a built-in feature of a featurephone. On my Cobra phone, for instance, I can dial numbers by pressing key B, plus a one-digit number, to dial any one of 10 frequently used numbers.

A good application is in the residential phone or small business phone where there is no PBX with this capability. Local operating companies—telephone operating companies—are even offering speed dialing to residential areas, giving you the option of memorizing about 10 numbers. This is done in much the same way as in the PBX. Using a portion of the memory in the telco switch, the numbers are programmed in through the dial on the phone. Thereafter, a simple key sequence calls the number desired.

Auto-dialers can be either Touch-Tone® or rotary, and they're electronically controlled and made by a number of different manufacturers. Some companies have manufactured a Touch-Tone® dial that replaces the standard dial in a telephone and has the capability of speed dialing. The dial itself has a RAM chip on it to memorize 10 different numbers. Some of the first auto-dialers used a type of card reader in which the number you want dialed would be blacked out in little boxes. The card would be inserted into a reader, and the number dialed.

Another method used a side magnetic tape on which the impulses would be stored. These impulses would be scanned and read, and used to open up a relay contact to imitate the pulse of a rotary dial.

The most popular unit now uses a programmed memory. It usually has a battery that will hold memory if the commercial power is lost. These are either available in rotary or DTMF (Touch-Tone®) versions.

7.3.4 Digression: Ring Detectors and Tape Recorders

Digression: ring detectors

The next three devices use a ring detector that allows the device to be triggered by the 100 or so volts of AC ring voltage on the line. Possibly, for describing these units, it might be good to bring up some methods by which ring-detect circuitry works, before describing these three items.

Ring-detect circuitry is used in a line card in a KSU, the trunk card in a PBX, or the bell in your phone. Of course, in your phone, instead of closing a switch or activating a circuit, the ring voltage rings the bell in your phone. In the other circuits, the ring voltage generally is used to close a switch or activate a solid-state circuit. Usually, when this is done, the circuit activated puts a "loop" or a "ground" onto the line (or trunk) to seize the line and connect the call. The KSU would ring the multiline phone or keyphone, and the PBX would alert the attendant console or accept digits for DID dialing.

Often, telephones are located in noisy areas, and can't be heard. There are devices that can be triggered on the ring signal, and then used to control a loud bell or power horn, for these noisy areas. Just as easily, the device can control a light or a flashing bulb, alerting the people in the area that a call is coming in. This is nothing more than a relay that trips on an incoming ring signal and controls the device. One difference in this type of circuit is that the line isn't actually answered, and that the ringing goes on. A circuit that could be used for this purpose is shown in Figure 7-8.

Second digression: tape recorders

Since the next two devices use tape recorders, showing the idea of how a tape recorder works, and showing uses of the different recorders, would be a good thing to bring up here—how the signal is transformed into the magnetic variations on the tape, and played back. A lot of people just don't understand the "magic" behind a tape recorder. This will help you later to understand a floppy disk recorder or a

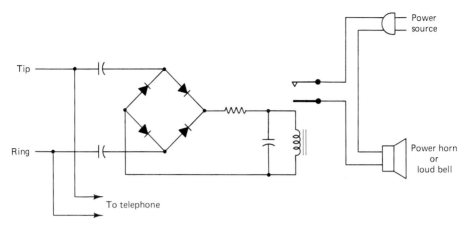

Figure 7-8 Simple ring-detect circuit.

hard-disk unit. I once was explaining to one of the techs how a tape recorder worked. He didn't seem to have any trouble understanding how it recorded, but understanding how it erased just didn't seem to make sense to him. I don't quite undertand that. We'll make sure to explain how to erase a tape, as well as how to record on it.

Recording.   Magnetic tape is a plastic ribbon coated with a material that can be permanently magnetized by a nearby magnetic field. This material is usually ferric oxide (iron oxide), but can be one of several other metals or metallic oxides. A tape recorder acts on a piece of demagnetized tape by exposing it to an electrically controlled magnetic field from the record head of the tape recorder.

The **record head** is an electromagnet activated by current running through a coil. It is something like a horseshoe magnet with a coil of wire wound around it, but with the U shape of the magnet bent almost into a circle, and with a very small gap between the ends of the U. Electrical current through the coil produces a magnetic field in this gap, with very little stray magnetism elsewhere around the U. The magnetic tape that passes the gap does not magnetize in proportion to the current through the coil; it is *nonlinear*. If it is premagnetized, or the recording is done with a certain level of background magnetism always in place, the magnetism on the tape becomes larger and smaller in about the same way that the current becomes larger and smaller; it is more *linear*. This background magnetism is called **biasing**, and it is done by adding a DC currrent to the audio current signal being recorded. For efficient utilization of the tape's response to both positive and negative polarities of current, **bias current** is switched back and forth from one polarity to the other at a high frequency (often 30,000 to 75,000 times a second or more). This **bias frequency** must be much higher than the audio frequency being recorded. It is an ultrasonic frequency that is beyond the range of human hearing. This technique helps further to reduce *nonlinearity* in the tape's response to the audio recording current signal. Biasing makes the design of tape recorders somewhat complicated, but results in far better quality in the sound of the recording.

The magnetic signal thus recorded on the tape is a composite of alternating regions of magnetic polarization, each with a field strength proportional to the incoming audio signal plus the momentary bias level.

Playback.   The magnetic fields recorded on a strip of tape are played back simply by moving the tape past the gap in a head, like the gap on the record head. In some cases, the same head is used for both recording and playback. In other cases, there is a separate **playback head**.

When a changing magnetic field from the moving tape sets up a changing magnetic field in the gap, the coil wound around the head develops an electric voltage. The amount of voltage depends on how rapidly the magnetic field is changing. Circuits that amplify the electric signal from the playback head's coil also filter out the bias frequency. The signal that comes back is proportional to how fast the magnetic field on the tape changed, rather than the absolute value of the magnetic field. If the audio signal were recorded directly without bias, the playback signal would be the rate-of-change wave instead of the audio wave, or as electronics engineers say, it would be a **differentiated wave**. With the bias frequency added, however, the amount of change between a positive and a negative alternation of the bias signal (with audio added) is proportional to the audio level itself. Once the bias frequency is filtered out, the resulting signal is very nearly identical to the original audio wave. This is another good reason why the bias method of recording is used in magnetic tape recorders.

Erase.   The best recording results are obtained when the record head magnetizes a tape that was previously unmagnetized. To demagnetize a tape, an **erase head** is used. It is similar in nature to the record head, but has a wider gap and a larger current passing through its coil.

A high-frequency alternating current (usually the same frequency used for bias) is passed through this erase head. This alternating current sets up an alternating magnetic field near the tape. As the tape passes by the gap, it is exposed to a large alternating magnetic field that gradually tapers off as the tape leaves the vicinity of the gap. As the alternating magnetism applied to the tape decreases, the amount of magnetization in the tape also decreases. When the amount of alternating magnetism tapers off toward zero, the remaining magnetism in the tape is also very near zero. Thus each portion of the tape passing the erase head, and leaving its vicinity, ends up with nearly zero magnetism. This is now a blank tape and can be recorded on with maximum efficiency.

Digital Recording.   In some methods of digital data recording, the bits of binary data are recorded using **saturation recording**. No effort is made, in these methods, to make the tape behave in a linear fashion. The pulses used to record are simply "overdriven" onto the record head, and saturate the tape with 100% magnetization. Various methods of playback and identification of a 1 or a 0 binary level are used, but in all these saturation recording methods, there is no bias frequency used, and none is needed. Saturation recording techniques are also used for floppy-disk and hard-disk magnetic data recording. In each of these cases, and with most digital tapes as well, the current through the record head is always at some saturation level, either "high" or "low" logic being represented by a strong current. If there are no unmagnetized spaces between magnetized regions on the magnetic recording medium, no erase cycle is needed, as the head saturates over any data already present. Schemes that use pulses of magnetism with unmagnetized states in between are called **return-to-zero** (RZ) recording methods. In any case where RZ recording is done, the tape must be erased before being reused for new data.

7.3.5 Message Announcer (Intercept Recorder)

Another item that can either be connected to a station line or a telephone line, or possibly to a line that might be accessed by any caller in a PBX system, is called a **message announcer**. This unit is used to answer an incoming line, and it has a recorded message on it such as "I'm sorry your call cannot be answered. The office is closed," or "The films that will be showing tonight at the Odeon theatre are. . . ." Typically, a magnetic tape recorder is used for this purpose, although recently, digital memory, combined with voice digitizers and undigitizers, have been used. As with other devices using semiconductor digital memory, these devices should have battery backup to avoid loss of all recorded messages if power fails. Another name for this item is **intercept recorder**. It gets that name from uses in the telephone company's central office, where it's used to do things like intercepting calls to nonexistent numbers or improperly dialed numbers, or perhaps a number that's out of order. When you dial the weather, you're dialing into an intercept announcer.

7.3.6 Telephone Answering Machine

Although it's unlikely that you'll find message announcers in many homes, the telephone answering machine is first cousin to the message announcer, and is a popular item in homes everywhere. It not only delivers a message, but gives the caller the ability to record a message. Some of these are almost as simple as the tape recorder itself; others are more complicated and are made for easy transcribing of messages by a secretary. There's a place to plug in a headphone, and usually a foot pedal to turn the tape on and off to make it possible to transcribe the messages for the people. In more sophisticated units, you often have the option of using the Touch-Tone® dial to stop, rewind, or fast-forward through the message you are recording. This enables you not only to record a message, but to edit it. This unit would have to have some kind of register to recognize DTMF tones, decode those tones, and use those tones to rewind, erase, or what-have-you. Still other units are able to answer an incoming call, ask questions, pause, wait for answers, trigger off the amplitude of the incoming voice signal to keep the recorder running, and then after the voice stops, perhaps stop and ask another question. Using this kind of controlled dialog (one that's programmable) is the next best thing to having a person answer your telephone inquiries. Later you would probably call your callers back with all the answers to their inquiries.

These devices could either be shared by all users in a system, or could be used by a person to answer one line only. Voice message units that are connected to PBXs now are even an integral part of some PBXs. In these systems there is no need for stand-alone recording devices, and many of these optional devices that were developed before switching systems (PBXs) had these capabilities.

7.3.7 Call Diverter

Another type of stand-alone item is a **call diverter**. Calls coming into the unit would be diverted to another number that's programmed into the unit. You could use something like this at an office, so that anyone calling the number of that office would trigger the unit. It would dial to a different number and then connect the call through. This duplicates the call-forwarding feature built into modern PBXs and available to local telco subscribers for an added fee. Some PBXs will only forward a number within the PBX's system itself. With others, you can set your phone to forward even to an outside number. There are delays, of course, encountered in triggering the diverter and in its dialing the number out.

7.3.8 Restriction Dialing Unit

We were talking about a speed dial, which actually replaces the Touch-Tone® dial on a phone. Another device that replaces the dial on a phone is a restriction dialing unit. This could be set so that only 10 digits can be dialed, which makes it difficult to dial long distance, where often more than 10 digits have to be dialed. Again, this special dial mimics features that are found on PBXs.

To determine the programming of these dials, there's usually a "dipswitch" on the back of the dial which you can program to set the maximum number of digits that may be dialed out of the phone, or to prevent dialing of numbers whose first digit is a 1 or a 0. Of course, you have to take the phone apart to reach the dipswitch, but someone who knows how to set the switches could always defeat this restriction. This type of limited dialing ability is generally known as **toll restriction.**

7.4 DATA TERMINALS

There are a great number of terminals manufactured by different makers, having various features that are unique to each manufacturers' machines. They have several things in common, however. The most standard feature possessed by all terminals is a keyboard. The standard (Sholes) typewriter keyboard arrangement is adapted to the terminal (although there are variations on key arrangement, such as the Dvorak keyboard). In terminals, and on typewriters, for that matter, there are character keys and function keys. The **character keys'** purpose is self-evident; the "S" types an "S," the "Q" types a "Q," and so forth. The **function keys** on a typewriter handle functions as common as shifting the characters on the keyboard from lowercase to capitals, or as exotic as switching from the black ribbon to the red ribbon. On the terminal, the control keys change the function or mode of operation in a similar way. Function or control keys all have one thing in common: they don't produce characters (don't type anything) when they are pressed. Some function keys on a terminal even control the unit on the distant end; for instance, the carriage return key will make the distant terminal—if it is printing on paper—move the print element or the paper so that the next character will be placed at the left-hand edge of a line and printing will proceed from there. A linefeed key will roll the paper up so that the printing will take place on the next line down on the sheet of paper. Terminals with video screens respond with similar actions, except that in the place of a type element, a screen has a block, underline, or marker called a *cursor* that shows where the next character will be printed. Most teleprinters combine linefeed and carriage return in a single key called *enter* or *return*. An indicator found on all typewriters is a bell that rings when you approach the end of a line. In the data terminal, this bell or tone indicator can be operated by the person calling you as well as by your own actions.

7.4.1 Keyboards

Keyboards are the primary input device to a terminal at the person/machine interface. The output device at the person/machine interface is usually an **alphanumeric display.** The word "alphanumeric" is a contraction of "alphabetic" and "numeric" and represents any print display. These displays come in two forms, hardcopy and readouts. **Hardcopy** displays produce printed sheets with words, numbers, and perhaps even pictures. **Readouts** display the same information in a transient form on some sort of readout panel, such as a TV screen, LED panel, or LCD segmented display.

7.4.2 Hardcopy (Printers)

Hardcopy for printing terminals is usually thermal or needle matrix printing. In both of these methods, print is formed of an array of dots. It doesn't look as good as the print from a typewriter, but it is possible to do it much faster than a typewriter can, and the mechanism is much less expensive than a mechanism for printing fully formed "letter-quality" type. **Needle printers** print by striking an inked ribbon with electromagnetically driven "needles" which make the individual dots. Almost any kind of "plain paper" can be used in a needle printer. **Thermal printers,** on the other hand, print on chemically treated paper using small heating elements which cause the chemical coating on the paper to change color. Only thermally sensitized paper can be used in thermal printers (usually blue or black), and the printouts are still sensitive to heat after being printed. (Don't lay your printouts from a thermal printer on the radiator!)

7.4.3 Readouts (Displays)

Readouts display print on a screen, again using dot matrix characters. The screens display the dots that form the characters by manipulating light in some way. TV screens, also called video or CRT displays, use precise timing of electrical pulses to control the brightness of a spot as it scans across the surface of a TV picture tube (a cathode ray tube, or CRT). By controlling the exact times at which the spot is bright and dim, dot-matrix print can be formed on the screen; since the scan pattern (the **raster**) is repeated in exactly the same amount of time every scan, the position *where* a dot appears on the screen can be controlled by *when* the beam (a beam of electrons, called a cathode ray) is "gated" to produce maximum and minimum levels of brightness. For instance, most CRT displays used in video display terminals (VDTs) scan a picture in 1/60 of a second (16.667 milliseconds). Suppose that you gate the electron beam ON for a very brief time, at 1/120 of a second after the beginning of each scan (8.333 milliseconds). The scan would be halfway down the screen every time you gated the beam ON, and there would be a bright spot precisely at the center of the screen. It's a long way from placing a dot in the center of the screen to placing dots that form all the letters of a 200-word page in places all over the screen, but it's just a matter of timing a more complicated pattern of ON and OFF cycles. The basic principle—exact timing to match the action of the raster scan—is still the same.

7.4.4 Machine/Machine Communication

In talking about keyboards and displays, we have considered the human/machine interface between the terminal and its operator. There is also, and equally important, a machine/machine interface between your terminal and the one at the distant end. These communicate to one another in digital codes, the most common of which is **ASCII** (the American Standard Code for Information Interchange). In this code, 7 bits, or digital states of ON and OFF, are used to represent the characters of the alphabet, the numbers, and punctuation marks. Other codes are used, too, but generally they are transmitted serially, as ASCII is, one bit at a time, with an entire character, letter, number, or whatever being transmitted at consecutive moments in time. Some terminals produced by IBM use **EBCDIC** (Expanded BCD Interchange Code), an eight-level code with an extra bit for parity instead of ASCII, since it is the code used in large IBM computers. Another code is **Baudot**, a five-level (5-bit) code used on older Teletype (TTY) terminals. TTY terminals that use this code are no longer being produced by Teletype.

7.4.5 Modems

In an analog telecommunications system, the signaling is designed primarily to communicate voice. A **modem** is used to convert digital data into a form that can easily be carried on analog lines. It performs the task of converting serial data bits into audio-frequency signals, and vice versa. The modem derives its name from the fact that when it is transmitting audio frequencies, it is converting the 1 and 0 voltages from (usually) +12 volts and −12 volts into two different frequencies, **mark** and **space.** This is called **frequency modulation,** or in the case of two frequencies, **frequency shift keying** (FSK). The modem gets the "mod" end of its name from its **modulation** of voltage levels into frequencies when it is transmitting. When mark and space frequencies are received, they are converted by the modem into the 1 and

0 voltages, a process called **demodulation.** The modem gets the "dem" part from its demodulation of the frequency signal. Two frequencies are used to represent the *high* (1) and *low* (0) state of digital logic. Usually, the mark, or 1, state is represented by a lower frequency, and the space, or 0, by a higher frequency. The words "simplex." "half-duplex," and "full duplex" refer to transmission and reception of the digital signal. **Simplex** means "one way only," with one end of the telecommunication link functioning as a transmitter only and the other end functioning as a receiver only; it is almost never used. **Half-duplex** is transmission in two directions on a single line, alternately—this is necessary if both ends of the digital link use the same 0 frequency (a space frequency) and the same 1 frequency (a mark frequency). In **full-duplex** data communications, one pair of frequencies is used for marks and spaces from one end of the line, while the other end uses a different pair of frequencies for its marks and spaces. In full-duplex communication, **answer** and **originate** refer to the "calling" party and the "called" party. The terminal doing the "calling" is originating the communication, and the "called" terminal is answering. Frequencies of 1070 and 1270 Hertz are used to represent the mark and space for answer, while 2025 and 2225 Hertz represent the mark and space for originate. The **baud rate,** or rate at which data mark and space signals are sent, is usually 300 per second (300 baud). The unit of data transmission speed is also called **bps** (bits per second). This term is being used to supersede the baud, and is defined in a slightly different way, but at present the two terms are used interchangeably in telecommunications jargon. In full-duplex communication, information is transmitted both ways at once on the same line, by separating the "northbound" signal from the "southbound" signal using different frequencies.

7.4.6 Host Computer

Generally, terminals do not talk to other terminals very often. Most of the time, they are dialing into a dial-up line that connects to a **host computer.** The host computer is generally in the answer mode. It receives communications from terminals throughout the communications network, and may relay information from one terminal to another. If the host computer is attached to the public network via a modem, it can be reached by anyone with a terminal who knows the correct telephone numbers, and the "passwords" that get by the computer's security protection. (If you saw the movie "War Games," you probably know what problems the WOPR had because it was connected to the public network.)

One use of modems is to connect terminals together through the public network or through any analog network that works primarily with voice transmission. In that kind of system, every terminal and host computer is connected to a modem, and thence to the system.

There are a growing number of systems that use digital signaling to transmit *everything,* whether voice or data, by using PCM (pulse-code modulation) or delta modulation. In **pulse-code modulation** the analog signal or voice signal is sampled, and a digital code is sent out that indicates the amplitude of the sample. This is called the **pulse code,** and through it, the original analog signal can be reconstructed, at least the parts that were sampled. In **delta modulation,** the digital code indicates whether the sample was larger or smaller than the preceding sample. As with PCM, the samples must be taken at a fast enough rate to keep up with changes in the signal, or the reconstructed waveform will not be accurate. For voice information, samples may be taken 8000 or 12,000 times per second. With the ordinary voice bandwidth of the telephone set from 300 to 3000 Hertz, samples taken at 8000 or 12,000 Hertz can reproduce the voice reasonably well.

The terminal or host computer, or any digital equipment hooked into this system, does not need its signals to be modulated and demodulated. A voice station, however, needs a **codec** to "talk" over these lines. "Codec" is contracted from "coder" and "decoder." It performs the task of sampling and "coding" the voice signal being transmitted, and of reconstructing the analog signal from the digital codes that are received.

7.4.7 Protocols

Since the digital station equipment in this kind of system is transmitting on digital transmission lines, you might think that no interfacing at all is necessary. You would be wrong. The line and the terminal might employ different protocols to transmit their data. What's a protocol, you ask?

A typical terminal might have a plug with 25 pins, some of which carry signals such as the following:

1. **Transmitted data.** Bits are going out from terminal.
2. **Received data.** Bits are coming in to terminal.
3. **Ready to receive/terminal busy.** This signal indicates whether a terminal is busy or waiting for data.
4. **Request to send.** This signal indicates that your terminal may transmit to the distant end any data that it has.

These signals use four of the pins on the plug. Two of the pins have the actual data traveling on them; the remaining two pins are handshake signals that keep the data from overflowing the capacity of the terminal or host computer to handle it. In this protocol, the Ready to Receive signal should be activated before the machine at the distant end sends the terminal more "receive data." The machine at the distant end should wait for this signal to become active. Since this involves waiting, taking turns, and being polite, this kind of handshaking is called **protocol.** The Request to Send is the other guy's Data Terminal Ready. When the machine at the distant end is ready to receive, *your* machine must be informed so that you don't transmit while that machine is busy. By the same token, your Transmitted Data is the other person's Received Data, and vice versa. What your machine has to say is what the other machine has to listen to, and so on.

These plugs are obviously intended to connect to other plugs just like them. Originally, the protocol just defined was intended for direct connection of one device to another through these plugs. We have, however, chosen to connect the devices indirectly, through the telecommunications system.

The telecommunications link connecting the digital devices has only one path for signals, not four. What do we do? We have to multiplex the four signals onto one path. A data terminal interface between the devices and the link must place each signal in turn on the line, providing (possibly) information telling which signal it is. This is called time-division multiplexing (TDM). On the transmitting end the data is multiplexed, and on the receiving end the data is demultiplexed. You would expect telecommunications people to call this thing a "muldem," but they didn't. Its shortest name is **mux/demux,** and they often refer to the thing by its unabridged name, **multiplexer/demultiplexer,** in spite of its awkwardness. The protocol used on the digital line is defined by when the Transmitted Data, Received Data, Ready to Receive, and Request to Send signals are placed on the line, and what codes accompany them, if any.

A modem would not be necessary in an RS-232-based local-area network (LAN), where the terminals and the network are compatible. The digital stations (terminals) in such a system would contain their own interfaces, and externally, stations would just be connected together with wires.

Data terminal equipment can have the data terminal interface (to a digital line) or modem (to an analog line) built in. If this equipment is made for a specific system, it may have no data terminal interface, but this function may be designed (or programmed) into the equipment itself. A separate data terminal interface would always be needed for converting from one digital format or speed to another. The most common example is RS-232 to any transmission system; there's always some kind of data terminal interface to convert between the two protocols, as the multiplexers do in the example mentioned earlier.

7.4.8 Automatic Data Collection Unit

An automatic data collection unit is a centrally located processor that picks up data from lots of remote terminals. It commonly uses the public network and incorporates an automatic speed dialer to "poll" different stations, requesting that information be updated from each. For example, a network of automatic weather-recording stations could be connected to the public network at scattered sites throughout the country. A central processor with an automatic data collection unit would use its speed dialer to poll each station in turn. The station's modem would answer the call and dump its information into the central data base through the public network. A system like this runs automatically, and people aren't involved in taking the measurements. Although the transmission link in this example was an ordinary dial-up phone line, digital lines could easily be used for the same purpose.

REVIEW QUESTIONS

1. What types of station equipment have become standardized, and what types are still in the process of development?
2. What are some of the methods by which standards become established?
3. What are the major parts of a single-line telephone, and what are they used for?
4. What are the advantages of the DTMF over the rotary dial?
5. Which is more nearly binary signaling, rotary or DTMF?
6. What is another name for multiline phones?
7. Are more pairs of wires used to connect the KSU to a mechanical keyphone or electronic keyphone?
8. Where did the author see a "banana phone" displayed?
9. Would a featurephone be likely to contain a microprocessor to do the things that an electronic phone relies on the PBX to do for it?
10. Is digital or analog signaling used between an electronic phone and the PBX?
11. What information is exchanged between an electronic phone and the PBX?
12. What is a speakerphone? What has its operation got to do with *half-duplex* operation? How does this avoid feedback?
13. What different devices use the magnetic recording principle?
14. What is a bias frequency?
15. What is the difference between a record head, a playback head, and an erase head? Are these all present on a *floppy* or a *hard disk?*
16. What part of a terminal performs an analogous function to the dial on a telephone?
17. Name one advantage, and one disadvantage, of a CRT readout compared to a hardcopy printer.
18. What is the most common code used to transfer information between terminals in the United States?
19. How does a modem use FSK (frequency shift keying) to communicate in full duplex on a single line?
20. What is meant by answer mode and originate mode in reference to two-way modem communications?
21. What is codec an abbreviation for? What is accomplished when the codec is converting voice information for transmission?
22. What is meant by the terms protocol and handshake?

8

Transmission Media

The goal of Chapters 8 to 10 is to describe the path by which information is communicated from one place to another in a telecommunications network, and in what form information is carried along that path. We will refer to the physical paths that carry information as **media.** The information itself is converted to, and carried in, various forms we will call **transmission forms** (commonly referred to as "the signal"). The actual **transmission** is, for example, a phone conversation or a screen full of text to be displayed on a video display terminal. When the transmission form is carried from one place to another, it consists of two parts. The first part is the information itself, and the second part is the signaling that makes it possible for the voice or data to reach its intended destination. In other words, the medium carries the information and signaling in a transmission form that is compatible with that medium.

Our objective in this chapter is to tell you about various transmission media. In Chapters 9 and 10 we discuss how transmission forms and signaling are used. In previous chapters, our goal in describing transmissions and transmission links was to tell you what they are, but here our goal is to tell you how they work.

In Chapter 7 we discussed station equipment, which creates and propagates the signals that we transmit in a telecommunications system. In this chapter we look at the paths the signals travel, and the signals themselves; in later chapters, we look at the switching network that is responsible for making the proper connections to link one station to another.

8.1 MEDIA CHARACTERISTICS

All signals used in telecommunication travel along electrical conductors at some point in their travels. There are different varieties of signal paths that use conductors, but they all have certain characteristics in common.

8.1.1. Resistance

Conductors used for telecommunications all have resistance. Electrical current travels through wires in the form of free electrons. These are the negatively charged outer particles in the atoms of metal that make up the conductor. They are loosely held in the vicinity of the atom, but can move freely through the metal of the wire when a force is applied to them. A type of friction hampers the passage of electrons through the wire. Like fluid friction encountered by water flowing in a pipe, the friction electrons encounter limits their rate of flow to a speed that depends on how much force or pressure is applied to them. We call this friction **resistance.**

The fan-blade analogy

This resistance "friction" arises in the following way: Imagine that an atom is like a fan with rapidly whirling blades. The electrons in the outer orbitals of each atom in the metal conductor are the blades of the fan. If a free electron "bumps into" the atom, it will be bounced off like a marble thrown into the fan. There is space between the blades, but you are unlikely to throw the marble through it.

Now, slow down the fan so that the electrons sometimes get through the blades, and sometimes are hit and driven in unpredictable directions. Now suppose that a "flow" of electron "marbles" is taking place through a conductor "tunnel" filled with spinning fans. An electric field in the conductor pushes the electrons along, in much the same way that gravity would if our conductor "tunnel" were tilted downhill. The electrons start rolling downhill through the conductor, picking up speed as they go along, in accelerated motion. Sooner or later, every electron gets whacked by a spinning fan blade in the form of a bound electron orbiting in some atom's outer shell. It then bounces off in an unpredictable direction, usually not downhill. Then the electron has to be slowed down by the electric field "gravity" and starts rolling downhill again.

In this way, electrons flowing "downhill" through a conductor are victimized by countless collisions with atoms. They never do get to pick up very much speed, since they are constantly being bounced backward by collisions, and as a result, they go from one end of the conductor to the other at an average speed that depends on how much speed they pick up between collisions.

8.1.2 Effect of Changing Materials

Now imagine that we change the fans. If they have more blades, or are whirling more rapidly, the rate at which electrons pass through the conductor changes. What are the fans with more or faster blades? In one case they might be atoms of a different element. That means a different metal or different type of conductor. Depending on the electron structure of the shells in which the bound electrons orbit, each atom could act like a fan with more or fewer blades, going at greater or lesser speeds.

The material that is used in the conductor determines how much resistance it has.

8.1.3 Effect of Changing Temperature

Another reason that the electrons might have an easier time getting through a conductor is its *temperature*. Remember what we said would happen if we slowed down the fan blades?

An atom's orbitals are like the blades of a fan. The speed with which they wobble around depends on the temperature of the material the atom is in. If we cool down a piece of conductor metal, it loses energy. Its atoms do not bounce around as much, and electrons are likely to pass between atoms without being hit as often. This means that they can reach a higher speed between collisions, on the average, than electrons in a warmer material. They flow at a faster rate with the same force applied—we say that the material has less resistance. Some materials, when cooled sufficiently, have no resistance at all—this is called **superconduction**—but the temperatures are so low that this technology is unlikely to see any practical use in telecommunications until a high-temperature or room-temperature superconductor is found.

The temperature of a conductor determines how much resistance it has.

8.1.4 Conductor Size

As you may recall from Chapter 2, the size of the conductor determines its resistance in two important ways:

1. **The longer it is, the more resistance it has.**
2. **The broader its diameter is, the less resistance it has.**

Table 8-1 shows the resistance of copper conductors of various diameters at two temperatures for 1000-foot lengths of wire. Wire sizes are commonly referred to as **wire gauge** and are identified with numbers that get larger as the wire gets smaller.

TABLE 8-1   WIRE GAUGE AND RESISTANCE CHART

Wire gauge (AWG No.)	Diameter (inches)	Resistance at 20°C per 1000 feet (Ohms)	Resistance at 0°C per 1000 feet (Ohms)
26	0.016	40.81	37.60
22	0.025	16.14	14.87
18	0.044	6.38	5.88
12	0.081	1.59	1.46

Ohm's law

In Chapter 2, we also mentioned Ohm's law, which describes *how much* current (electrical charge rate of flow) will pass through a conductor with a certain electromotive force (voltage) applied and a certain amount of resistance.

8.1.5 Skin Effect

Electrons are negatively charged. They repel other electrons, which are also negatively charged. Since electrons push each other apart, you might expect them to arrange themselves inside the wire so as to be as far from each other as possible. They do, and this means that they are clustered at the outside edges of the conductor. When the charges travel mostly on the outer surface of the conductor, instead of throughout its thickness, this reduces the effectiveness of the conductor for carrying current. The effect produced by this tendency is called the **skin effect**, and it is more intense at high frequencies than at low frequencies. At high frequencies, the electrical fields in wires can be more intense than when a DC current flows in the same wire. The more intense electric field causes greater concentrations of electrons to build up in places, and their repulsion increases. This results in even less of the wire being usable for electron flow. The skin effect thus produced results in wires that have higher resistance at high frequency than at low frequency. You will remember that this result comes from the fact that a narrower conducter is harder to get through than a wide one—it has more resistance—and the wires which conduct only through their outside "skin" are in many ways like a narrower conductor, since the inside is no longer used for conduction. **At low frequencies, like voice frequencies, the resistance of a wire is less than at high frequencies, like microwave frequencies.**

Note: Microwave waveguide is not solid conductor for this very reason. The outer conductor of a waveguide is like a solid conductor that has had its middle cut out, making it a hollow tube. If you don't use the center of the conductor, why bother having it?

8.1.6 Power Loss due to Resistance

The primary effect of resistance on a signal traveling through a conductor is to consume power from the signal as it passes down the length of the conductor. The amount of power used up depends on the conductor's resistance and the amount of current passing through the conductor. An electrical current passing through a conductor is composed of moving charges. As these charges do work getting through the resistance in a length of conductor material, they lose energy. The amount of energy lost by each unit of charge is called the *voltage drop*, as mentioned in Chapter 3, and the power used up in the length of wire is the voltage drop multiplied by the amount of current flowing. Since the voltage drop depends on resistance:

$$(\text{Ohm's law}) \quad E = I \times R$$

and the power is related to voltage and current:

$$P = I \times E$$

it turns out that the power of a signal passing through a conductor is used up at the rate

$$P = I \times (I \times R)$$

which can also be written

$$P = I^2 R$$

What all this means is that a signal current traveling down a wire loses strength if the wire has any resistance at all, and wires used in ordinary telecommunication *always* have some resistance. As a result, the longer a wire, the more strength is lost from the signal as the signal travels greater and greater distances. This was described in detail in Chapter 2. In summary, the primary effect of resistance is greater loss of signal strength as the signal is transmitted over longer conductors (since the resistance in the longer conductor is larger).

8.1.7 Capacitance

The effect of charge storage, or capacitance, between two conductors was discussed in Chapter 3. Its effect on AC signal strength was described. The effect of AC attenuation on signal strength is somewhat like that of resistance, but varies according to what frequency of AC transmission is used (resistive signal attenuation does not). With every signal path, there must be two conductors—to and from the *load*—to make a complete circuit. The conductors have different voltages whenever a signal is present, and there is *always* some capacitance between the two conductors.

8.1.8 Frequency Response

The strength of an AC signal is weakened as it is transmitted down lengths of conductor. The capacitive reactance of the signal path attenuates the signal progressively as longer conductors are used, but also as higher frequencies are used, signal strength is depleted more rapidly. This means that higher-frequency signals don't get as far down a capacitive line as do low-frequency signals.

8.1.9 Loading Coils

In Chapter 3 we noted that coils of wire called *inductors* can counteract the effect of capacitive AC attenuation. This allows signals to be sent longer distances without losing the high-frequency part of the signal. At least, the high-frequency part of the signal isn't lost at a faster rate than the low-frequency part, although all parts are still attenuated by resistance.

8.1.10 Crosstalk

In Chapter 3 we brought forth the idea of a **transformer**. In the transformer, signals in a wire called the **primary** were able to induce, or develop, signals in another unconnected wire called the **secondary**. This effect is called **mutual inductance**. This effect allows a signal from one wire to pass into, or **induce**, a signal in another wire that is not electrically connected to the first. Multipair cables with multiple pairs of wires traveling long distances alongside each other comprise "accidental transformers," with mutual inductance between the wires carrying the primary signal and other wires in the same cable, which we can think of as secondaries. This mutually induced voltage can result in signals of significant strength appearing in wires that are not supposed to be carrying that signal [see Figure 8-1 (a)]. When this happens, the stray signals that "get into" the unconnected wires are called **crosstalk**. Another method of getting a signal into an unintended set of wires happens when the conductors actually make contact and are connected. This is called a **short circuit**, but we will discuss that when we discuss insulation and insulators.

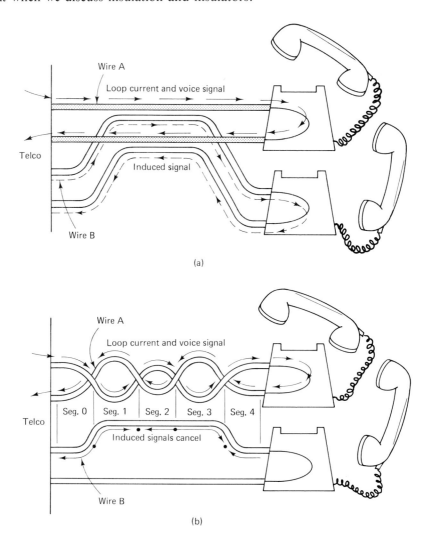

(a)

(b)

Figure 8-1 (a) Crosstalk. (b) Twisted-pair.

To prevent the mutual inductance of a wire from developing a "secondary" signal in its neighbor, both wires in the loop are twisted together. This is called a **twisted pair**, and is able to prevent crosstalk (due to mutual induction) in the following way. In Figure 8-1 (b) a twisted pair is carrying current to and from a receiver. The twisted pair runs alongside a single wire called B. In the part of wire B identified as segment 1, current in the adjacent part of the twisted pair is flowing toward the receiver. In segment 2, a little farther along, the adjacent part of the twisted pair is a wire carrying current back away from the receiver. In segment 3, the adjacent wire carries current *into* the receiver, and in segment 4, *outward* from the receiver, and so on. The straight wire called B is subject to induced forces that are trying to both push and pull on the charges in the wire. The pushes and pulls cancel out, and the net effect of the mutual inductance from the twisted pair to wire B cancels out.

8.1.11 Balanced and Unbalanced Lines

Another source of crosstalk that might be considered apart from mutual inductance is the capacitance between two wires. If two conductors share a common electrical point, they can develop voltages relative to one another, and this causes charges to be moved around in one conductor by the electrical potential difference of another nearby conductor. We saw in Chapter 3 that this is called capacitance. To prevent this from happening, it is necessary that separate loops or pairs must not share a common electrical point.

In Figure 8-2 (a), two loops share a common connection to an electrical point called **ground**. In the two wire loops, one wire of each loop is connected to this point. The other wire has some voltage with reference to the ground, which depends on the signal it is carrying. In wire A, the signal is becoming more positive. In wire B, which is alongside wire A, that positive electrical field is forcing positive charges out of wire B, even though there is no signal applied to wire B's loop. The grounded conductor of loop B is still connected to ground, and it does *not* become more positive. Suddenly, wire B of the loop is positive, and the other wire of the loop isn't, and a current begins to flow through the receiver. As a result, there *is* a signal in receiver B, and any other signal put into wire B's loop will merely be added on to that signal. If someone is talking on the wire B loop, they will hear the conversation of the wire A loop between words.

This is also **crosstalk**, and it is a problem that is solved by **balancing** the wires in loops A and B. This balancing amounts to isolating loop A from loop B by disconnecting one or both loops from the ground (or any electrical point that they might be connected to in common).

A **balanced line** (loop) now has wires with no reference point, as shown in Figure 8-2(b). In this case, loop A has one of its wires grounded, but neither wire in loop B is grounded. Wire A has no voltage with respect to wire B. Since there is no point in loop B that connects to the ground, there is no way that wire B can be more positive or less positive than wire A.

I know that this is a hard point to understand. If wire A is more positive than

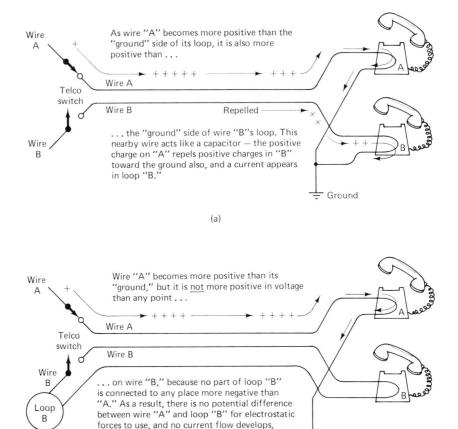

Wire
A
+

As wire "A" becomes more positive than the
"ground" side of its loop, it is also more
positive than . . .

+ + + + + ———————→ + + +

Wire A

Telco
switch

Wire B   Repelled ————————→

. . . the "ground" side of wire "B"'s loop. This
nearby wire acts like a capacitor — the positive
charge on "A" repels positive charges in "B"
toward the ground also, and a current appears
in loop "B."

Wire
B

A

+ +
+ +

B

Ground

(a)

Wire
A
+

Wire "A" becomes more positive than its
"ground," but it is <u>not</u> more positive in voltage
than any point . . .

+ + + + ———————→ + + + +

Wire A

Telco
switch

Wire B

Wire
B

Loop
B

(balanced)

. . . on wire "B," because no part of loop "B"
is connected to any place more negative than
"A." As a result, there is no potential difference
between wire "A" and loop "B" for electrostatic
forces to use, and no current flow develops,
because positive charges in "B" are not driven to
ground by the positive "A"-to-ground voltage.

A

B

Ground

(b)

Figure 8-2   (a) Two lines with a common ground. (b) Two balanced lines.

it was *before*, don't all the other wires in Figure 8–2(b) "see" the increase in its positive electric field? Of course they do! That's not the point, though.

If the wires in loop B are unconnected to any ground electrical point, *both* wires of the loop become more positive at the same time, under the influence of wire A's electric field. Neither wire in loop B is more positive than the other wire in loop B. Electrically, we say that the voltage in the two wires of loop B is **balanced**. No current will flow through the receiver in loop B, because neither wire is more positive than the other. The only currents that will flow in the receiver of loop B are the ones put into that receiver directly. (*Note:* Current does not flow through a load connected to two wires unless one is *more* positive than the other. As long as they are equally positive, electric charges don't have any "downhill" direction in which to flow.)

Now, that's pretty hairy, but what it comes down to is that the wires running alongside one another in the cable of Figure 8–2(b) don't transfer a signal from loop

A to loop B by capacitive charge induction unless the two loops share a common connection somewhere. As long as the two loops are electrically isolated from each other, everything is okay.

In reality, you can't *ever* get every loop in a cable to be *completely* isolated from all the other loops. Weak coupling between loops is always a possibility, and when that happens, there will be a certain degree of crosstalk. Poor connections or a bad splice between wires in a cable can also give rise to crosstalk, because the resistance of the splice drops voltages on one side of the loop that are not balanced on the other side of the loop (unless an identical bad splice exists in the same place on the other wire of the loop—an unlikely situation).

8.1.12 Twisted-Wire Pairs

The most obvious way to carry an electrical signal is by passing it through a conductor. Since electric current is the circulation of electric charge around a complete circuit, two wires are needed to carry a signal, one for the "outbound" current and one for the "returning" current. Communication on wire **pairs** is the oldest and simplest method of electrical transmission. As we discussed in talking about crosstalk, it is better to transmit signals using a twisted pair than with straight wires. Each wire is covered with a layer of insulation (to keep it from touching the other wire and making an electrical contact). The two wires of a loop are placed alongside each other and wound together into a double helix, as we saw in Figure 8–1. In a cable carrying many pairs, each pair is twisted first, then the twisted pairs are laid alongside each other and the bundle is enclosed in another layer of protective material. The bundle of twisted pairs may even be twisted together before being bound in a protective jacket. Even so, sometimes, improperly twisted pairs will produce crosstalk between unconnected loops in a telecommunication cable.

Inside installations—that is, connections inside a private telecommunications system—usually use 24- or 26-gauge copper wire. This wire is about the thickness of a sewing needle. The wire is wrapped with a layer of insulation somewhere between the thickness of a sheet of paper and a sheet of cardboard, depending on what material is used for insulation.

8.2 NONCONDUCTORS AND INSULATORS

Insulation is the name for the coating or wrapping you find around copper wire in ordinary wiring. Commonly, solid wires are coated with a layer of rubber or plastic which is a nonconductor, and through which electric current cannot pass. Older wiring was wrapped with cloth and cloth/rubber composites, and some woven fiberglass cloth wrappings are used for insulation today. The major characteristic that makes insulators useful is the fact that they are made of materials that contain few free electrons. Insulators contain very low amounts of free-charge carriers of any kind. The atoms or molecules of insulators hold on to the electrons in the outer orbitals with greater tenacity than the atoms or molecules of an electrical conductor.

There is also little tendency for atoms of insulating materials to become charged themselves.

If an atom or molecule becomes electrically charged by losing or gaining an electron, the atom is called an **ion**. Substances that contain ions can carry electric current even if there are no free electrons, so to be a good insulator, a nonconductor must have few free electrons and also be difficult to **ionize** (hard to form electrically unbalanced atoms or molecules).

Why do wires have to be coated with insulating jackets? The obvious reason is safety. Where large or dangerous voltages exist, insulation keeps current from flowing through the wire into people or other creatures foolish enough to touch a "live" conductor.

The second reason is to prevent short circuits. If two wires make electrical contact, current will flow between them to equalize electric charges or voltages on the two wires. Often, before charges or voltages can be equalized, the wires are heated enough to damage the signal sources, or cause a fire. Since fires are a safety matter, this is really a case of the first reason, safety, as well as an electrical problem. When two wires carrying telecommunications channels make electrical contact, the channels become mixed together—we mentioned this before in describing crosstalk.

The third reason is to protect the conductor from contact with a corrosive material, or with water, which usually contains enough dissolved ions to be a good electrical conductor. In the case of water, the moisture is actually a second conductor in a short circuit, so this is really a case of the second reason, as well as protection of the conductor material.

8.2.1 Color-Coded Wires

The fourth reason for using insulation is not electrical, exactly. Wires are often coated with insulation in different colors to aid in identifying each conductor in a cable or bundle with many conductors. In telephone wiring, cables often contain so many conductors that no one could keep track of them if they were all the same color. In fact, even if they were all solid colors—red, green, and so on—in a 50-conductor cable, there are not enough shades of color to tell apart 50 wires. Most people could not tell 50 different shades of colors apart, so the wires are striped with two different colors. There is a code to the colors (a *color code*). Ten different colors are used for the wires and their stripes. Usually, there are only 25 combinations, for the 25 pairs used in 25-pair cables. In this system, 5 of the 10 colors are used for color A and the remaining 5 colors are used for color B. Each pair of wires combines a pair of colors, one from color A and one from color B. Each wire in the pair has both colors; in one case, the body of the wire is color A with a color B stripe, and in the other, the body is color B with a color A stripe. Since 25 combinations of this sort are possible, cables wtih more than 25 pairs use some color combinations twice. To tell them apart, the first 25 pairs are wrapped in a blue ribbon within the cable, and the second 25 pairs wrapped in an orange ribbon, repeating the color B code used for the wires in the color of the binding ribbon. For cables over 100 pairs, two-colored ribbons are used to group each bundle of 100 pairs. The scheme used for color-coding 25 pairs of wires is given in Table 8–2.

TABLE 8–2   COLOR CODING FOR 25 WIRE PAIRS

Color A	Color B	Pair number	Wire colors	Body stripe
White	Blue	1	White-blue	Blue-white
Red	Orange	2	White-orange	Orange-white
Black	Green	3	White-green	Green-white
Yellow	Brown	4	White-brown	Brown-white
Violet	Slate	5	White-slate	Slate-white
		6	Red-blue	Blue-red
		7	Red-orange	Orange-red
		8	Red-green	Green-red
		9	Red-brown	Brown-red
		10	Red-slate	Slate-red
		11	Black-blue	Blue-black
		12	Black-orange	Orange-black
		13	Black-green	Green-black
		14	Black-brown	Brown-black
		15	Black-slate	Slate-black
		16	Yellow-blue	Blue-yellow
		17	Yellow-orange	Orange-yellow
		18	Yellow-green	Green-yellow
		19	Yellow-brown	Brown-yellow
		20	Yellow-slate	Slate-yellow
		21	Violet-blue	Blue-violet
		22	Violet-orange	Orange-violet
		23	Violet-green	Green-violet
		24	Violet-brown	Brown-violet
		25	Violet-slate	Slate-violet

8.3 COAXIAL CABLE AND OTHER SHIELDED CABLES

"Coaxial" means "having the same axis." The axis of a wire or conductor is a line running through its center and down the length of the wire. In **coaxial cable,** both conductors are centered around the same line. There is an inner wire at the center, and an outer conductor, which is like a hollow tube centered around the inner conductor. Usually, the two conductors are separated by a space filled with a flexible insulator. Figure 8–3 shows a cutaway view of several types of coaxial cable.

Coaxial cable, "coax" for short, is used instead of twisted-pair conductors when the frequencies being transmitted are high enough that most of the energy in the signal is radiated from a twisted pair as a radio signal. This causes loss of energy. At frequencies between audio and a few megaHertz (between 10,000 and a few million cycles

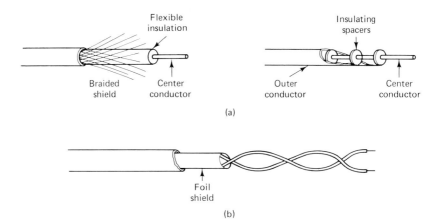

Figure 8-3 Cutaway diagrams of several types of shielded cables. (a) co-axial and (b) twisted-pair.

per second), a simple conductor begins to "broadcast" significant amounts of its electrical signal as electromagnetic radio waves. This energy is lost from the conductor, and is much like energy loss through a conductor with resistance, except that this energy interferes with radios and TV sets, whereas the heat radiated by a resistor does not.

The outer conductor of a coax cable is usually grounded. As a result of this, the outermost conductor has a constant voltage and does not radiate electromagnetic waves. The inner conductor cannot produce any electromagnetic signals that will penetrate through the outer conductor, so the coaxial cable does not lose as much energy from radio-frequency radiation as a twisted pair would. This effect of minimizing RF (radio-frequency) radiation by surrounding a conductor with another grounded conductor is called **RF shielding,** or simply **shielding.** Essentially, the unshielded conductor acts as a broadcast antenna, and loses power along its length as that power is broadcast into the space around the wire. In reverse, shielding can also be used to keep a wire from *picking up* a signal (like a receiving antenna) by using the same approach. A shielded enclosure around a device that might otherwise receive stray RF radiation is called a **Faraday cage.**

Shielded twisted pairs are used for frequencies in the low-megaHertz range, and work excellently. As long as there is shielding, the RF radiation is effectively stopped at these low-end radio frequencies. For higher frequencies, coaxial cable, which is just a special case of a shielded conductor, is better (loses less power through radiation). With coaxial cable, signals may be transmitted at frequencies into the gigaHertz (billions of cycles per second).

Other advantages of coaxial cable and shielded cable (compared to unshielded twisted pairs) are:

1. The shock hazard is reduced. (If the outer conductor is exposed, it is grounded and will not give a shock to anyone touching it.)

2. There is no problem if the outer conductor is exposed and makes contact with earth ground through water or any other conductor.

3. Ground reference—since the signal carries its own ground reference at the transmitting end and receiving end (see the discussion of ground start in Section 10.1.6), this assures that all the things at the transmitting end and receiving end have signals referenced to the same potential.

8.4 WAVEGUIDE

Waveguide is a name for various special conductors used to carry microwave frequency signals (see Figure 8-4). Microwave frequencies are in the range 1 gigaHertz to a few hundred gigaHertz (from 1 billion cycles per second to a few hundred billion). At these very high frequencies, signals traveling down a conductor aren't currents going back and forth through wires, anymore; they're electromagnetic waves that are "trapped" inside the waveguide. The wavelengths are much shorter than the length of the conductor. In the case of most waveguides, the diameter of the waveguide (a sort of hollow conducting tube) is related to the wavelength of the microwaves being transmitted, so that the microwaves "resonate" inside the cavity within the waveguide and are propagated without much loss or radiation of power. For instance, if the frequency 30 gigaHertz is used, the waves radiated by a broadcast antenna would be 1 centimeter long (for further information, see Section 8.5, where we discuss "broadcast". If the signal at 30 gigaHertz is sent down a waveguide instead, the signal traveling inside the cavity of the waveguide is still 1 centimeter in wavelength. If the cavity inside the waveguide is exactly 1 centimeter across, a **standing wave** or resonating condition is set up. The waves can now vibrate transversely (across the width of the waveguide) as they propagate down the tube. This is the normal way electromagnetic waves travel [see Figure 8-5(a)], and in the case of waves traveling inside a waveguide, the standing-wave condition inside the tube permits a strong microwave signal to exist inside the tube, and travel down its full length, without penetrating to the outside of the conductor [Figure 8-5(b)].

Suppose that the wavelength of the microwaves is 1.5 centimeters (you would get this if the frequency were 20 gigaHertz instead of 30) [Figure 8-5(c)]. In that case, if you could get 1.5-centimeter standing waves into a 1-centimeter-wide waveguide, half a centimeter of the wave would penetrate outside the conductor and radiate into

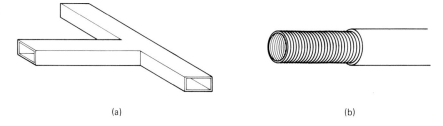

(a)   (b)

Figure 8-4   Rectangular (a) and helical (b) waveguides.

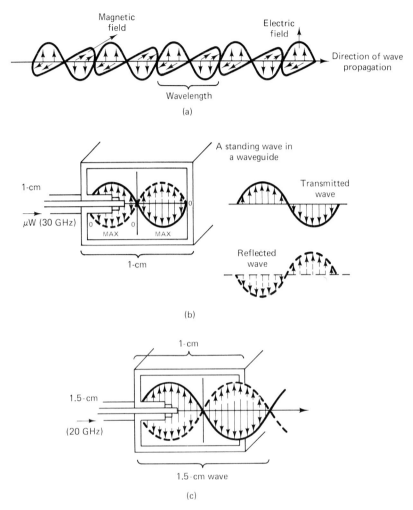

Figure 8-5 (a) Normal propagation of electromagnetic waves. (b) Combined (standing) wave. (c) Diagram shows no standing waves because reflection takes place at the wrong point on the wave.

the space surrounding it. In fact, you could not even get the 1.5-centimeter waves into the waveguide in the first place. Most of the wave's energy would be lost where you tried to stuff it into the waveguide, and the rest would be lost rapidly.

But we used 1-centimeter waves, which fit into the waveguide just fine, and we don't have that problem. Since none of the wave penetrates outside the conductor, it doesn't radiate its energy away.

Note: Standard waveguide is usually $\frac{1}{2}$ wavelength wide, instead of a full wavelength, but the points of the above argument are not changed.

8.5 THE WIRELESS MEDIUM

We've just spent a great deal of time looking at ways to prevent electric signals from radiating away from one set of wires and being picked up by other wires a distance away. In describing shielded cable and coaxial cable, we found that at higher frequencies, electrical signals have a greater tendency to radiate energy into space. What if we *want* a signal to go from one place to another without wires? In that case, we would have a **radio broadcast,** and the wires would be called **antennas**. What advantages are there to this sort of transmission?

First, the cost of setting up a link of this type may be considerably less than for a wire or coax link. If you want to set up a small-volume, private, dedicated link between two places across town, you have to string wires from one place to the other if you use a conductor. There must be poles from which to suspend the wires, or a trench dug to lay an underground cable, and somebody has to purchase the right-of-way across the property through which the wires will travel. If this weren't bad enough, the cable or wires themselves are not cheap.

On the other hand, inexpensive solid-state radio transmitters are available today at prices which make their use extremely attractive in cases such as the one just mentioned. The second reason you might want to use wireless transmission for your telecommunications needs is feasibility. In some places, it's just plain impossible to string wires from where you are to where you want to talk to. Cars, boats, and air and spacecraft are natural candidates for wireless communication.

Another growing reason for using wireless telecommunication is convenience. The wireless telephone, although limited in range, frees you to take the phone wherever you go in the house or office, instead of forcing you to run to the phone at *its* request. The convenience of having a telephone in your car was once a luxury limited to the very wealthy. The technology to make this available to everyone is now in place, and it is just a matter of time (and public demand) before this becomes a universally acceptable adjunct to the family car. As with that other great trend for the future, picturephone, the **cellular radio** concept may not catch on with the public. Perhaps, like teleconferencing, cellular radio may find a more limited field of customers than is presently anticipated by the phone company. Then again, maybe the mobile phone in the car will replace the cigarette lighter and everyone will want to have one. We discuss cellular radio in more detail in Section 8.5.2.

The ubiquitous "beeper" pocket pager, the dispatch radio used by taxicab or delivery trucks, and the police-fire band of two-way radio are "public utility" uses of wireless transmission. The question is, however, are they *telecommunications?* Since police-fire and dispatch radio is so limited in capacity, it would not be fair to call them telecommunications systems, but the beeper pocket pager is very close to a wireless communication network. It is, of course, a one-way channel of communication—and you might think of the pager as an adjunct to the ordinary telephone system—sort of a long-distance way of ringing you on the telephone. With its capacity for selectively informing one person from among thousands, the pager is sort of a "simplex" telecommunications link. Some of them even have the capacity for receiving messages, such as which number to call when you reach a phone.

Most people do not think of wireless transmission as a part of the telecommunications system. The specialized carriers, such as MCI (Microwave Communications, Inc.), use microwave links and satellite transmission as well as existing land lines to convey their signals. Microwave transmission can be made *directional*. Certain types of microwave antenna can be used to focus the broadcast signal into a narrow beam, which can then be directed to a specific receiver. In circumstances such as this, separate frequencies for each broadcast transmitter are not necessary. You could use the same frequency from dozens of different transmitters, each aimed at a specific receiver, and even have the microwave beams intersect, without any chance of interference or crosstalk. Like flashlight beams, two microwave beams can cross through the same space without interfering with each other. Figure 8–6 illustrates some aspects of microwave transmission.

We mentioned that a wire carrying a high-frequency signal would radiate energy as electromagnetic waves, and that this energy would be lost from the signal travel-

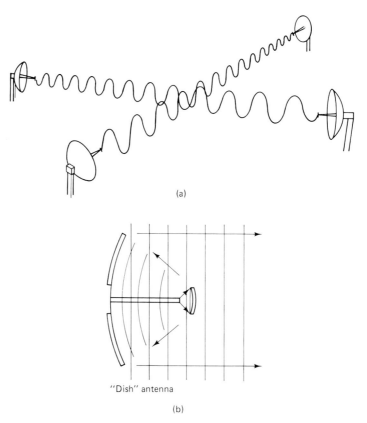

(a)

"Dish" antenna

(b)

Figure 8-6 Some aspects of microwave transmission: microwave beams can intersect without crosstalk (a); microwaves can be reflected and focussed like light waves (b).

ing down the wire. If we want the radiation of energy from the wire to be a large part or all of the signal, we have to configure our wire deliberately to radiate RF energy with a high efficiency. Such a specially configured wire is called a **broadcast antenna**. In cases where the antenna radiates electromagnetic energy in all directions, it is called **omnidirectional**. A stub of wire of the right length (usually one-half the wavelength of the radio wave) is a sufficient omnidirectional antenna. To focus or confine the beam into a narrow path, an **array of antennas** may be arranged so that their radio waves reinforce each other in one direction and cancel in another. At microwave frequencies, radio waves behave much like visible light. They can be reflected off conducting surfaces, and focused the way a flashlight beam is focused by the shaped reflector behind the flashlight bulb [Figure 8-6(a)]. The shape of a **dish antenna** used to focus reflected microwaves is identical to the shape of the flashlight reflector, although larger in size. Antenna arrays and dish antennas are both **directional antennas**.

Of course, satellite transmission and reception is by wireless means. You have probably seen satellite receiver dishes for the reception of television broadcast from geosynchronous communications satellites. It is clear from the shape of these dishes that the signals can be focused like light, and from the discussion above, you have probably already concluded that these satellite signals are microwave.

What is a **geosynchronous communications satellite?** The word "geosynchronous" refers to the fact that the satellite orbits around the earth in the same time it takes the earth to rotate underneath the satellite. If the satellite is placed in an orbit 23,600 miles above the earth, traveling west to east, it will remain above a fixed point on the equator (provided that it is in an equatorial orbit). With the aid of small rocket thrusters "stationkeeping" the satellite in its proper location, the satellite can be used as a *very tall* microwave tower.

8.5.1 Propagation of Signals without Wires

How is the signal carried from a transmitting antenna to a receiving antenna? If there is no conductor for electric current connecting the two ends of the wireless link, what conveys the message? Scientists thought for a long time that there was a medium called "the ether" that was "jiggled" by electric charges and carried light waves, radio waves, and microwaves as vibrations of this ether material. It was assumed that light waves and radio waves were produced by electric forces that acted on the atoms of the materials light passed through. A vibrating charge on a wire would electrically push and pull on the atoms of the air around the wire, and they would push on their neighboring atoms, and so on, propagating a wave of radio energy outward from the wire (the antenna). That was fine, except that there were electromagnetic vibrations that arrived (light) from long distances through spaces where people knew that there was no air.

To handle that problem, scientists invented the idea of a sort of medium that filled the empty space between the earth and the sun, and the spaces between atoms, and they called this material "the ether." They imagined that it was like stiff Jello with atoms embedded in it here and there, and when something electric or magnetic happened, it jiggled the Jello at one end and the waves traveled outward from there. By the end of the nineteenth century it was clear that this explanation would not work. Several experiments proved that ether couldn't be like any familiar material, and one in particular, the Michaelson-Morley experiment, proved (to people who could interpret the results) that ether couldn't exist at all.

Today, we describe electromagnetic waves in the following way. An electric field surrounds a charged object and extends out into the space around it like rays of sunlight. The force of electric attraction or repulsion acts on other charged objects along these lines of force (the rays we mentioned). If the charged object is moved, its field is warped, and a wave of realignment spreads outward in the electric field from the point where the charge was moved, as shown in Figure 8-7.

A magnet has a similar field of magnetic lines of force surrounding it and acts along these lines on other magnetized objects. If you move the magnet, a similar thing happens to the field. Now moving electric charges are always accompanied by magnetic fields, so when a charged object somewhere, like an electron in a wire, is "jiggled," the distortion in the pattern of magnetic and electric field lines around the object spreads out as an electromagnetic force. The idea that these waves exist and that they can carry off energy from the object that is being "wiggled" was first predicted by a Scotsman named James Clerk Maxwell in 1860, and first observed

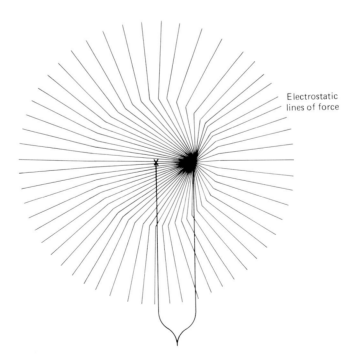

Electrostatic lines of force

Figure 8-7   A charge at rest is suddenly moved to the right at a constant speed of ½ c. This causes a wave of warped or distorted electric field lines.

in the laboratory by Heinrich Hertz in 1886. Long-term broadcast of radio waves took about 20 more years to reach a commercial stage.

What is the electromagnetic signal like? A distortion or disturbance in the web of electric and magnetic force lines spreads out like a spherical bubble from the point of origin. At a point on the surface of this bubble, if the electric field lines swing up and down, the magnetic field lines swing left and right, and the whole wave travels outward along rays or straight lines that originate at the place where the wiggling electric charge is. Figure 8-7 shows how a part of this expanding, spherical wavefront might look.

So the medium that carries signals without wires is the web of electromagnetic field lines that penetrate every bit of space—even empty space—surrounding a charged or magnetic object. Disturbances in that medium travel through empty space with a speed of 300,000 kilometers per second. Electromagnetic waves which travel through space that is not empty—especially light waves—may travel more slowly than this, as they are absorbed by atoms, then reemitted later in the same direction after a certain time delay.

When these waves or distortions are produced by an antenna, it is a stub of wire containing charges that are wiggled by a broadcast transmitter. This is just a high-speed electrical oscillator that first applies a force pushing electrons into the wire stub (antenna) and then pulls the extra electrons back out, repeating this thousands, millions, or billions of times a second.

How can a current be pushed into a wire that sticks out into space and is an open circuit? Remember our discussion of capacitance and capacitors in Chapter 3? Well, this antenna wire is something like one plate of a capacitor. It's possible to "compress" more electrons into the wire than it contains normally, and build up a "pressure" (an electric negative charge and a negative voltage). When the pressure is released, the electrons flow back out the wire until a balance is reached, and the wire is uncharged. It happens that current in the wire does not start flowing everywhere along the length of the wire at the same time. The effect of starting a current flowing in the wire spreads down the wire at the speed of light (300,000 kilometers per second). If the wire is long enough, the current in the wire has been reversed, and charges are flowing out of the wire, before the faraway end of that wire even knows the charges were put in. If the length of the wire is just right (see Figure 8-8), no current has to flow into or out of the far end of the wire even though your transmitter is making AC current flow at the near end.

What is a "just right" length for an antenna? Well, the oscillator is producing AC "waves" at a particular frequency, and at 300,000 kilometers per second, the electrical signal produced at the wave's beginning goes a certain distance through space before the end of the cycle. This distance is called a **wavelength,** and the length of a simple antenna should be a quarter or a half of this length. In Figure 8-8 we are looking at a quarter-wave antenna. When the antenna is attached to its oscillator, a standing wave is set up where the AC current is maximum at the end of the wire attached to the oscillator (what we called the "near end") and *zero* at the other end of the wire segment (what we called the "far end").

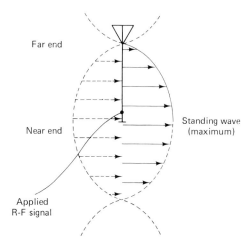

Top of antenna (0 volts)
(reflection point)
(minimum node of standing wave)

Far end

Near end

Standing wave
(maximum)

Applied
R-F signal

Figure 8-8 Quarter-wave antenna-node
of standing wave (minimum AC current
and voltage) is at far end of antenna.

8.5.2 Cellular Radio

Cellular radio has found its way into telecommunications by way of car phones. It allocates many frequencies to small areas in a way that avoids crowding. In each little area, or cell, you could have 100 frequencies. If those 100 frequencies were used in a more-powerful, city-wide broadcast area, for instance, there would be only 100 channels of transmission in the entire city. With a city covered by 10 cells, each having small transmitters for the 100 frequencies, someone in one cell could be using frequency 5, and elsewhere in the city, mobile phones in other cells could be using frequency 5 as well, without conflict. Effectively, we have 1000 channels instead of 100, by isolating the users into cells.

As shown in Figure 8-9, Smedley Wolverton III is driving his car, talking on the car radio, through the outskirts of greater Boswash. As his car passes out of cell A and into cell B, the transmitting equipment in cell A detects the fact that Wolverton's car is entering cell B. The transmitter at A passes control to cell B, which will allocate a frequency for Wolverton's transmission. This will not necessarily be the frequency Wolverton was on in cell A, but Wolverton's car transceiver will be signaled to switch to the new frequency, and his telephone conversation will be picked up by the cell B transmitting equipment. All this will take place automatically without interrupting the conversation.

Clearly, some kind of computerized control similar to a telephone switching system is at work here. Unlike a central office switch, this system must be able to change connections *during* the conversation. If Wolverton were to drive into a cell where all the frequencies were allocated, a blocking situation, similar to blocking

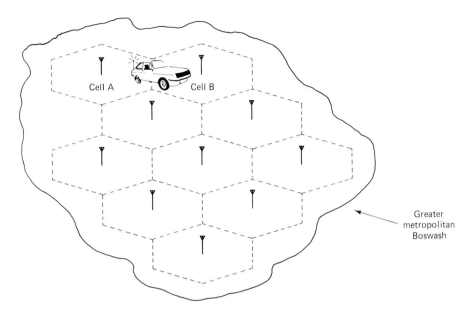

Figure 8-9   Cellular radio grid.

in an overloaded telephone switch, would occur. However, since this is going on during a conversation, not when dialing, it would appear to Wolverton that he had been suddenly disconnected without warning. To avoid this, the number of available frequencies in any cell should exceed the number of users likely to be using their mobile phones within the cell. More frequencies, or smaller cells, will have the same effect.

8.6 FIBER-OPTIC LIGHT PIPE

At higher and higher frequencies, the transmission of electromagnetic (radio) waves looks more and more like the propagation of light. In fact, light is just a very high frequency wave of the same sort as radio waves, so why not use light as the transmitting medium? For broadcast in the open air, this would not be a very good idea. Clouds, small birds, rain, and dust in the air all get in the way of light beams. The same problems (to a smaller degree) occur in microwave links, and the best solution is to use coaxial cables and waveguides (which do not allow things to block the beam and can go around corners). Light can be transmitted through a sort of waveguide called a *light pipe*. Its working principle is **total internal reflection,** which is something that happens at the boundary between two dissimilar transparent materials. The speed of light in glass or silica is slower than its speed in air (about two-thirds as fast in silica as air). When light traveling through silica strikes air, if it hits the boundary between the silica and the air at a shallow angle, *all* the light that strikes the boundary is reflected back into the silica; *none* of it ends up as a beam in the air [Figure 8-10(a)].

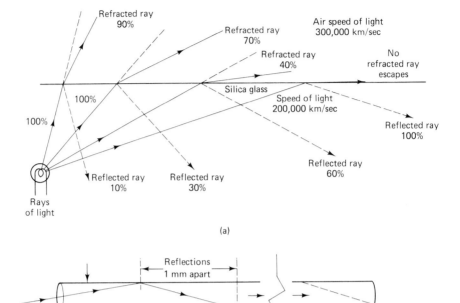

(a)

(b)

Figure 8-10   Total internal reflection.

This is a better reflector than the best metal mirror that can be made. A pure silver shiny-metal reflector reflects about 93% of the light that strikes it and absorbs the remaining 7%. **Total internal reflection** is necessary to "trap" a beam of light in a light pipe [see Figure 8–10(b)]. Suppose that we have a silica fiber 0.1 millimeter in diameter and 1 kilometer long. If the light loss at each reflection is 1 part per million, and the light strikes the silica-air boundary at a shallow angle so that it rebounds and strikes the other edge 1 millimeter down the fiber, there will be 1,000,000 reflections along the kilometer of silica. At 1 part in a million loss per "bounce," there will be only 36% of the light left at the end of a kilometer. If 10 parts per million is lost at each bounce, only 45 millionths of the original light makes it to the far end of the fiber. Real mirrors lose 7%, at best. After 200 reflections (2 meters), less than half a millionth of the original light would remain. At the end of a kilometer, nothing measurable would remain. Fiber light guides that use total internal reflection lose less than half the signal's power in 6 kilometers. What loss there is happens because the silica is not perfectly transparent and has impurities in it. At the reflections, however, there *must* be 100% reflection—or at least 99.9999% (with more nines than we'd like to count)—otherwise practically nothing would arrive at the far end of the fiber.

The silica used in optical fibers has exceptional transparency. To carry signals more than a kilometer, the clarity of the silica must be incredible. For example, if you were to stack layers of ordinary window glass until you had a "window" a foot thick to look through, you would hardly be able to see through it at all—a "window" a yard thick would be impossible to see through—and yet the silica glass in optical fibers is clear enough that we can see through a window *miles* in thickness!

Light frequencies are around 10^{15} Hertz. This covers the range from infrared to visible light. Frequencies like this (10,000 times microwave frequencies) suggest a correspondingly higher capacity for multiple channels carried on a single fiber. Since the fibers are thin, and each thin fiber is quite flexible, a bundle of thousands of fibers can be made into a fiber-optic cable with flexibility as good as stranded-copper wire cables. One advantage of fiber-optic cables is that only one strand is needed for each "circuit," copper-conductor cables need two wires for each line.

8.6.1 Advantages of Fiber Optics

As mentioned briefly above, to communicate information, silica fiber does not require a "loop" with two conductors. A single path can carry information in two directions at once (full duplex) and can carry many different colors of light at once (frequency-division multiplex). In addition to this, silica glass is chemically quite inert and not subject to attack by corrosive chemicals that would destroy a copper conductor. It is also lighter than a comparable metal wire.

No insulation is needed around each fiber, to prevent crosstalk to other fibers. There is no shock hazard from contacting fiber optics cables carrying a signal. Similarly, if a fiber-optic cable should become waterlogged, there is no chance that the signals would "short out" and be lost through the water. Finally, the raw material for making silica glass is sand. Strategic supplies of sand are unlikely to be cut off by any political catastrophes, as is the case with copper, tin, and other metals available from a limited number of sources in the world.

8.6.2 Disadvantages of Fiber Optics

Disadvantages to the use of fiber optics as compared to copper include the fact that it takes more energy to process sand into usable silica fibers than to process copper. Copper is more expensive than ordinary glass because it is more rare than sand, not because of the energy it takes to smelt it from the ore. The cost of energy is the reason for ultrapure silica fiber being more expensive than copper wire. Interconnection or termination of connections between fiber cables is an immature technology. It will probably never be as simple to connect optical conductors as it is to connect electrical connectors. Certainly, it is not now—but new developments in fiber-interconnection technology are taking place all the time.

Another disadvantage that accrues to light, as any electromagnetic energy, is that it must be converted into an electrical signal at the distant end, to operate most telephone equipment, and the electrical signal of most phone equipment must be converted into an optical (light) signal. This is accomplished by devices called **transducers,** but in any event, it is not as direct as dealing with electrical signals everywhere. Work is proceeding on telephones that will convert sound directly into light energy, and vice versa, without any intermediate electrical signal, or electrical power requirements.

A device called a **transphasor,** devised at Herriott-Watt University in Edinburgh, amplifies and switches pulses of laser light at speeds of nearly 1 trillion per second with no electrical equipment at all. Mentioned as elements of a proposed optical computer in the February issue of 1983 *Scientific American*, this may also be the all-optical telephone-repeater element of the future as well, for fiber-optic telecommunications systems.

REVIEW QUESTIONS

1. Do all signals used in telecommunication travel along electrical wires?
2. Electrical resistance generally decreases with colder temperatures. Does the electrical resistance ever disappear altogether in any electrical conductors?
3. Is electrical power dissipated in the form of heat by a wire's resistance? Must there be current in the loop?
4. What causes higher frequencies to be attenuated more than lower frequencies in twisted-pair conductors?
5. What is one cause of crosstalk?
6. Why is it necessary to use balanced lines?
7. For what reasons are wires in a cable coated with an insulating material?
8. Describe the body-and-stripe insulator color code on the nineteenth pair of conductors in a 25-pair telephone cable.
9. A T1 span uses a data transmission rate of 1.544 Mbps. What kind of cable is used as a link for T1 to prevent this signal from being radiated in the commercial broadcast band?
10. What kind of transmission forms are carried by waveguide?
11. What happens when two microwave beams cross?
12. What is an electromagnetic wave? Are both light and radio broadcast examples?
13. What is the ideal length for an antenna? How is it related to the frequency of the signal?
14. Could cellular radio be used for video as well as voice? What changes might have to be made?
15. What medium is used to carry light-wave or infrared transmissions?
16. What are the advantages and disadvantages of using silica strands for fiber optics compared to other media?

9

Transmission Forms

9.1 ANALOG AND DIGITAL SIGNALS

The unmodified electrical signal produced by the human voice spoken into a microphone is an analog signal. If the voice is modulated onto a **carrier signal** by the AM or FM methods normally used for radio broadcast, it is also an analog signal. Commercial radio broadcast is, of course, analog in nature. What is the characteristic of these signals that makes them analog?

If you show an analog signal on a piece of graph paper, one that plots the signal's value over an interval of time, as shown in Figure 9-1(a), the analog signal will show a smooth variation from each minimum to each maximum throughout the span of time. Digital signals plotted in the same way tend to show only two values, that occurr no matter how long you watch the signal [see Figure 9-1(b)]. The values in between never appear. We say that an analog signal is a *continuous* variation of some physical quantity to communicate information, since any of the values between the maximum and the minimum might appear. We say, on the other hand, that a digital signal is *discrete*. It doesn't take on all values between its maximum and minimum. The maximum and minimum values are all there are. This is called a **binary** system of transmission. There are digital codes where three, or four, or more discrete levels are used, but they are not used for telecommunication purposes.

In the definition above, we "hedged" a bit about what kind of values we were plotting on our graph paper. The brightness of a light, the voltage or current of an electrical signal, or the frequency or amplitude of an oscillator's waveform might all be used as legitimate ways to indicate information. Any electrical or electromagnetic quantity that can be varied for this purpose (carrying information) is **modulated** by

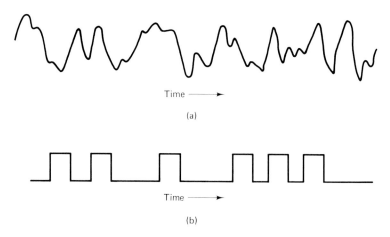

Time ─────▶

(a)

Time ─────▶

(b)

Figure 9-1   (a) Analog signal. (b) Digital signal.

the information that varies it. **Modulation** of an electrical or electromagnetic quantity may be either digital or analog, depending on how the quantity is varied.

Once you have a device that measures the level of the electrical quantity you're modulating—a measuring instrument that measures the level of the quantity that you're varying to transmit a signal—you have a **demodulator.** This is the instrument that receives the signal and identifies the information that was modulated onto it.

The electrical quantity you choose to vary when you transmit information (for instance, the brightness of a light or the frequency of a radio wave) is called the **carrier** for the transmission. The carrier itself might be any variable electrical or electromagnetic commodity that you can control. We have already stated that the type of control is called **modulation**. The information you **modulate** onto the **carrier** may be either digital or analog. For instance, suppose that you have a wave generator producing sine-wave variations in voltage, as depicted in Figure 9-2(a). The height of the sine wave is the difference between its most negative voltage and its most positive voltage. In Figure 9-2(a), this wave height is called the **amplitude** of the voltage wave, and it is comparable to the brightness of a light or the loudness of a sound. If you want to vary the amplitude of the signal coming out of this oscillator, you have **amplitude modulation,** or AM transmission. Suppose that you attach a telegraph key to the oscillator, and you connect the oscillator to a broadcast antenna through the telegraph key. Now, when you close the contacts by pushing the telegraph key, there is a wave of a certain amplitude on the antenna, and its energy is radiated in the form of radio waves. When you release the telegraph key, the contacts open, and no wave reaches the antenna, so no radio waves are radiated. The amplitude of the radiated radio waves has two levels, ON and OFF, where one condition results in a wave with a certain amplitude, while the other condition results in a wave with no amplitude at all (no wave). This is a *digital* transmission: the schematic and wave are represented in Figure 9-2(b). Although most people use telegraph keys to transmit Morse code, this arrangement could transmit binary code just as easily. In this case,

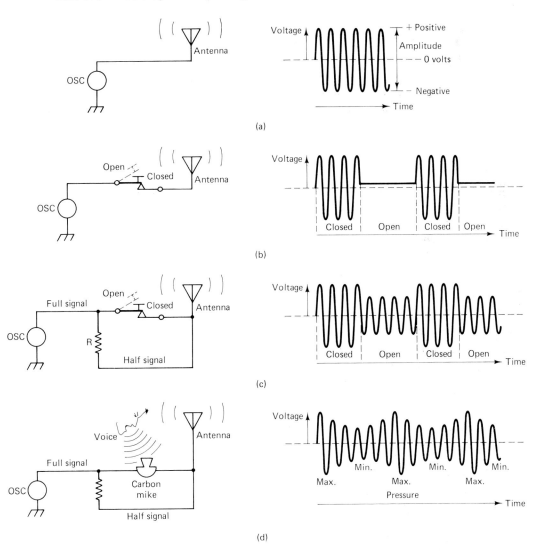

Figure 9-2   Amplitude modulation. (a) Wave generator producing sine-wave variations in voltage. (b) Schematic and wave of AM digital transmission. (c) Digital AM signal modulated by telegraph key, at 50% modulation. (d) Analog AM signal modulated by carbon microphone, also at 50% modulation.

we're modulating the carrier rather extremely: it's either there or it's not. Another way to transmit the digital signal from the telegraph key would use the key to cut the carrier amplitude to the antenna in half, rather than cutting it off completely, as shown in the schematic and waveform of Figure 9-2(c). Part of the signal could be arriving at the antenna all the time, and closing the contacts on the telegraph key could connect the rest of the signal to vary the amplitude.

Now suppose that instead of the telegraph key in Figure 9–2(c), we used a carbon microphone, the kind found in a telephone transmitter. A voice or sound would vary the amount of amplitude arriving at the antenna between half and full amplitude, depending on the sound pressure on the carbon granules in the microphone. This is shown in Figure 9–2(d), where the carrier is modulated by an *analog* signal rather than a *digital* one, as in Figure 9–2(c). Notice that the only difference between the two waveforms is that in Figure 9–2(d), the wave goes from half to full amplitude smoothly. In Figure 9–2(c) the wave amplitude jumps from half to full and back again, without having any other value of amplitude. The maximum and minimum amplitudes are the only ones that appear. In Figure 9–2(d) the maximum and minimum amplitudes, and all the ones in between, appear as the voice wave compresses and releases the carbon granules in the microphone.

We say that the telegraph key is being used to transmit *digital* information, and the microphone is being used to transmit *analog* (voice) information. In both cases, we've got the same carrier wave, but the information modulated onto it is digital in one case and analog in the other. The carrier wave itself is a sine wave. It varies smoothly between its most positive voltage and its most negative voltage, so the carrier itself is an analog wave. But it is not the voltage of the carrier we are modulating, it is the amplitude. This means that although the carrier wave is, of itself, an analog variation in *voltage*, the information modulated onto its amplitude may be either analog or digital.

To illustrate this point, we've drawn a graph that shows the signals in Figure 9–2(c) and (d) a different way. In Figure 9–3(a), we see a graph of the amplitude of the wave from the telegraph key over the same span of time that the voltage was shown in Figure 9–2(c). In Figure 9–3(b) we see the *amplitude* of the wave from the carbon microphone, instead of the voltage of the carrier wave. Looking at the amplitude of the carrier wave, instead of its voltage, we see the information that's being transmitted rather than all those little "wiggles" in the voltage that just tell you the oscillator is turning over. It's clear that Figure 9–3(a) has only two levels and is binary digital information, whereas Figure 9–3(b) is an analog voice wave. In Figure 9–3(a), although the **carrier** wave was analog, the **modulation** we see is digital. We say that the graphs plotted in Figure 9–2(c) and (d) are voltage versus time graphs of the carrier wave, but the graphs plotted in Figure 9–3 show the **modula-**

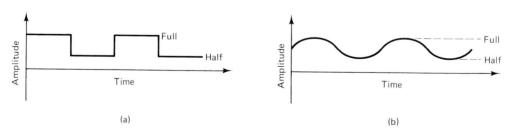

Figure 9–3 Voltage versus time graphs. Amplitude of waveform telegraph key (a) and from carbon microphone (b).

tion envelope which is the information modulated onto the carrier by the telegraph key or microphone. The modulation envelope is the actual information; the carrier wave is merely a way to carry that information. Other ways to carry information do not require antennas. Wires could carry the AM signal we saw in Figures 9–2 and 9–3. If we modulated the brightness of a light, the light could carry the transmission, without an antenna, and if we wanted to, we could modulate the amplitude, frequency, or phase of ultrasonic sound waves, and use a speaker to transmit, where the electromagnetic waves of Figures 9–2 and 9–3 use an antenna. This is done, for instance, in most remote-control TV channel changers, where digital information modulated onto an ultrasonic sound carrier wave is used to select channels.

9.2 FREQUENCY AND BANDWIDTH

We have already defined **frequency** in Chapter 2, and done a number of things since then which use the frequency of an electrical signal to accomplish various things. In this section we briefly recap on what frequency is, then detail what frequencies are used for various things and which frequencies arise as the result of various processes. Then, after describing the natural range of frequencies that appear in nature, or are used by humankind for various purposes, we go into how much information can be carried on different portions of the frequency spectrum. The concept of **bandwidth** is tied in very closely with how much information can be transmitted and how fast it can be transmitted. This concept is then used to explain the advantages and disadvantages of transmission at various frequencies.

9.2.1 Frequency

Certain events that occur in nature are periodic or cyclic. The growing of plants, regulated by sunlight, is cyclic, and we call that cycle of sunshine and darkness "a day." The seasons are cyclic, repeating in an interval called "a year." In natural and humanmade systems, cyclic, repeating processes have an interval between repetitions, called the **period,** and an rate of repeating, called the **frequency.** Frequency is always expressed as the number of repetitions of a recurring event in a standard interval of time. If a phonograph record is identified as one that rotates at $33\frac{1}{3}$ revolutions per minute, that number is a frequency. Most electrical frequencies are measured in cycles (repetitions) per second. When the standard unit of time is the second, the standard unit of frequency is the Hertz. If a fly's wings flap at 100 beats per second, we can say they flap at 100 Hertz. (This unit is named after Heinrich Hertz, mentioned earlier as the first person to identify electromagnetic waves with a laboratory experiment—not the inventor of the rent-a-car concept.)

Frequencies arise in electrical, acoustical, optical, hydraulic, and mechanical systems. Sound is a periodic variation in air pressure. Pressure waves originate from whatever is making the sound, and have a frequency that depends on something mechanically vibrating against the air and transferring energy to it. The lowest-frequency sounds a human ear can hear are around 20 cycles per second (20 Hertz).

Below that frequency, mechanical vibrations can be felt but not heard. These are called **subsonics.** Pressure waves in air travel on average about 1000 feet (300 meters) per second. A source of 20-Hertz waves is making new pressure waves every 1/20th of a second. That is the same as 50 milliseconds, or 50 thousandths of a second. In 50 milliseconds, each pressure wave travels 50 feet (15 meters) at the speed of sound, 1000 feet per second. We arrive at this conclusion from the fact that

$$\text{distance} = \text{rate} \times \text{time}$$

The pressure waves spreading out from a 20-Hertz source are 50 feet apart, traveling at 1000 feet per second (15 meters apart, traveling at 300 meters per second). We call this 50-foot, or 15-meter, separation between pressure waves the **wavelength.** Sound waves 15 meters long are much larger than a human ear and are difficult to hear. As the sound waves get smaller and closer to the size of a human ear, they become easier to hear. What sound waves are smaller? If we increase the frequency of sound waves, we decrease their wavelength. For instance, if sound waves are produced with a frequency of 2000 Hertz, 2000 of them will be produced in a second. Since that means that 2000 waves will be produced in the time it takes sound to travel 1000 feet (300 meters), each wave will take up ½ foot of wavelength (15 centimeters), and 2000 of them will fit neatly into a 1000-foot (300-meter) space. A frequency 10 times larger, 20,000 Hertz, has a wavelength 10 times smaller. The eardrum of a human ear is about this size (0.6 inch or 1.5 centimeters). Frequencies much higher than this cannot be heard, because the wavelength is too short to affect the whole eardrum at once. Part of the eardrum is being pushed by one pressure wave while another part of the eardrum is being pulled by the low-pressure part of the same wave; the net result is that the eardrum doesn't move enough to notice (that is, you can't hear anything). This limits the range of human hearing to sounds between 20 and 20,000 Hertz. Practical communication—voice that you can understand, for instance—can take place in a much narrower range. Telephone communications can be restricted to sounds between 300 and 3000 Hertz, and it is still easy to understand the speaker clearly.

The concepts of **frequency, period** and **wavelength** apply to electrical signals traveling in wires and electromagnetic signals traveling through space just as they do to sound waves. Electric current in a wire is made up of charges (free electrons) that move when an electric field is applied. The effects of this electric field move much faster than the charges travel down the wire. This is due to a *shock-wave* effect, that works like as follows. Electrons are pushed into a portion of wire. They push the electrons already there in that part of the wire and begin to move them over. Those electrons push on their neighbors farther down the wire, and so on, so that an entire column of moving electrons is stacking up electrons ahead of itself, and the column is getting longer. The chain effect travels down to the end of the wire very rapidly. The time between pushing a bit of electric charge into one end of a wire and having the same amount of charge pop out the other end appears instantaneous, but, of course, it is not.

The shock-wave effect travels from one end of the wire to the other at speeds near the speed of light (300,000 kilometers per second, or 186,000 miles per second). In space, the electric a magnetic fields of an electromagnetic wave, such as a radio broadcast, travel with the speed of light. This means that a radio signal with a carrier wave frequency of 186,000 Hertz would produce waves one mile long, and a frequency of 300,000 Hertz would produce electromagnetic waves with a wavelength of 1 kilometer. A 20-Hertz electromagnetic wave would have a wavelength of 9300 miles (15,000 kilometers), compared to the 50-foot (15-meter) wavelength of acoustic (sound) waves. The clear difference between electromagnetic and sound waves is the speed with which they travel. For a given frequency, electromagnetic waves will have a wavelength 1,000,000 times as long as sound waves, because they travel 1,000,000 times as far in the same period of time.

The only difference between electric frequencies used in wires and electromagnetic frequencies used for wireless broadcast is whether they are allowed to radiate or are confined to a conductor. Cables used for telephony may convey information modulated onto carrier frequencies, just as radio does. Figure 9-4 shows how various acoustic and electromagnetic frequencies are produced, what they are used for, and what they are called.

When you modulate information onto a carrier wave, there is a limit to how fast you can change the carrier wave you are modulating. Figures 9-2(c) and (d) and 9-3 indicate how a carrier wave's amplitude might be modulated with digital or analog information. Most **detectors**—the devices that pick up a modulated signal—need several cycles to estblish what the amplitude, frequency, or phase of the carrier wave is, before you begin to change it. If you change the carrier wave too quickly, the detector will be unable to "keep up" and may fail completely to detect the information you put on the carrier wave. As a result of this, it is possible to vary the carrier wave (to modulate it) only at speeds that are less than the frequency of the carrier wave itself. This means that for a carrier wave with a slow frequency, information can only be transmitted at a slow rate—in fact, a rate less than the frequency of the carrier—while a higher-frequency carrier can be modulated with more information in the same amount of time.

9.2.2 Bandwidth

We are going to use the word **bandwidth** to represent a range of frequencies between a low frequency and a higher one. For example, the **voice bandwidth** used in telephony is the group of frequencies that will pass through ordinary telephone equipment. The lowest frequency generally accepted by telephone equipment is 300 Hertz, and the highest that is able to get through all telephone equipment is about 3000 Hertz. This defines a bandwidth of 2700 Hertz—the difference between 3000 and 300 Hertz—as the bandwidth of telephone equipment. Another example of a **band** is the range of frequencies picked up by an ordinary AM radio receiver. (AM, by the way, represents

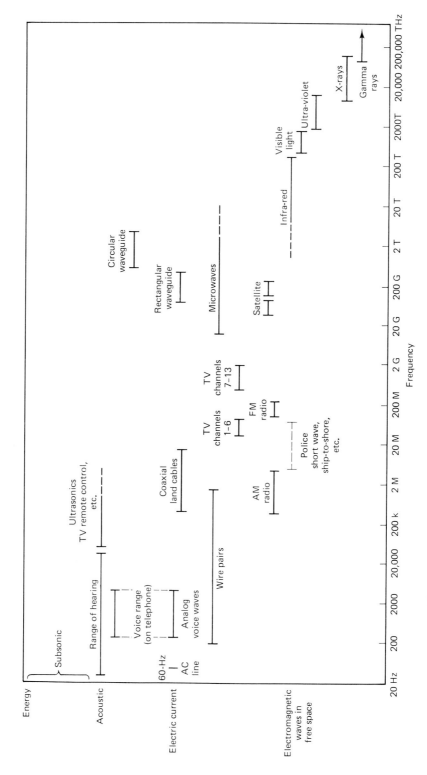

Figure 9-4 Production, use, and types of various acoustic and electromagnetic frequencies.

amplitude modulation, which we just saw in Figures 9-2 and 9-3.) The AM bandwidth is defined by where you can tune on the dial of your radio and receive AM radio stations. This range on the dial is between 500,000 and 1,650,000 Hertz. The bandwidth of the AM band is 1,150,000 Hertz, from the lowest to the highest frequency of electromagnetic radiation received by your radio. Visible light, as you will recall, is an extremely high frequency of the same kind of electromagnetic energy as radio. Human beings can see light between 400,000,000,000,000 Hertz (or 400 teraHertz) and 750,000,000,000,000 (or 750 teraHertz). Within this bandwidth are all the colors we can see. Now, let's imagine a stoplight. On the stoplight there are three filters in front of the bulbs. One filter (the red one) passes frequencies between 400 and 460 teraHertz. The amber filter in the center passes light whose frequency is between 490 and 510 teraHertz. The green filter passes light between 525 and 555 teraHertz. We see each bandwidth of light frequencies as a different color.

This idea of splitting a particular frequency band of energy into smaller bands, in this case, ones we identify as "red" or "yellow" or "green," also applies to sound energy. As we divide light's frequencies into colors—red, orange, yellow, green, blue, indigo, violet—we divide sounds into notes—do, re, mi, fa, sol, la, ti—making up names for individual bands of frequencies we can distinguish from one another.

Each color is a **band**. A color is not just one frequency. Colors we recognize as shades of "red" might be anything between 400 and 460 teraHertz. As the frequency gets higher, the "red" we see shades off toward orange. In the case of the stoplight, we have three different channels of information—stop, caution, and go—that are modulated onto different-frequency carrier waves. The stoplight is a transmitter that broadcasts information in three different frequency bands. Since each channel carries information in a "yes/no" form (light on or off), the lights in a stoplight are actually transmitting digital information.

The amount of information that may be modulated onto a channel—in the case of digital information, this is called the **data rate** or **baud rate**—depends on the bandwidth that is available to carry the information. For instance, let's say that you use the AM radio band to transmit information. The bandwidth—from 500 to 1650 kiloHertz—is 1150 kiloHertz. Now suppose that each channel of information you wish to transmit contains data at a rate of 5000 bits per second (5 kilobaud), or audio information containing frequencies between 0 and 5000 Hertz (5 kiloHertz). For each channel, we would need 5 kiloHertz of "space" within the bandwidth. We can't let the channels overlap or we won't be able to tune each one in separately, so we put them right next to each other across the AM frequency band. In that case, we can "cut apart" the AM band—1150 kiloHertz—into 230 "strips," each 5 kiloHertz wide. That means that we can transmit 230 channels of 5 kiloHertz audio, or 5 kilobaud digital, on the AM radio band. That's true only if everything works perfectly. In the real world, where things slip out of adjustment all the time, far fewer than 230 channels are used in the AM radio band. This situation is described more thoroughly in a later section of this chapter devoted to amplitude modulation.

Now, suppose that there is a medium or link that will carry only frequencies in the AM radio broadcast band. With such a link, you could never transmit more than 230 channels of 5-kiloHertz data or voice. A medium or link with a larger bandwidth could transmit more channels, or an equal number of channels with a higher rate of information (modulation). For instance, a coaxial cable can handle frequencies from DC (0 Hertz) to at least 1 gigaHertz (1 million kiloHertz). In theory, you could modulate 200,000 "single-sideband, suppressed-carrier" voice channels 5 kiloHertz wide onto this cable. In practice, there are a lot of reasons why this wouldn't work. It is still possible, though, to get several thousand voice channels onto this cable. Even with everything working according to theory, however, you could never get more than 200,000 of these 5-kiloHertz channels onto the cable.

Instead of a coaxial cable, suppose that you used a fiber-optics lightpipe, capable of carrying visible light and infrared. The frequencies carried by such a lightpipe might be between 300 teraHertz (infrared) and 800 teraHertz (near-ultraviolet). (Note: 1 teraHertz = 1,000,000,000 kiloHertz.) How many 5 kiloHertz voice channels could the fiber-optics medium carry?

First, we need to know the bandwidth:

$$\text{bandwidth} = \text{highest frequency} - \text{lowest frequency}$$

$$= 800,000,000,000 \text{ kHz} - 300,000,000,000 \text{ kHz}$$

$$= 500,000,000,000 \text{ kHz}$$

Next, we need to see how many 5-kiloHertz voice channels will fit into this bandwidth of 500,000,000,000 kiloHertz:

$$\text{channels} = \frac{500,000,000,000 \text{ kHz (bandwidth of medium)}}{5 \text{ kHz (bandwidth of one voice channel)}}$$

$$= 100,000,000,000 \text{ (5-kHz channels)}$$

That's 100 billion voice channels! Since there aren't 100 billion people in the world, it unlikely that we would ever need all those channels, nor would we be likely to build a receiver for all of them.

We can conclude from this that fiber optics has enough bandwidth for all the channels we would ever need, whereas untwisted twin-lead copper conductors (whose bandwidth is about equal to the AM band) would carry no more than a few hundred, and probably only a few dozen, channels. For a comparative representation of the bandwidth capabilities of various media, see Figure 9–5.

9.3 FREQUENCY-DIVISION MULTIPLEX

In Section 9.2.2 we looked at the example of a stoplight transmitting three channels of digital information in three different frequency bands (colors). This is the way in which all **carrier transmissions** allow multiple channels of information to be carried on one medium. When more than one channel of information is carried on a single conductor, we say that we have multiplexed several channels onto the same conductor. Multiplexing can be done in several ways, but the one that was developed first is still the most widely used in radio and television broadcast. In the example

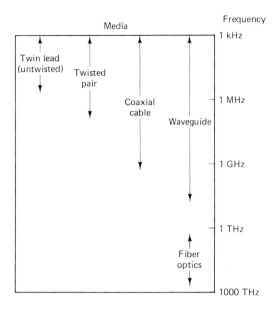

Figure 9-5 Comparative representation of bandwidth capabilities of various media.

of the stoplight, each channel's information—stop, caution, or go—was modulated onto a different frequency band. The same carrier—light—was used in every case, but some of the light was red, amber, or green accordingly. Commercial radio and television transmission separates voice channels in the same way. Although we don't perceive various radio frequencies as colors (in fact, we don't perceive them at all), each channel of the commercial televison broadcast band, for instance, is modulated onto a band of frequencies that the other channels don't use. To separate the channels, a receiver whose antenna is receiving all the different frequencies at once must filter out all but the desired frequency, much as the red filter on the stoplight blocks light frequencies that are not red. To tune in television channel 2, you need a filter that blocks all frequencies that are not channel 2. When you switch channels, you are merely connecting different filters into the receiver circuit. In the case of fiber optics, where each carrier frequency actually is a different color of light, multicolored light can be carried down the fiber and separated into different channels at the receiving end by colored filters. This frequency separation method of tuning in different channels from a multichannel signal is called **frequency-division multiplex** (FDM). Whenever multiple channels are modulated onto a separate carrier frequencies, we have FDM.

9.4 AMPLITUDE MODULATION

Amplitude modulation has already been described in the first part of this chapter. We saw, in Figures 9-2 and 9-3, what the AM modulation of a carrier signal looked like for digital and analog information. The concepts of carrier, modulation, and modulation envelope were introduced there.

There were a few things about the amplitude-modulated signal that we didn't explain. One is the reason why we can transmit only 230 5-kiloHertz voice channels in the AM band from 500 to 1650 kiloHertz. It seems logical that if all we have changed in the carrier wave is its amplitude (in other words, its wave height), then all of that carrier wave, whether tall or short, should have the same frequency. For instance, if we voice-modulate the amplitude of a 1000-kiloHertz carrier wave at a rate of 5 kiloHertz, isn't the wave's frequency still 1000 kiloHertz everywhere? Why shouldn't we be able to put another channel closer than 5 kiloHertz from the first? Why shouldn't we be able to put channels as close together as we want?

The answer is that by changing the amplitude of a wave, we create additional frequencies, even if the waves have the same wavelength all the time. Figure 9–6 shows the effects of modulating a carrier wave by changing its amplitude.

Figure 9–6(a) shows a **frequency spectrum** of the unmodulated carrier wave. This is a graph showing how much energy is being radiated at all frequencies between 990 and 1010 kiloHertz. The carrier wave is oscillating at 1000 kiloHertz. Energy is mostly radiated at 1000 kiloHertz, with a little on each side because no oscillator is perfectly steady.

Figure 9–6(b) shows what happens when we modulate the carrier (1000 kiloHertz)

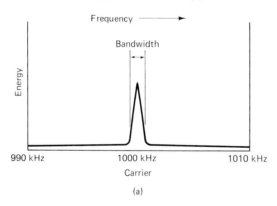

(a)

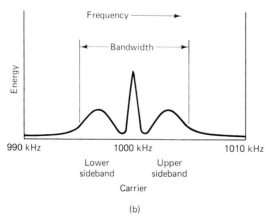

(b)

Figure 9–6 Effects of modulating carrier wave by changing amplitude. (a) Frequency spectrum of unmodulated carrier wave. (b) Carrier modulated with voice signal.

with a voice signal that contains frequencies up to 5 kiloHertz. The energy that appears on either side of the carrier frequency depends on the energy in the voice modulation. If the voice signal contains a large amount of energy at 2 kiloHertz (a piercing high-soprano note, for instance), there will be a lot of energy radiated at 1000 kiloHertz plus 2 (1002 kiloHertz) and 1000 kiloHertz minus 2 (998 kiloHertz). These additional energy peaks on the spectrum are called **sidebands.** The higher frequency (1002 kiloHertz) is called the **upper sideband,** and the lower frequency (998 kiloHertz) is called the **lower sideband.**

This is why we can't put channels closer together than the audio bandwidth of each voice channel. In fact, with two sidebands generated by the same voice signal, we should have radio stations twice as far apart as the audio bandwidth of the sounds they are transmitting. It happens that AM radio broadcast (commercial) stations *do* use an audio bandwidth of 5 kiloHertz, and thus they require 10 kiloHertz or more between carrier frequencies to make it possible for receivers to tune in each station separately.

The bandwidth of the radio station's transmission is now the carrier wave's bandwidth, plus the sidebands, which are each as large as the frequency range of the original voice signal modulated onto the carrier.

If the information modulated onto the carrier appears as two sidebands, the radio transmission takes up twice as much frequency space as the original audio sounds did. It appears that all the information in the voice signal appears twice in the radio transmission. Can we get by with transmitting just one of the sidebands? If the information isn't a part of the carrier wave, but appears at frequencies above and below the carrier, why not just transmit one sideband without the carrier? This is the principle of **single sideband suppressed-carrier** (SSB-SC) **transmission.** It actually *does* work, and that was the method we had in mind when we said that it was just possible, theoretically, to transmit 230 channels of 5-kiloHertz information on the 1150-kiloHertz bandwidth assigned to AM radio.

Figure 9-7 shows varying amounts of modulation. The waves marked 25, 50, 75, and 100% show the carrier varied by increasing amounts of amplitude change. At 100% the bottom of the modulation envelope is low enough that the amplitude of the carrier is zero. Since there is no way to reduce the amplitude of a carrier to less than zero, this is the largest possible percent of modulation. Attempts to modulate the wave with larger amplitude changes merely result in the information becoming garbled and unintelligible.

9.5 FREQUENCY MODULATION

There are three characteristics of wave phenomena that we can modulate to carry information. We have already discussed AM, the modulation of a wave's amplitude. Now we take up the modulation of a wave's frequency. This is called **frequency modulation** (FM).

Like AM modulation, the most familiar use of FM modulation is radio broadcast. The FM band of radio transmission is located between 88 and 108 megaHertz.

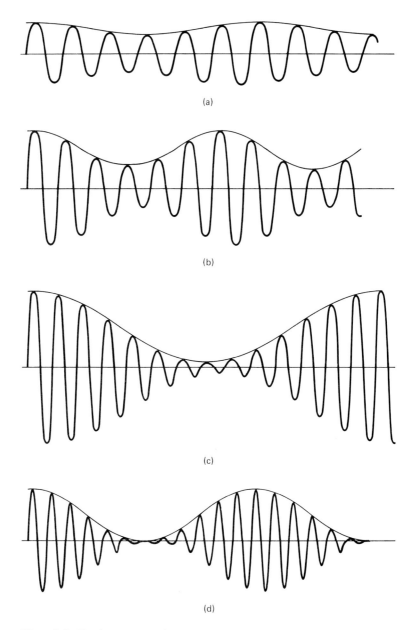

Figure 9-7 Varying amounts of amplitude modulation: 25% (a), 50% (b), 75% (c), and 100% (d).

When used for modulation of an electric current on a wire rather than transmission by wireless broadcast, FM modulation can be used in any frequency band. Frequencies around 2000 Hertz are often used for digital data transmission via modem on ordinary voice telephone lines. Although modems transmit only digital information, frequency-modulated onto telephone lines, FM radio is used for voice and music (analog) information.

9.5.1 Frequency-Shift Keying

Of the two types of information modulated by FM onto the carrier frequency, digital data is probably the easiest to understand. When there are only two levels in the information transmitted—off and on—there are only two frequencies. For the higher level of the digital signal—the "on" signal—the frequency of the carrier wave is shifted to a higher frequency. For the lower level of the digital signal—the "off" signal—the frequency of the carrier wave is shifted to a lower frequency. When there are only two frequency shifts like this, in fact only one frequency shift, between the higher and the lower frequencies, the modulation is called **frequency-shift keying** (FSK) rather than FM. The name derives from the fact that FSK was originally used for radio telegraphy, and a telegraph key was used to shift the frequency of the modulated wave.

Figure 9–8 shows a carrier wave that has been modulated by frequency-shift keying. In Figure 9–8(a), you can see the digital wave that is used to modulate the

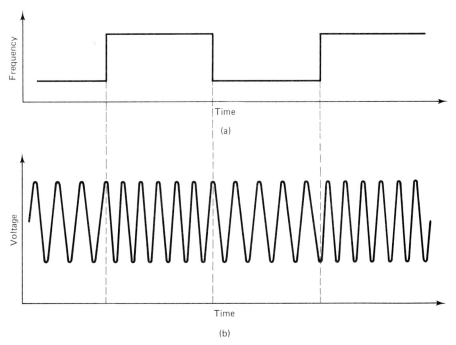

Figure 9–8   Carrier wave modulated by frequency-shift keying. (a) Digital wave used to modulate frequency. (b) Modulated wave.

frequency. This is the **modulation envelope,** a graph of the wave's frequency plotted over an interval of time. In Figure 9–8(b), the modulated wave itself is shown. Notice that the wave is the same height—the same **amplitude**—everywhere throughout the interval of time. It is the spacing between waves—the **frequency**—that changes. The close spacing, or short wavelength, is a higher frequency and represents the "on" state of the digital signal. The open spacing, or long wavelength, is the lower frequency and represents the "off" state of the digital signal.

9.5.2 Frequency Modulation (Analog)

In FSK transmission, the frequency of the carrier wave "jumps" between two frequencies. If an analog signal, such as a voice wave, is used instead of a digital signal, the frequency of the carrier wave varies smoothly between its highest and lowest values. This is similar to AM modulation, except it is the wave's frequency that varies instead of the amplitude. Instead of frequency-shift keying, this is simply called **frequency modulation.**

FM modulation is different from AM modulation in another important way: If the information that is modulated onto the carrier wave is voice or music, it affects the carrier wave frequency according to its loudness. The louder the audio information is, the greater the changes are in the carrier wave's frequency. In AM modulation, the loudness of the audio didn't affect the bandwidth of the AM channel. Only the frequency of the audio could change the width of the sidebands. Greater levels of loudness would put more energy into each sideband but would not affect its location. In FM modulation, the bandwidth of the channel depends on the loudness of the analog or digital signal. The frequencies used in the information we modulate have some effect on the appearance of the sidebands, but it is the loudness of the signal that determines how far out the sidebands extend from the main carrier-wave frequency (the bandwidth). For this reason, most FM radio stations employ a **compressor** in their transmitting apparatus. The compressor makes sure that the volume, or loudness, of the audio signal stays within certain limits. The station also uses a filter that passes only a certain bandwidth of frequencies before the transmission is amplified and broadcast from the station's antenna. This compression and bandpass filtering ensures that FM transmissions will not be buried in interference from stations at nearby FM frequencies. FM broadcast stations are usually allocated a bandwidth of 75 kiloHertz for each station. This means that if the compresors and bandpass filters are working correctly, that FM radio broadcast stations could be spaced every 75 kiloHertz across the FM dial. With a bandwidth of 20 megaHertz, this means that about 266 FM stations could fit into the FM band, spaced 75 kiloHertz apart. In reality, there are not that many FM stations in any one locality. FM receivers are not generally built with such precision as to be able to separate stations as close together as 75 kiloHertz apart. A tuner with that precision would be able to separate frequencies less than 1/10 of 1% apart on the frequency spectrum. That kind of precision is very expensive and not likely to be found in a mass-produced

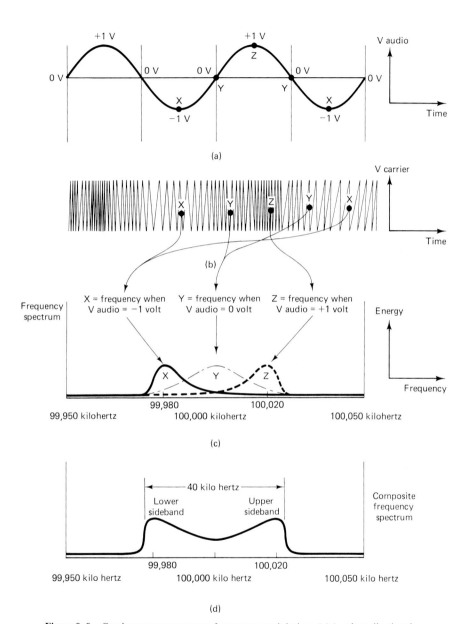

Figure 9-9 Carrier wave response to frequency modulation. (a) 1-volt audio signal. (b) FM modulated carrier waveform. (c) Frequency spectrum of carrier wave. (d) Composite frequency graph.

device such as an FM radio, so most FM stations broadcast from frequencies several hundred kiloHertz from their nearest neighbor.

As an example of FM bandwidth and frequency, suppose that a 1000-Hertz tone is modulated onto a carrier wave by an FM transmitter whose carrier frequency is 100,000 kiloHertz. Further, let's suppose that the frequency of the carrier wave deviates 20 kiloHertz every time the audio signal goes up or down 1 volt. If the audio signal goes positive 1 volt and negative 1 volt at each end of the wave, the carrier wave's frequency will change from 100,000 + 20 = 100,020 kiloHertz to 100,000 − 20 = 99,980 kiloHertz, and this will happen 1000 times a second as the audio wave varies from + to − volts.

In Figure 9-9(c), we can see this effect. The frequency of the carrier wave deviates back and forth from X to Y to Z, and back again, 1000 times a second. Figure 9-9(a) shows the 1-volt (amplitude) audio signal. Figure 9-9(b) shows how the carrier wave responds to FM modulation (not to scale). Figure 9-9(c) shows how the frequency spectrum of the carrier wave looks at various moments, X, Y, and Z. Figure 9-9(d) shows a composite frequency graph, compiled over the course of several cycles, showing how the peak of the frequency spectrum, "sloshing" back and forth 1000 times a second, averages out. In this figure it becomes apparent where the bandwidth and sidebands of the transmission come from.

What would happen if the 1000-Hertz 1-volt (amplitude) audio wave changed frequency (to 2000 Hertz)? The energy peak's frequency would slosh back and forth—would **deviate**—twice as often in a second, but would still go to the same places. The composite frequency spectrum, bandwidth, and sidebands would not look any different from Figure 9-9(d).

Now suppose that the 1000-Hertz 1-volt wave's *amplitude* got 50% bigger (to $+1\frac{1}{2}$ volts and $-1\frac{1}{2}$ volts). The energy peak would now deviate 50% farther (30 kiloHertz) on either side, and the bandwidth of Figure 9-9 (d) would become 60 kiloHertz instead of 40.

Earlier we mentioned that although FM radio broadcasts are limited to frequencies between 88 and 108 megaHertz (88,000 and 108,000 kiloHertz), when FM is used for current conducted on a wire, any carrier-wave frequencies you want can be used. This does not change the basic "how it works" of FM modulation, however, and all the principles we explained here for radio broadcast work equally well for signals carried on wires.

9.6 PHASE-SHIFT MODULATION

Of the three types of modulation discussed so far, varying the carrier wave's phase is probably the least straightforward, and hardest to understand. Since this method is used almost exclusively for digital data, the examples we will use are exclusively digital.

To start with, what is **phase**? In Figure 9-10(a) we can see three segments of a wave with the same amplitude and frequency. The only difference between these

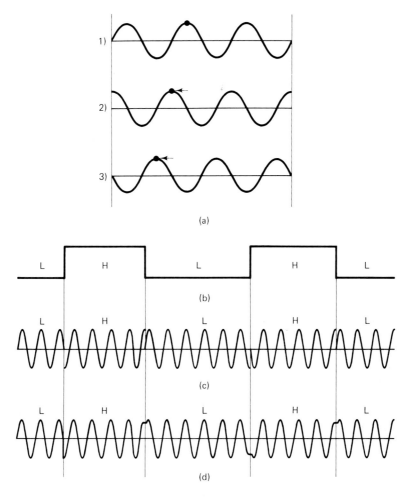

Figure 9-10   Carrier wave phase modulation. (a) Three segments of a wave with same amplitude and frequency. (b) Digital signal. (c) Carrier wave after modulation. (d) "Smooth" version of wave as produced by a real PSM transmitter.

segments is the phase, which we can relate to the point at which the wave starts. Each of the three pictures shows a wave with the same height and wavelength. Picture 2 is like picture 1 "slid over" a quarter of a wavelength. Every place on picture 2 is a quarter-wavelength ahead of picture 1. Another way to look at this is to suppose that the oscillator that makes wave 2 started a fraction of a second earlier than the oscillator that makes wave 1, and wave 2 constantly stays ahead of wave 1 by the same amount all along. Picture 3 shows the wave slid even farther ahead—half a wavelength ahead of picture 1—so that its positive peaks appear where picture 1's negative peaks do. You could see this two ways: picture 3's wave is picture 1's wave upside down, or picture 1's wave slid over half a wavelength. In each case, the wave has the same amplitude and frequency, but has been shifted over to a different *phase*.

In Figure 9–10(b) and (c), a digital signal is used to modulate a carrier wave's phase. Figure 9–10(b) shows the digital signal. Figure 9–10(c) shows the carrier wave after it is modulated. Where the digital signal changes from L to H, the carrier wave shifts ahead a quarter of a wavelength. Throughout the H portion of the digital signal, the carrier remains at the same frequency and amplitude but a quarter wavelength ahead in phase. When the digital signal changes from H to L, the carrier wave drops back a quarter wavelength, to its former phase, but keeps the same amplitude and frequency as before.

The phase shifts in Figure 9–10(c) would probably be hard to produce using a real oscillator, so we have provided a smoother versin of that wave in Figure 9–10(d), like you would get from a real PSM transmitter. The phase shifts between the H and the L parts of the modulation envelope give this method of modulation the name **phase-shift modulation** (PSM). Even smoothed out, the shifts between H and L phases are evident.

In AM, FM, or PSM, some device has to be able to vary the amplitude, frequency, or phase of the carrier wave. Conversely, when the carrier wave is received, some device has to be able to detect changes in amplitude, frequency, or phase. The gizmo that makes the changes we call a **modulator;** the gizmo that detects the changes is called a **detector** or a **demodulator.**

9.7 TIME-DIVISION MULTIPLEX

Multiplexing was discussed earlier in this chapter when we saw what frequency-division multiplex (FDM) was. To reiterate briefly, let's say that the objective of multiplexing is to put more than one channel—more than one stream of information—onto a single conductor or signal path.

In many cases we have assumed that the way to do this was to put each channel of information onto a separate carrier frequency by some form of modulation and then transmit all the frequencies at once, separating out the frequencies (and thus the channels) at the receiving end. That was the principle of FDM.

There is, however, another way to include many channels of information on a single signal path, and separate the channels out at the receiver. Imagine that a radio station wants to reach audiences that speak English, Polish, Lithuanian, and Spanish. They might have all four announcers speak at the same time, but that probably would not work too well. Instead, they would probably have the Polish-language broadcast from 8 A.M. to noon, the Lithuanian broadcast from noon to 4 P.M., the Spanish broadcast from 4 to 8 P.M., and English the rest of the time. The pattern would repeat every day, so each group of listeners would know what time to tune in their particular broadcast. 8 A.M. would always be Polish time, noon would always be Lithuanian time, and so on. In this example, we see how four languages can be **time-division multiplexed** onto one radio station. In telecommunications we would like to do this without having to wait 4 hours for our channel's turn to come up, so we switch channels a bit faster.

In a PBX, the TDM bus samples a voltage at one channel in less than 1 microsecond, encodes that number, moves on to another channel, repeats the same thing, and manages to sample all the channels, returning to the first in less than a ten-thousandth of a second. That means that the samples for each channel are taken, and the code for each sample sent on, at a rate much faster than that at which the voice signal changes. Since the highest frequency in a normal voice communication is 3000 Hertz, samples taken at a much higher rate, say, 12,000 Hertz, can reconstruct the original waveform with some degree of accuracy, even for the highest voice frequencies.

Imagine that there are three voice channels multiplexed onto a TDM bus. (That is the conductor carrying all the channels' information at once.) Every third sample is the sample for channel 1. As long as there is circuitry that places a sample from channel 1 onto the bus in coded form at the transmitting end every third sample, and a circuit that routes every third sample out to a channel 1 receiver, we won't have to concern ourselves with how that is done (we look at that in Chapter 11).

In Figure 9–11(a), three channels with different waveforms are sampled in a 1–2–3 pattern. Each sample taken from channel 1's waveform, for instance, is marked with a dot on waveform 1. The same thing is shown for channels 2 and 3. In Figure 9–11(b) we see the samples put together into a single waveform. The TDM three-channel composite wave, Figure 9–11(b), is made up of the samples from every channel, and in consequence, it doesn't look like anything familiar. (Actually, many systems transmit the samples as sequences of digital code instead of transmitting the voltage of the sample as an analog electrical signal, but that doesn't change the idea of what's being done here.) Figure 9–11(c) shows how the samples would reconstruct the waveforms from channels 1, 2, and 3 if each channel's sample was pulled out of the composite wave and sent to some destination at every third time interval. It is probably clear that channel 1, with a higher frequency, doesn't "look as good" as channels 2 and 3, but its basic amplitude and frequency (and phase) can still be ascertained from the "reconstructed channel 1" waveform. If the frequency in channel 1 were higher, or if it were sampled fewer times than this, it might become difficult, or impossible, to tell the frequency, amplitude, and phase of the signal at channel 1 from its reconstructed wave. With our hypothetical sample rate of 12,000 samples per second, mentioned earlier, the highest voice frequency, 3000 Hertz, would look just like channel 1's reconstructed waveform when it is **demultiplexed.** In the receiver, there will be filter circuitry that will "connect the dots" for waveforms such as reconstructed waves 1, 2, and 3, and the smoothed-out waves will resemble the incoming information in channels 1, 2, and 3 quite closely.

9.8 PULSE-AMPLITUDE MODULATION

The example of Figure 9–11, where the samples of each channel's voltage are sent along the TDM bus as analog levels, already uses **pulse-amplitude modulation** (PAM). Each channel is sampled, and the sample is turned into a pulse whose height is related to the voltage of the sample. The multiplexing puts the pulses together from the three PAM channels into one composite wave. PAM itself is just sampling each

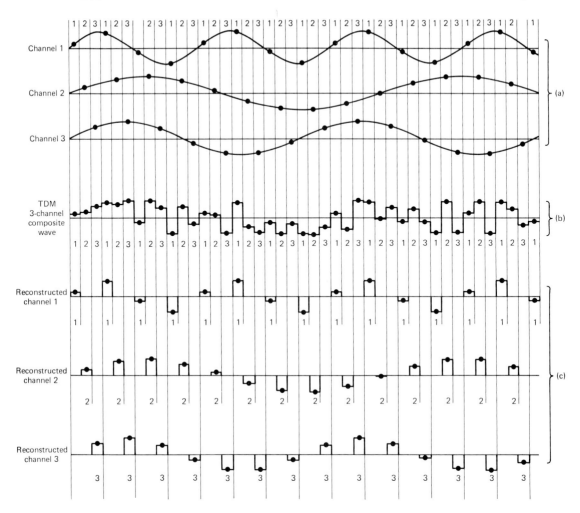

Figure 9–11 Three voice channels multiplexed onto TDM bus. (a) Three channels with different waveforms. (b) Samples put together in single waveform. (c) Reconstructed channels.

channel and turning the sample into a pulse. The "reconstructed" channel 1, 2, and 3 waveforms in Figure 9–11(c) are just the PAM version of the original channel 1, 2, and 3 waveforms.

PAM is almost always used in systems where channels are being multiplexed together, but PAM is not the same thing as multiplexing, of course.

9.9 PULSE-CODE MODULATION

If we convert the samples of Figure 9–11(b) into digital code by an A/D (analog-to-digital) converter, and send the codes instead of the analog pulses, we are using **pulse-**

code modulation (PCM). Digitizing a pulse or sample involves turning its voltage into a binary code. In Figure 9–12(a) a waveform is sampled, and the voltage in each sample period is converted into digital code, which is sent out on a four-wire bus. This arrangement of sending each digital number out on a group of wires so that all 4 bits are sent out at once is called **parallel data transmission.** In Figure 9–12(b), an extra stage called a **UART** (universal asynchronous receiver/transmitter) puts the bits of each coded number onto a single wire one after another, a method called **serial data transmission.** For short distances, such as inside a PBX cabinet, the parallel arrangement is more efficient, but not usable over long distances. Longer-distance links, and "outside" connections such as a T1 span, would use a single shielded pair and usually employ serial transmission. Transmissions to satellite repeaters are broadcast using PCM modulated microwaves.

9.10 DELTA MODULATION

A variation of pulse-code modulation is **delta modulation.** Instead of sending the entire code every time a sample is taken, the delta signal—one bit—merely indicates whether the level has increased or decreased. If a channel is being delta-modulated,

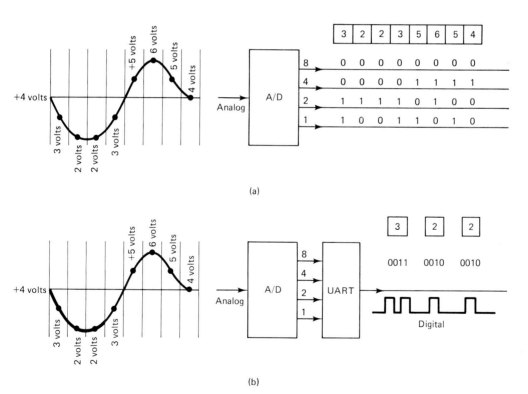

Figure 9–12   (a) Parallel data transmission. (b) Serial data transmission.

and its level has increased since the last time it was sampled, the delta signal will be a 1. If the level has decreased, the delta signal will be a 0.

For delta modulation to be effective and accurate, sampling should take place at a higher rate than for PCM. There is no difference between serial and parallel transmission when only one bit of data is involved, as with delta modulation.

9.11 SIMPLEX, HALF-DUPLEX, AND FULL DUPLEX

The terms "simplex," "half-duplex," and "duplex" refer to different ways in which signals can travel on various links. **Simplex** communication is one-way communication. Information is sent from a source to a destination and does not return in the opposite direction. An example of this is the ubiquitous paging "beeper." It is a one-way channel entirely, since you cannot beep back.

Duplex communication means two-way transfer of signals. There are two methods of duplex communication, half-duplex and full duplex. **Half-Duplex** communication is like a telephone conversation. It is two-way communication, but each person takes a turn and the communication is only one way at a time, with the direction alternating. **Full-duplex** communication is two-way, both at the same time. This involves either two pairs of wires, one for sending and one for receiving, or some form of TDM or FDM multiplexing.

By comparison with something more familiar—traffic—simplex communication is like traffic on a one-way street. Half-duplex communication is like a reversible lane on an expressway, which is inbound to the downtown area in the morning and outbound from downtown in the evening. Full-duplex communication is like the expressway itself, with inbound and outbound lanes carrying traffic in both directions separately and simultaneously.

REVIEW QUESTIONS

1. Can transmission forms be modulated by either an analog or a digital quantity?

2. In a system that varies the brightness of a light to either of two discrete levels, is the source of modulation analog or digital? What is the carrier that is being modulated?

3. Can the quantity of information transmitted can be increased by increasing either bandwidth or frequency?

4. What is the approximate wavelength of a 500-Hertz tone (sound waves)? Assume standard temperature and pressure.

5. What is the typical bandwidth of human hearing? What is the bandwidth of typical voice-grade telephone transmission? What is the bandwidth of light we can see?

6. As you speak into the mouthpiece of a telephone, electrical waves are set up in the wires with the same frequency as the sound waves. Which wave *goes faster,* the sound wave or the electrical signal?

7. What types of transmission forms are commonly frequency-division multiplexed?

8. Draw three amplitude-modulated sine waves with a 0, 50, and 100% modulation. Assume that a "slow" sine wave is used to modulate the amplitude of a "fast" sine wave in these examples. Is it possible to have a wave modulated *over* 100%?

9. Name four ways to modulate a signal onto a carrier.

10. Is time-division multiplexing used in broadcast radio?

11. Can pulse-amplitude-modulated signals be multiplexed onto a single carrier? Is PAM essentially digital or analog?

12. What kind of code is used in pulse-code modulation? Is the transmission of this code serial or parallel?

13. If a codec uses delta modulation to transmit digitized voice, what is the main advantage over pulse-code modulation?

14. From among the examples we discussed in this chapter, select one each where information is transmitted in: (a) simplex form; (b) half-duplex form; (c) full-duplex form.

10

Signaling

Signaling guides, directs, and supervises the flow of information. The media that carry information and signaling determine what form that information and signaling may take. For example, signals sent on copper wires would be transmitted in electrical form, signals on a fiber-optic light pipe would be optical in form, and wireless transmission would use radio waves. To paraphrase Marshall McLuhan, the medium may not be the message, but it does limit the forms in which it can be conveyed. Signaling may take place in analog form—like Touch-Tone—on a link that carries voice transmissions in an analog form. As distinguished from voice or digital information, signaling conveys facts such as where the information is to be sent, when it is to be sent, and who is going to pay for the use of services and equipment. Signaling sometimes conveys these facts to the PBX, to the phone company's central office, and especially, to the listener, in human-interpretable forms such as dial tone, ring-back tone, and busy signal. On most links, the signaling and the transmission are going over the same path (e.g., DTMF and voice both go over the same copper wire). For example, on a tie line—full-duplex four-wire transmission—the signaling would be superimposed on the potential levels of the two pairs. This is called DX signaling. This is often converted to E&M signaling. **Supervision** is another name for the handshaking that keeps track of who's talking, who's listening, when the connection has been established or when it should be disconnected, and keeps track of billing for a call.

Three types of signaling that are used in telecommunications: analog (or AC) signaling, DC signaling, and digital signaling. **Analog signaling** (AC signaling) involves the use of tones or sounds that are transmitted in the same way, and on the same wires, as ordinary voice transmissions. Examples are the dial tone, busy signal, and DTMF (Touch-Tone®) signaling for dialing digits of a telephone number. These are carried as part of the audio signal and can be heard on the telephone. For the most part, such signals are *supposed* to be heard on the telephone—as in the case of the dial tone, and the busy signal, which are, after all, signals for the human user and not intended to control automatic equipment. There are also classes of signals that are *not* intended for human users, but automatic equipment, instead. Some of these signals are **out of band**—that is, they are tones whose frequency is beyond or above the range of human hearing. They cannot be heard by the user of the telephone but can still control automatic equipment perfectly well.

DC signaling takes advantage of the fact that the voice signal in a telephone is direct current. Variations or fluctuations in the value of the current produce the sound at the receiver, but the current itself always flows in the same direction. Reversing the direction of the current, or altering its average value, is a method of signaling. It is not intended for the human user and is generally inaudible to a human being. Fluctuations in a negative current or a positive current sound the same at the receiver.

Digital signaling is used where digital data are being transmitted instead of analog voice signals. The signaling information bits appear "between" blocks of digitized voice transmission data, or blocks of digital information originating at a terminal or computer. As with some other methods mentioned, digital signaling is for automatic equipment and is not intended to be "heard" by the human user at the receiver.

Often, the signaling information, especially that not intended for human ears, is passed through a separate pair of wires (a "signal pair") instead of the wires used for the analog voice transmission (the "voice pair").

10.1 SIGNALING OVER TELEPHONE TRANSMISSION LINKS

10.1.1 Advisory Tones

When you pick up your phone to make a call, the switch that will make the connection must be told that you want to make a call. You cannot do this with words, because the switch doesn't have the "smarts" to understand your spoken language (at least, not yet). Instead, a change in the electrical condition of your phone must take place in a way that the switch can "understand." You do this by merely picking up the handset, an action that completes a circuit and allows a "loop current" to flow in the line. This is your signal to the switch that you want to dial out. Now, the switch has to tell you when it has a register ready to accept your dial pulses. Since the switch doesn't have the smarts to say, in spoken words, "Okay, buddy, it's your quarter—go ahead and dial," the switch uses an electrical signal instead. This signal is the **dial tone.** You've heard this bunchteen zillion times in your life, so we will not bother to describe it here. If the phone system isn't ready when you are, you won't hear a dial tone until it's ready to accept the digits you'll dial.

The next signal you would encounter would be the ringing tone or busy signal. The **ringing tone,** of course, is supposed to sound like a distant echo of the telephone ringing on the distant end of your call (in fact, the rings you hear may have no connection with the number or rate of rings at the distant end). The **busy signal** is, of course, the switch's way of notifying you that the circuit to which you want to connect is already in use. Clearly, you are aware of all these signals in the system, and it is necessary for you to understand them to be able to use the system appropriately. When your call "goes through," however, there is one more signal that occurs in the system, of which you are unaware, as you were unaware of sending a "request for service" signal by picking up the handset. That is a signal from the called-party's side of the system, which indicates when the phone is answered, and you see the effects of this signal when you receive the bill at the end of the month.

10.1.2 Telco Signaling Methods

There are several types of signaling used by the public network that are "transparent" to the users (that means that you won't hear this signaling on your receiver). These signals are used for *supervision* (to tell how long, and to where, the call is connected, for billing purposes), and for *addressing* (to tell how to connect a call based on the numbers dialed by the caller). These methods are:

1. **SF signaling.** This is a 2600-Hertz tone that is present when the phone is on-hook, and absent when it's not. This tone is used for supervision of the call by the telco, and you do not hear it.

2. **Two-frequency signaling.** Two-frequency signaling is used for both supervision and addressing. It is pulsed from one place in the system to another, using frequencies of 2040 Hertz for the binary 0 and 2400 Hertz for the binary 1, to make numbers using binary code. Although this is similar to the two frequencies placed on the voice line by a digital telephone modem, these two tones are only used internally by the telco, and are never heard at the receiver.

3. **Multifrequency signaling.** This is the telco's internal version of DTMF. It uses five or six frequencies, two at a time, to handle addressing, supervision, and control operations within the system. It is only used on trunks, and is not **supposed** to be accessible through ordinary telephone transmitters.

4. **DX signaling.** DX derives its name from "duplex" and involves signaling in two directions. It is a form of DC signaling superimposed over the wire pairs. The voltage potential difference between the transmit and receive pairs of wires carriers supervision information about whether phones are off-hook or on-hook. DX signals also carry addressing information in the form of rotary dial pulses encoded into DC signaling.

Note: Since these types of signaling are used exclusively by the telcos, you won't be working with them if you are working with private telecommunications equipment. Since, for instance, there are places where DX signaling is converted to E&M signaling (used inside a private branch exchange), it is important for the person working within the private telecommunications system to know what DX signaling is, even though it is not used within the private system.

10.1.3 Central Office Lines

When you pick up your telephone handset the hookswitch completes a current loop. A current of 20 to 40 milliAmperes begins to flow, and this current is your electrical signal to the switch that you want a dial tone. When you pick up a phone that is ringing, you also start a loop current, but in this case, your signal to the switch does two things. First, it signals the switch to stop ringing the phone. (What's that? You thought small elves inside the phone disconnected that for you?) Second, it signals the phone company that the call has "gone through," and they start billing the call to the account of the person who dialed you. This "answer supervision" signal is what tells the billing equipment to start clicking off the dollars and cents.

10.1.4 Central Office Trunks

Central office trunks are links between private switching systems and the local central office. Local trunks, WATS (wide-area telephone service) trunks, FX (foreign exchange) trunks, OCC (other common carriers), and MABS (metropolitan area business service) trunks are examples of central office trunks. These names don't reflect anything about the type of signaling these trunks use, only how the customers will be billed and the ownership of part (or all) of the path being used. It also reflects whether the trunk will be used for incoming calls only, for outgoing calls only, or for calls going both ways. While your home telephone can call out or receive calls, this is not necessarily the case with all trunks. 800 numbers, for instance, can only **receive** calls. Certain types of WATS lines (called outWATS) can only call out. The "number" of the WATS line, in this case, is not even "dialable" from outside; it's just an identification number.

10.1.5 Loop-Start Central Office Trunks

The only difference between a loop-start central office line (see above) and a central office trunk is the type of equipment attached to it. When a PBX puts a loop across the line, instead of a telephone, the central office doesn't know the difference, and functionally, there isn't any. Designating a "line" or a "trunk," in this case, refers only to whether the equipment attached is a station or a piece of switching equipment. Naming these facilities is a matter of convenience, depending on their use.

In Figure 10–1, we see a **loop-start trunk.** Figure 10–1(a) shows the line before a call is placed or received; the T (tip) and R (ring) leads have voltages of 0 volts (tip) and −48 volts (ring). In Figure 10–1(b), a load is attached across the leads (the PBX's trunk interface card); a current flows, signaling the central office with a request to dial. Figure 10–1(c) shows an incoming call; the ring-lead voltage adds to the −48 volts a 100-volt AC signal, pulsating at 20 Hertz. This is detected by the PBX as an incoming call. On an ordinary phone, this voltage would ring the bell (the ring signal). Finally, we see, in Figure 10–1(d), how the PBX answers the call by placing a 600-Ohm load across the leads, establishing a loop current that is

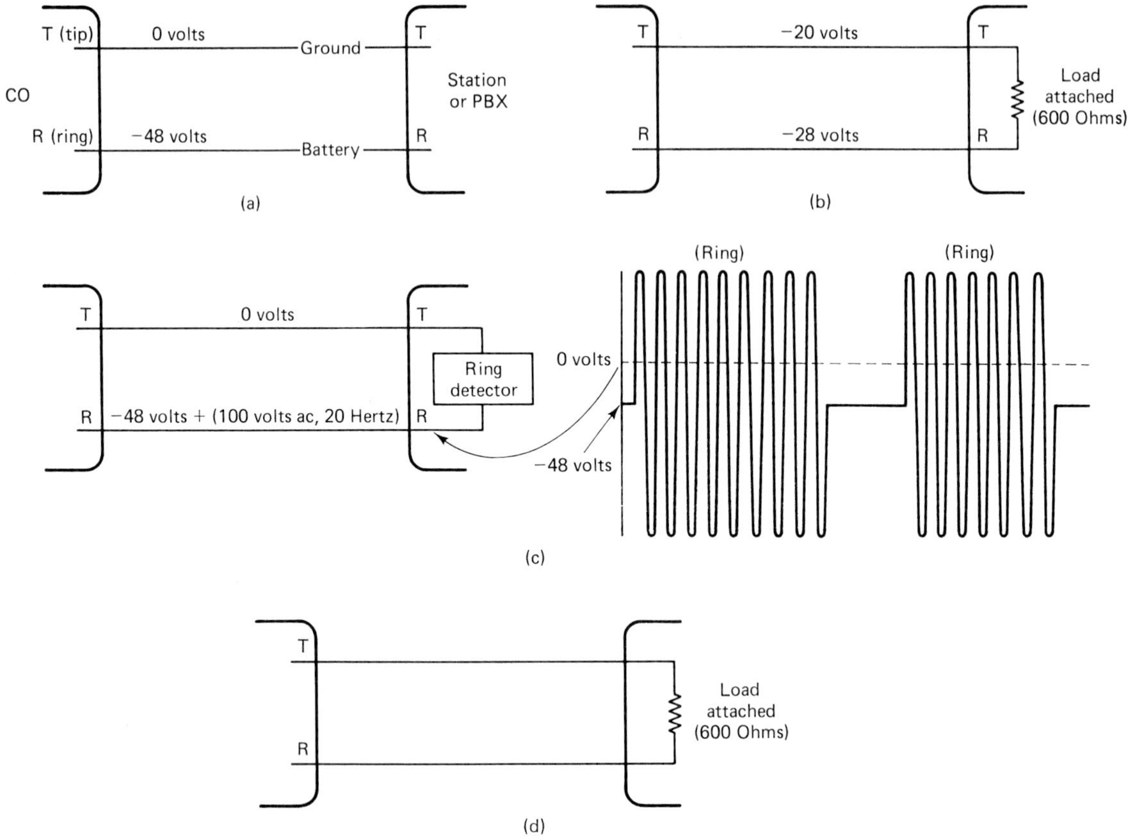

Figure 10-1 Loop-start trunk: idle (a); outgoing call (b); incoming call (c); incoming call answered (d).

equivalent to what you produce when you pick up your phone. All these figures are the same for a call to or from your home phone, except that a PBX makes the connections.

10.1.6 Ground-Start Central Office Trunks

Another method of signaling is **ground start.** Rather than establishing a current loop when the telephone goes off-hook, a ground start signals the system by applying a specific voltage level. To get a dial tone on a ground-start central office trunk, the PBX must momentarily ground the ring side of the line. The mere presence of current in a loop is not sufficient to identify a "request for service" from this type of trunk. After the ground start has been signaled, the PBX has to remove the ground and complete the loop to hold the circuit to send tones or rotary dial pulses. Anybody could supply this kind of trunk to a PBX—a central office from the telco or any other common carrier—ground start might even be used for something besides a PBX, such as a pay phone.

Perhaps you saw the motion picture "War Games." In one scene, the teenage computer whiz was at a pay phone booth where he wanted to place a call. Lacking money, he cast about for an alternative way to get into the telephone system. He found a pull-tab from an aluminum can, unscrewed the mouthpiece of the telephone, and used the metal pull-tab to connect the chassis of the phone to a point inside the handset. Immediately, he was rewarded with a dial tone and placed his telephone call.

What did he do? Would it work?

Apparently, he assumed that the pay phone was attached to a ground-start central office trunk. Using the pull-tab (a metal conductor), he connected the ring side of the line to a ground (the metal chassis of the telephone itself) directly, which is presumably what happens when you drop the correct change into the coin box.

For him, it worked. However, your chances, should you try the same thing, are not very good. On an older phone, perhaps, in an out-of-the-way location (as depicted in the movie), it might work if you know where the contact for the ring side of the line comes out inside the handset. At present, pay phones generally use other methods of signaling. You should be warned that doing this is *fraud,* and it is a federal offense. You should be further warned that they wouldn't show this in a movie if you could still do it.

Reminder: Although we mentioned this in Chapter 1, it is worth repeating: The name "ring" for one of the two wires of a telephone line has nothing to do with the fact that a telephone rings a bell to signal its user. It is so named because the metal contact on a phone plug, to which this wire is connected, is shaped like a ring, while the other wire is connected through the hole in the ring to the tip—the contact at the tip of the plug. (If the inventor of the telephone had been named Alexander Graham Buzzer, we wouldn't have any of this confusion.)

In Figure 10–2, we see a ground-start central office trunk. In Figure 10–2(a), the trunk is idle; the T (tip) lead connection to the central office is open (not connected), and the R (ring) lead connection to the PBX is open. Figure 10–2(b) and (c) shows an outgoing call; the PBX places a momentary ground on the ring lead, then the central office "sees" the ground and returns a ground on the tip lead. When this ground arrives at the PBX, it connects a load to the tip and ring leads, resulting in a loop current. In Figure 10–2(d) the central office completes the connection and is ready to receive dial pulses. For an incoming call, the tip lead from the central office is connected to ground, while the ring lead receives the 100-volt AC ring voltage superimposed on its − 48-volt level.

Advantages of ground start over loop start

Since a PBX isn't a human telephone user, it can't "hear" the dial tone, so a PBX connected to a loop-start trunk will just have to "guess" at when the CO is ready. With ground start, the returned ground on the tip lead can be recognized by the PBX, and it knows when the CO is ready. This ensures that the PBX won't begin dialing until the CO is ready.

When a call comes in to a loop-start trunk, all that's on the line—to let the PBX know that an incoming call wants to use the line—is the ring voltage signal.

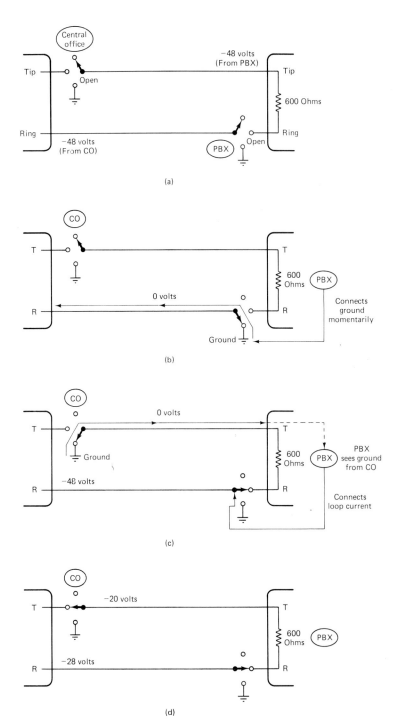

Figure 10-2 Ground-start central office trunk: idle (a); outgoing call (b); outgoing call connects ground to T(tip) (c); completed connection (d).

That is an interrupted signal, with "dead time" between rings. Between rings, the PBX doesn't have any signal to identify the trunk as an "active" one with a call trying to get in. It might seize that trunk and try to use it for an outgoing call. By contrast, the ground-start trunk has a ground on the tip lead, which lasts during the ringing and between rings as well. This identifies the trunk as "busy" so it won't be seized for an outgoing call by the PBX. This problem is called "glare" or sometimes, a "head-on" collision.

Loop-start systems don't give a PBX any positive way to tell when the far end hangs up. Ground-start systems do. When the distant end hangs up, the central office gives a ground-start trunk an open on the tip lead—the idle condition in Figure 10-2(a)—this informs the PBX that the call was disconnected.

10.1.7 Signaling over DID Trunks

Direct inward dialing (DID) is a scheme to reduce the number of wire pairs connecting the central office to a PBX for incoming calls. These circuits cannot be used for calls going out—other provisions must be made for outgoing calls. For example, suppose that Bonebreak Community Hospital wants to be "phonable" at any of 1000 numbers from 555-1000 to 555-1999. It's possible to connect 50 trunks to the central office from Bonebreak Community Hospital, but perhaps there are never more than 50 calls in progress at any one time. In that case it would make more sense to assign the "block" of numbers (555-1000 to 555-1999) to Bonebreak Community Hospital but connect only 50 **DID trunks.** Any calls coming in to 555-1000 through 555-1999 would be connected to one of the 50 trunks. The central office would "repeat" the last four digits dialed, and the PBX at Bonebreak Community Hospital would connect the incoming call to one of 1000 extensions.

To the PBX at Bonebreak, the DID trunk is somewhat like one of its own extensions. When the PBX "looks out" at the central office, it "sees" the DID trunk much like one of its own single-line phones. When someone picks up a single-line phone inside the PBX system, they can dial four digits to reach any extension. When the central office gets a call for a number within the block (555-1000 to 555-1999), it's programmed to "pick up" one of the 50 DID trunks. To the PBX, this is like one of the people inside the institution picking up the phone, except that the PBX does not respond to *this* request with a dial tone. Instead, the PBX gives the central office a **wink-start** signal. This "wink" signals the central office to repeat the last four digits dialed and the PBX acts on the four-digit number to ring the extension.

Why does the central office need a signal (wink start) from the PBX? This is a "handshake" that tells the central office when the PBX is ready to receive the last four digits. It's possible that the circuits involved in identifying a four-digit number (registers) are momentarily busy, but it won't be long before one of these registers becomes available. When it does, the PBX can receive the repeated digits from the central office, and it signals the central office to go ahead and send those digits. Human beings dialing from another extension would be given a dial tone at this time, because they haven't dialed a number yet, but the DID circuit at the central office doesn't need a dial tone, since it already "knows" the last four digits of the extension wanted. Instead, when the "wink" is received, it outpulses (DTMF or rotary dial pulses) the last four digits of the number previously dialed.

What is a wink-start signal? In this case, the wink-start signal is a reversal of polarity. For a quarter of a second, the PBX reverses the direction of current flow. Current is normally flowing because the central office DID circuit has placed a loop across the PBX line, and the PBX runs a current through it as though it were a phone that had just gone off-hook. In this sense, the DID trunk has signaled the PBX the same way that a telephone does—by **loop start**—with the presence of a current-carrying conductor across the line. The PBX must signal back with a "hi there" that shows it has a register ready. Since it is already passing current through the DID trunk's loop, a reversal of current is a good signal.

Answer supervision

When a central office DID trunk calls up a PBX and the extension answers, the PBX has to signal the central office that the extension has been picked up. Now the central office knows that the phone call has been answered, and can start billing (the person who called) accordingly.

How does the PBX signal the central office of this fact? It reverses the loop current (as it did during the wink), but this time leaves it reversed.

There are some connections that do not reverse the line current; for instance, if you dial Bonebreak Community Hospital and mistakenly dial a four-digit extension that's not being used (the extension does not exist), the PBX will connect you to an **intercept recorder,** which plays a recorded message stating: "The number you have dialed is not in service." You will not be billed for that call, because the intercept recorder does not reverse the line current and the central office does not recognize this as a call that has "gone through."

In Figure 10–3, we see the loop current flowing in a DID trunk as an incoming call is received. It's the same current whether it's flowing in on the tip lead or out

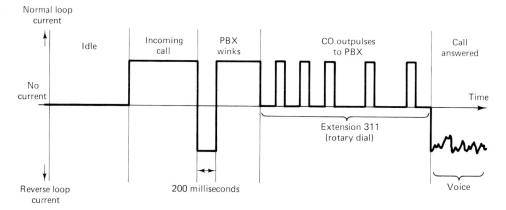

Figure 10–3 Flow of loop current in DID trunk as incoming call is received.

on the ring lead, so we haven't specified the lead on which the current is measured. At first we see the idle DID trunk; then the call coming in from the CO makes loop current flow in the normal direction. When the PBX is ready to receive the extension number from the CO, it "winks" (the current reverses for 200 milliseconds). The CO repeats the extension number it received—in this case, in the form of rotary dial pulses—and when the call is answered, the current reverses and stays reversed for the duration of the call. When the call is disconnected, the system returns to the idle condition (not shown).

10.2 SIGNALING OVER E&M TIE TRUNKS

Tie trunks are used to connect one PBX to another at remote location. These connections are often called **tie lines,** but since we defined a connection from one switch to another as a trunk, a tie line is actually a **tie trunk**.

Remember the Harts Company's tie line to Milwaukee, the one we mentioned in an earlier chapter? This was a private line that Harts leased from the local telco to connect the office in Chicago with its office in Milwaukee. Tie trunks might also be used to connect a PBX to an OCC (other common carrier), a company such as MCI, for long-distance calls.

The signaling over a tie trunk is usually E&M signaling. "E&M" stands for "ear and mouth"—because the E lead "listens" for incoming signals (ear) and the M lead is the one on which outgoing signals are "spoken" (mouth). We use the terms "ear" and "mouth" figuratively rather than literally here, because these leads actually have nothing to do with the voice part of the transmission and do not receive or deliver signals to the earpiece and mouthpiece of the handset. They are used only to signal between the switches. Figure 10–4 shows the "give and take" on an E&M link.

Let's identify two PBXs and look at some of the signaling that goes on between them over a tie trunk. Martha, in Washington, wants to talk to George, in Mount Vernon. A tie line links the two. Martha dials 7, the access code in her PBX

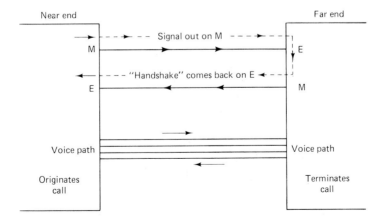

Figure 10–4   E&M interface (null-tie).

for the tie trunk. She receives a dial tone from the distant PBX. Then she dials George's extension number, and when George answers the phone, she can ask him about those items on his expense account that were puzzling her.

What is really happening during this transaction? There are two wires, used for signaling only, called the E lead and the M lead, effectively connecting George's PBX to Martha's. There will also be leads for the voice path, either two wires or four wires, depending on how the tie line is designed. The M lead is for sending signals, and the E lead is for receiving them. Martha's M lead (which is sending) communicates with George's E (receiving) lead, and vice versa.

When Martha dials 7, Martha's PBX puts a signal on its M lead (which is George's E lead), and requests a dial tone from George's PBX. George's PBX sends back the dial tone. This occurs on the voice leads, not the E&M leads. George's PBX "sees" the signal on its E lead, and in response to this, it puts "battery" on its M lead (that is, the voltage of the -48-volt battery that is used in the telephone system). Martha's PBX receives the signal from George's PBX (a handshake) that acknowledges receipt of Martha's signal and that the distant PBX has a register available (this serves the same purpose as the quarter-second "wink" on a DID trunk).

Like the wink signal from the PBX to a DID trunk, the battery voltage placed on George's M lead tells Martha to go ahead and send the extension number. Since Martha is not a trunk, the signal coming to her PBX's E lead means nothing to her. She needs a "human" signal—the dial tone—to tell her that it is time to dial the extension numbers. She dials the digits of George's extension, hears a ringing tone while George's phone rings, and waits for George to pick up the phone.

10.2.1 E&M Interface

The **E&M interface** is one of the most common for signaling between a switching system and the transmission facility that will connect it to another switching system. At the distant end of the transmission medium, another facility will probably be interfaced to the distant PBX with an E&M interface also. This is shown in Figure 10-5.

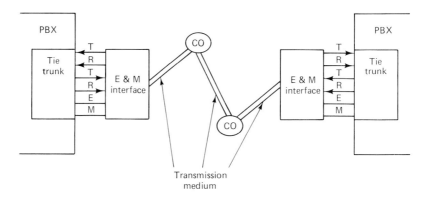

Figure 10-5 E&M interface between different mediums.

Why is the transmission link going through two central offices in the figure? Suppose that Martha's PBX is connected to George's by stretching wires from Washington to Mount Vernon. Somebody has to buy up all the right-of-way, put up poles, and string wires (and there's always the Potomac River to worry about!). This is very expensive, and impractical to the point of being impossible. The telcos, on the other hand, already have wire facilities in place that connect Washington to Mount Vernon. They will gladly lease some of that wire to connect Martha's PBX to George's PBX on a permanent basis. This is called "leasing out dedicated private-line service." Other common carriers besides the telco may also provide private-line service.

In Figure 10-5 a two-wire voice pair is used—identified as T and R (tip and ring) in the picture—and signaling takes place on the E and M lines. So T and R represent the voice pair, and E and M represent the signal pair. The E&M interface converts voice plus signaling into a standard or protocol that is compatible with the transmission medium. At each end the E&M interface is *co-located* (in the same switch room) with the PBX and converts the tie trunk's signaling between E&M and the transmission standard used by the link. A company might provide its own private line, using microwave, for example, as the medium between the two switches, instead of going through the public network. The E&M interface would still be used, but in this case it would connect the switch to the microwave transceiver.

10.2.2 Addressing

When Martha dials George, how do the numbers get sent on the E&M tie trunk? If the PBX sends out rotary dial pulses, they will go out the M lead, but a DTMF (Touch-Tone®) number will go out on the voice leads. The voice leads are used for transmitting signals that vary their amplitude or frequency over a wide range of values—analog information—while the E&M leads carry a limited number of levels—ground or battery—and are essentially digital. DTMF pulses, bearing a wide range of frequencies, are essentially analog, and thus go out on the voice leads, while rotary pulses, which are essentially digital in nature, use the E&M leads.

10.2.3 Signaling Arrangements

We are going to look at three methods of signaling that may be used by various E&M facilities. Immediate start, wink start, and delay dial accomplish the same thing using different signaling formats.

10.2.4 Immediate-Start E&M Signaling

The case we just discussed, with George and Martha, was an example of **immediate-start signaling.** Immediate-start signaling works like this. In Figure 10-6 we see the E&M leads "sitting idle" at step 0, before the call has been started. The calling (near-end) signals are shown at the top of the figure, and the called (far-end) signals are shown at the bottom of the figure.

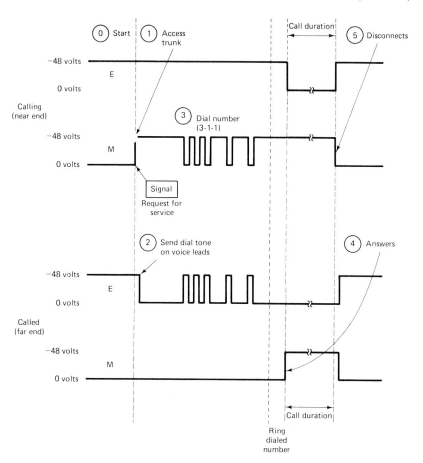

Figure 10-6   Immediate start E&M signaling.

Step 1: **Access trunk.** This happens when someone dials the access code for the tie trunk. The calling-end M lead goes from its idle state (0 volts) to its active state (-48 volts) to signal the far end. This causes the E lead at the far end to change from its idle state (-48 volts) to its active state (0 volts).

Step 2: **Send dial tone.** This is the far end's response to the signaling (change of state) on its E lead. It places a dial tone onto the voice leads (not shown in Figure 10-6).

Step 3: **Dial.** This is the rotary pulse code for the extension being dialed. It is sent from the near end's M lead as pulses which appear in inverted form on the far end's E lead. If the dial is a DTMF type, these will be tones sent on the voice leads and will *not* be placed on the M lead. When the dialed digits are received, the switch processes them and finds the extension being dialed. Once the extension is found, the far end rings that extension and sends a ringing tone to the near end through the voice leads (since these are not visible in Figure 10-6, the ringing is not indicated on the diagram).

Step 4: **Answers.** The called party answers, and the switch "cuts through" the voice connection to permit conversation. The far end signals this fact by taking its M lead, which has been idle (0 volts) all along, to an active state (–48 volts), which is maintained for the duration of the call.

Step 5: **Disconnects.** At the end of the call, either end can terminate the conversation by removing the "active" signal from its M lead. The other end will interpret this as a "hang-up" and will release the call. (In this case, the near end hung up first.)

10.2.5 Wink-Start E&M Signaling

There is an E&M signaling format similar to wink start on a DID trunk—in fact, it is also called wink start—but it doesn't involve the reversal of a loop current. Instead, a wink on the E or M lead (depending on which direction it's going) is a momentary change in the status of the signal (see Figure 10-7). For instance, if the E lead is normally at –48 volts (battery), it would go to ground (0 volts) for a quarter of a second during a wink and would switch permanently to ground when the called party answers.

Step 1: **Access trunk.** Access is provided to the tie trunk. The calling-end M lead goes from its idle state (0 volts) to its active state (–48 volts) to signal the far end. This causes the E lead at the far end to change from its idle state (–48 volts) to its active state (0 volts).

Step 2: **Register found.** The far end responds to the signaling (change of state) on its E lead by hunting for a register to receive the extension number. Wink start is typically used in systems with automatic dialing. No dial tone is sent to the near end, since the numbers have already been dialed. The automatic routing function of the calling PBX will repeat the numbers for the register of the called PBX as soon as one is available. In step 2 the register has just become available, and a "wink" is placed on the far end's M lead. When it appears on the near end's E lead, the dialing sequence will begin.

Step 3: **Dial.** In this case the dial is a DTMF type. These tones are sent on the voice leads, and are *not* seen in Figure 10-7. When the dialed digits are received, the switch processes them and finds the extension being dialed, ringing it and returning a ringing tone to the near end on its voice leads (not indicated on the diagram).

Step 4: **Answers.** The called party answers, and the switch "cuts through" the voice connection to permit converation. The far end signals this fact by taking

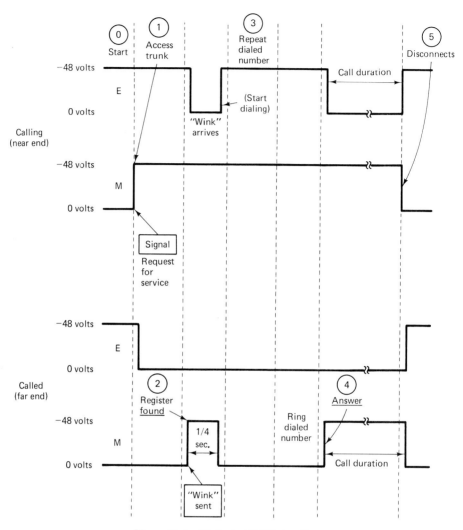

Figure 10-7 Wink start E&M signaling.

its M lead, which has been idle (0 volts) since the wink, to an active state (-48 volts), which is maintained for the duration of the call.

Step 5: **Disconnects.** At the end of the call, either end can terminate the conversation by removing the "active" signal from its M lead. The other end will interpret this as a "hang-up" and will release the call. (In this case, the near end hung up first.)

In all cases it should be kept in mind that the switch initiating the signal controls the state of its M lead and watches for a return signal on the E lead from the other switch it is "talking" to.

10.2.6 Delay-Dial E&M Signaling

In wink start, the wink occurs when the called PBX is ready to receive dialed digits. In **delay dial** (Figure 10–8) the status of the near end's E line is changed immediately when the called PBX recognizes a request for service, but does not change back until the equipment is attached (a register is found). The moment the E signal changes back, the calling PBX can start dialing, whereas in wink start, dialing can begin only after the wink is over. Both wink start and delay dial are necessary during automatic dialing with a "sender," as when digits are going to be repeated by the PBX, much

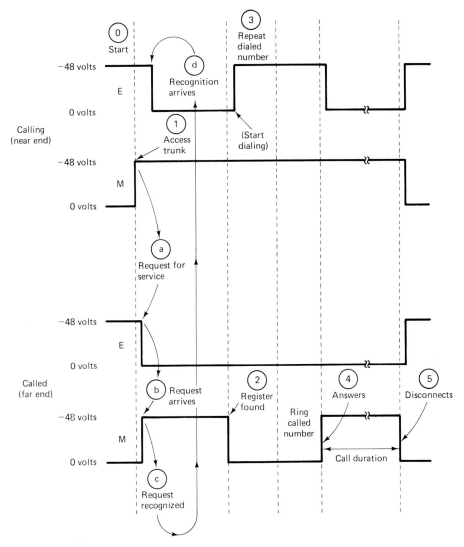

Figure 10–8   Delay dial E&M signaling.

as the CO repeats digits in a DID circuit. In Figure 10–8 the **access trunk** leads to the "recognition arrives" in the form of an active signal on the E lead. This signal remains active until step 2, when the register at the far end is located, and the M lead is made active. Step 3 follows immediately as soon as the near end's E lead responds, and dialing begins. Steps 4 and 5 proceed as in Figures 10–6 and 10–7, except that the far-end phone was hung up first this time.

10.3 DIGITAL SIGNALING AND TRANSMISSION

Technically, what follows is not, strictly, signaling. The signaling schemes we talked about are used to convey a limited number of types of information between a station and a switch, or between one switch and another. E&M signaling formats (immediate start, wink start, or delay dial), for instance, convey whether a telephone is on-hook or off-hook, while digital formats identify the actual information being "talked" between the stations as well as signaling. This is something like having both voice and DTMF tones handled in the same analog format.

10.3.1 RS-232

RS-232 is a system of signaling used between digital systems. In the data communications area, RS-232 is the "de facto" standard way to connect data communications stuff to other data communications stuff. Any connection between one data communicating device and another, or even between a data terminal and some kind of box that interfaces the data to a nondata link, probably uses the RS-232 connection to make the hook-up. Since the purpose of telecommunications now is to carry voice or data between one point and another, it is very important for the telecommunications person to understand RS-232.

RS-232 is an interface for serial binary encoded data. Usually, RS-232 is used to connect between a data terminal or a computer and a telecommunications network. The equipment used is going to depend on whether the network uses digital or analog lines. If analog, a modem is used to convert each bit of binary into a tone (or frequency) that can be transmitted down an analog line without deterioration. If the lines are digital, the digital format of RS-232 will be transformed into some *other* serial data transmission format (each manufacturer wants to have its own) for use within the system. That doesn't mean the RS-232 (which is serial digital data itself) can't be put onto a line directly. In some cases, RS-232 is used for short distances, usually under 100 meters.

The connector universally used to connect RS-232 devices together is the **DB-25 plug.** Figure 10–9 shows the layout of a "male" and "female" DB-25 plug, so called because it has 25 pins. Don't worry! All 25 are seldom used. Usually, only eight or fewer are needed. The signals on these pins are used for four basic purposes:

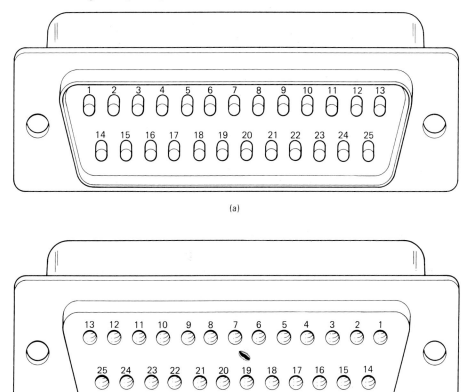

(a)

(b)

Figure 10-9   Layout of DB–25 plug. (a) DB–25 (RS–232) male connector. (b) DB–25 (RS–232) female connector.

Group	Purpose
I	Ground or common return
II	Data signals to transmit and receive bits
III	Control signals to show busy/ready status
IV	Timing signals for "synchronous" data (clock)

Since serial data is usually "asynchronous" (*not*-synchronous), group IV signals are seldom used.

The digital code most commonly transmitted on RS-232 serial data transmissions is ASCII. This is the American Standard Code for Information Interchange. It uses seven binary "bits" to represent each letter, number, punctuation mark, or control function found on a typewriter keyboard. Each ASCII character is transmitted one bit at a time (serial data). Before the first bit of an ASCII character is transmitted, a **start bit** precedes it, and possibly a **check bit,** if transmission of parity code is desired. The 7-bit ASCII character is transmitted in seven separate data bits, then a **stop bit,** or possibly two stop bits, are transmitted after the data bits.

In Figure 10–10(a), a **bit stream,** or timing diagram, shows the data transmission of an ASCII character, the letter A, using odd parity and one stop bit.

The first bit transmitted is a **start bit,** which is a 0 in this example. It is followed by an 8-bit ASCII character. In the figure, **odd parity** is used. That means that the check bit in each character is made 1 or 0, whatever is needed, so that the total bit count (per character) of 1's is 1, 3, 5, or 7 (an *odd* number). In this case the ASCII letter A contains two 1's, which is an even number; the check bit added is a 1, which brings the total bit count in the 8-bit ASCII character to 3, an odd number. The 8-bit ASCII character with odd parity is followed by a **stop bit,** which is a 1.

The reason for using parity is reliable transmission. It is a way to tell if the characters transmitted have lost or gained a wrong bit. If only odd-parity characters are transmitted, any character received with an even number of 1 bits is immediately *wrong.* The receiver can then request the transmitter to send that character again.

In Figure 10–10(b) we can see the timing of each bit transmitted. A pattern like this cannot be measured for its *frequency,* since the *high* and *low* pulses that make up an A are not all equal in length. Instead, the length of time each bit lasts—in this case, 0.833 milliseconds—is measured, and a rate is calculated for the number of such bits that can be transmitted in a second. You can fit 1200 pulses of 0.833-millisecond duration into a full second. We say that the **data transmission rate** is 1200 **baud** (the units of data transmission speed). The actual voltage is not switching on and off (from *high* to *low*) at anything like 1200 switchings per second in this figure, but that is the rate at which signaling elements—in this case, bits—are being transmitted, and that is what a **baud rate** is. In the figure, although 10 bits are transmitted, there are only two positive pulses, so the pulse rate is much less than the bit (baud) rate. Very few characters have even half as many pulses as bits; a character with all 1's would have no pulses or cycles at all because the voltage would always be *on.*

Another way to measure the speed of data transmission is to look at how many characters we could send in a second at this speed. Since the character we see in Figure 10–10 has 10 binary digits (bits), including stop and start bits, a character is transmitted for every 10 bits. If 1200 bits are transmitted per second (that's 120 10-bit groups), we are transmitting 120 **characters per second,** as long as the bits in consecutive characters last 8.33 milliseconds, like the ones in the figure.

The actual transmission of data is bipolar, non-return-to-zero format. In Figure 10–10(c) we see *positive* 12 volts used for binary 0 and *negative* 12 volts = binary 1. The two polarities of 12 volts are the "bipolar" voltages that represent the two binary states of logic. The "non-return-to-zero" part refers to the fact that only the positive

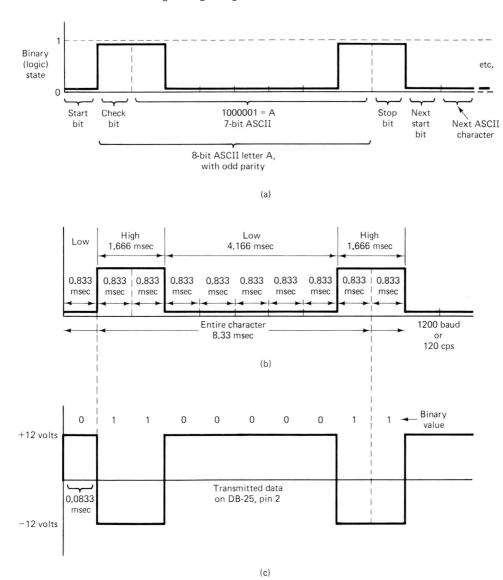

Figure 10-10 (a) Bit stream of ASCII letter A, using odd parity, one stop bit, 1200 baud. (b) Timing of transmitted bits. (c) Bipolar, non-return-to-zero format of data transmission.

and negative voltages are used for "legitimate" data. An open, nonconducting, or zero-volt state on the data line would indicate that no data were being transmitted.

Let's go back to the pins on the DB-25 plug and analyze what each is for. Sorted out by the four groups of functions, their signals are used as shown in Table 10-1.

TABLE 10-1 PIN NUMBERS FOR THE DB-25 PLUG

I Ground/return	II Data xmit/rcv	III Control signals	IV Clock
1, 7	2, 3, 14, 16[a]	4, 5, 6, 8, 11, 12, 13, 19, 20, 21, 22, 23, 25	15, 16,[a] 17, 18, 24

[a]Depending on whether equipment uses EIA standard pin assignment or Bell
208A, this may be either a "secondary received data" pin or a clock pulse on
a divided clock.

The most important signals on the DB-25 plug are:

Pin 1: Protective ground

Pin 7: Signal ground

The signal ground/return pin (7) is *absolutely* necessary. Without it, none of the other
signals would work. Whenever the other pins are made positive or negative, a current
must flow out of them, through the wires, to the circuit at the far end of the connector,
and back again, completing a current loop. This is the pin to which currents from
all the other pins return. They complete the current loop and are also at a zero-volt
potential. Pin 1 is also at a zero-volt potential and may be attached to the outer shield
of a shielded cable, used to prevent interference from stray external signals from
reaching the conductors inside. This protective ground is optional and is sometimes
strapped to pin 7.

Pin 2 (Transmitted Data) and pin 3 (Received Data) are the pins where the serial
data comes out (pin 2) and goes in (pin 3) to the terminal. One end's output is the
other end's input. The TD (transmitted data) at the near end is the RD (received data)
at the far end, and vice-versa.

For RS-232 data:

$$positive = binary\ 0$$

$$negative = binary\ 1$$

10.3.2 Control and Handshake Signals in RS-232

These are important signals that control the transfer of data but are not data
themselves:

Pin	Name	Symbol
4	Request to send	RTS
5	Clear to send	CTS
6	Data set ready	DSR
8	Data carrier detect	DCD
20	Data terminal ready	DTR
22	Ring indicator	RI

Control signals are active (ON) when they are positive. Some of these signals work together in pairs; one "answers" a "question" posed by the other:

<table>
<tr><td align="center">Request to Send
(near end)</td><td align="center">Clear to Send
(far end)</td></tr>
<tr><td>+ Can I send to you?
− Never mind . . .</td><td>+ Me? Sure, go ahead!
− Me? No way! I'm busy.</td></tr>
<tr><td align="center">Data Set Ready
(interface equipment)</td><td align="center">Data Terminal Ready
(terminal)</td></tr>
<tr><td>+ I'm ready to send (receive).
− I'm busy right now—wait!</td><td>+ I'm ready to receive (send).
− I'm busy right now—wait!</td></tr>
</table>

The second pair of signals could use some explanation: A **data set** is usually a modem or some other interface between RS-232 and another communication link. A **data terminal** is a piece of station equipment we have already discussed. Suppose that your terminal has a keyboard and prints on paper, and it connects to a telephone modem through an RS-232 connection. Let's illustrate the handshake that goes on between the data set and the data terminal via these two signals. We can imagine it in the form of a dialog something like this:

Checking data set ready

Terminal: My human has just typed the letter A, and I have digested it into RS-232 serial format. Before I send the serial bits to that modem over there, I better ask it if it's busy at the moment. "Excuse me—you, over there at the data set . . . ?"
Modem: Yes?
Terminal: I'd like to know, can you put this letter A onto the telephone line right now?

Data Set Ready—Negative

Modem: Sorry, right now, I'm receiving incoming bits from the distant end, and can't send your letter A.

Data Set Ready goes Positive

Modem: There! I'm finished receiving those bits. It's a letter Q from the distant end. Now I can take your letter A and transmit it.

Checking data terminal ready

Modem: Now I'm sending your A. By the way, if you're able to handle it, I'd like to send you the letter Q that I just got from the distant end.

Data Terminal Ready—Negative

Terminal: Oh, too bad—my typewriter is in the middle of printing the letter A right now. It's the one my human just typed; I've got to do it or he'll pound on my keyboard. I don't have a buffer where I can store your letter Q. Can you wait until I finish this A?

Modem: O.K. Let me know when you're ready, and I'll give you the Q to print next.

Data Terminal Ready goes Positive

Terminal: Hello? I just got done printing the A. Do you have the Q for me now?

Modem: Sure. Here it is.

And all of this is going on within a few hundredths of a second as two bytes of binary data are being exchanged!

Clock signals (used only with synchronous data)

You are not likely to encounter these unless synchronous transmission is used. Synchronous transmission does not depend on data bits always to be of constant duration. A **clock signal** identifies when each new bit starts:

Pin	Name	Symbol
15	Transmitter clock	TC
17	Receiver clock	RC

A transmitter clock is an output the near end uses to tell the distant end when it is starting to send another bit on the TD lead. The receiver clock is an input that is the near end's way of finding out when the distant end is sending another bit on the RD. The near end's transmitter clock is the distant end's receiver clock, and vice versa.

Other signals

These signals are not used consistently, and probably will not appear on most interfaces. They are provided without further explanation, on the off chance that you will encounter them in some equipment and need to know what they are called.

Pin	Name	Symbol
9	Positive test voltage (testing only)	
10	Negative test voltage (testing only)	
11	Equalizer mode (Bell modems only)	
12	Secondary data carrier detect (S)DCD	
13	Secondary clear to send (S)CTS	
14	Secondary transmitted data (S)TD	
16	Secondary received data (S)RD (Bell modems: divided clock transmitter)	
18	Divided clock receiver (Bell modems only)	
19	Secondary request to send (S)RTS	
21	Signal quality detect SQ	
23	Data rate selector	
24	External transmitter clock	
25	Unassigned, sometimes used as a "busy indicator"	

10.3.3 Null Modem for RS-232

A commonly used piece of equipment associated with RS-232 interfaces is called a **null modem** or **null modem cable**. The words ''null'' and ''modem'' imply that it is not a modem, and this is true. Most RS-232 equipment is documented as though they (manufacturers) expect you to attach it to the telephone system through a modem. A computer with an RS-232 interface might want to use a printing terminal with an RS-232 interface as a printer. As the manufacturers originally intended it to work, you would connect the computer to a telephone modem, call up the printer, and it would receive data through its telephone modem. But what if the printer is in the same room as the computer? It would seem wasteful to place a telephone call to yourself just to hook up the computer to the printer. A cable connecting the two RS-232 plugs would seem to be the answer. Would a straight extension cord do the job?

A cable that connects pin 1 to pin 1, pin 2 to pin 2, and so on, would not work. Both the near end and the far end (the computer and the printer) would be sending signals out of pin 2 (TD) and waiting for signals to come in at pin 3 (RD). The telephone company takes care of this when they put the data that goes in the near end's mouthpiece through to the distant end's earpiece. The near end's transmitted data becomes the far end's received data automatically, because the phone company transfers the signal from the near end's transmitter to the far end's receiver, and vice versa. Without a phone company in the middle, the cable will have to take care of these reversals itself.

The null modem cable reverses, or swaps connections, for the ''paired'' signals we just described. Figure 10-11 shows one arrangement for a null modem. The computer's transmitted data (TD) output goes to the received data (RD) input of the printing terminal. The computer's RS-232 ground is connected to the ground on the printing terminal, directly. ''Data Set Ready'' and ''Data Terminal Ready'' are swapped, to eliminate the need to wait for data conversion at a telephone line (there is no telephone line or modem to wait for now). That's basically all there is to a null modem.

The null modem is a handy way of connecting two pieces of station equipment via RS-232 without a modem and a telephone company. Before we leave this subject, take another look at Figure 10-4. It is a sort of null modem (a null tie trunk) for the E&M signaling and voice ordinarily communicated through a tie line. This hookup can be used to connect two PBXs with E&M signaling when they're in the same room, just like the null modem connects the RS-232 computer to the RS-232 printer in the same room. One modification that would have to be provided is an inverted M signal. The active and idle levels on the E lead are opposite to those on the M lead. Fortunately, most PBXs provide the option of inverted M, to make null modem operation easy. With the M lead polarity reversed, only wires and plugs are needed to connect two PBXs for direct E&M communication.

10.3.4 RS-449

Another popular interface, similar in nature to RS-232, is **RS-449.** Instead of a 25-pin plug, a 37-pin connector is used, similar to the DB-25. RS-449 is designed for higher-

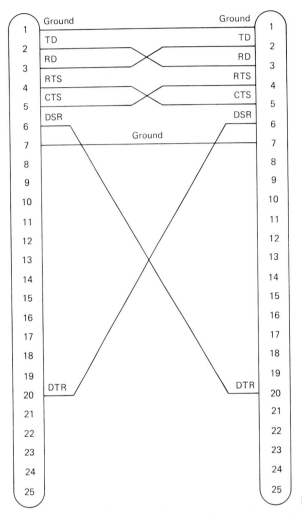

Figure 10-11 RS-232 null modem.

speed transmission. Each signal pin has its own return line, instead of a common ground return, and the signal pairs (signal, return) are balanced lines rather than a signal referenced to ground. Generally, this cable uses twisted pairs rather than the open-wire pairs of RS–232C, for each signal and its return.

10.3.5 T1 Span (D3 Format)

T1 is a digital transmission standard which uses binary code to convey both data and voice. It is the most common high-speed data format and is in widespread use for voice and data transmission; many PBXs and most microwave and satellite equipment are compatible with this standard.

In a **T1 span,** two twisted pairs, often identical to those used for voice, transmit and receive data at 1.544 Mbps (1,544,000 baud) and are sometimes referred to as a **T1 link.** T1 was developed by Bell for the transmission of 24 digitized voice channels. Several channels of data (at low baud rates) can be submultiplexed into one voice channel. Each pair, and the repeaters it contains, is used in only one direction.

Figure 10–12 shows how the 1.544-Mbps link is broken up into 24 64-bps channels. Data from each of the channels are transmitted as 8-bit **words.** Twenty-four of these words, one from each channel, make up a 193-bit **frame,** with 192 bits (24 × 8) of data and one **framing bit.** Twelve frames are organized into a **master frame,** and the pattern of 12 framing bits repeats with each master frame transmitted.

Although 64-kilobaud channels are theoretically possible, data transmitted in this system is usually transmitted in the form of 7-bit words, with the eighth bit used for signaling. In the sixth and twelfth frames, the eighth bit of all 24 words is used for in-band signaling. To keep things simple, only the seven most-significant bits are used for data transmission, even in the words where no signaling is included. That reduces the effective bandwidth of each channel to 56 kilobaud. For voice, all eight bits are used, except for those in the sixth and twelfth frames. In effect, voice transmission uses words $7\frac{5}{6}$ bits long.

The T1 interface is a balanced, bipolar, *return-to-zero* pulse stream. Unlike RS–232, the polarity—positive or negative—of the pulse voltage has nothing to do with binary 1 or 0 value. In fact, a 1 is a pulse regardless of whether it is positive or negative. The binary 0 is transmitted when no pulse appears. Bell requires that no more than 15 binary 0's be transmitted in a row; to maintain clock synchronization, pulses must be present at least once in every 16 bits transmitted. Also, there

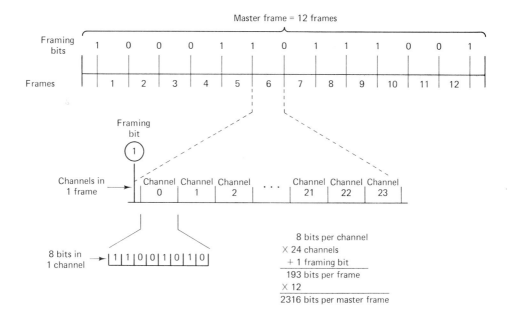

Figure 10–12   T1 data format.

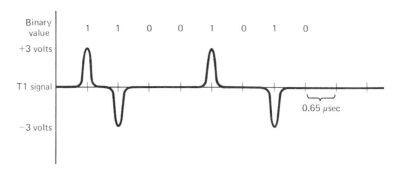

Figure 10-13   T1 bit stream.

should be at least three 1's every 24 bits. This is where the eighth bit (used every sixth frame for signaling) comes in handy for balancing the line voltage.

In Figure 10-13 we see a T1 bit stream, for the bits 11001010. As long as + and − pulses come along in equal numbers, reasonably close together, the average DC voltage of the bit stream is zero. Each pulse occurs in the middle of a time slot 0.65 microsecond wide. Pulses this far apart can occur at frequencies up to 1.544 megaHertz, which is the frequency of the T1 span.

T1 can be used between a PBX and the central office or other common carriers. It's used from PBX to PBX in somewhat the way tie lines are used, and it may be used in making up a **local-area network** (LAN). A T1 span without repeaters can carry a signal between 4000 and 12,000 feet (1 to 4 kilometers).

The need to switch voice from office to office and from city to city has been obvious for some time. It is now apparent that data—digital information—must also be switched from office to office and city to city. Since the paths are the same, why shouldn't the medium be the same? In the office, the PBX switches telephones now. Soon, most offices will have terminals as well as telephones. If the PBX has the capacity to handle data and voice, switching information between telephones or terminals, it seems logical to use that PBX switch to switch the data as well. If T1 is used, then with voice and data multiplexed onto the same lines, existing wiring and switching equipment can handle both kinds of information. This is already readily available in the PBX equipment of today's office. Eventually, it will have to become a readily accessible feature of the public network as well. The telephone network of today is becoming the digital/voice network of tomorrow.

REVIEW QUESTIONS

1. What is the purpose of signaling?
2. What is the difference between analog and DC signaling?
3. What are advisory tones, and why is this type of signaling needed?
4. What are four methods used by the telcos for signaling between offices?

5. What is the difference between a loop-start central office line and a loop-start central office trunk?

6. What is the meaning of ground in a ground-start trunk?

7. What sort of signaling do you give to your local telco when you take your telephone off-hook?

8. How can we reduce the chances of a glare condition arising?

9. Can DID trunks be used for both incoming and outgoing calls?

10. On a wink-start DID trunk, does the *wink* originate at the DID or at the central office? What does this signal convey?

11. Why is answer supervision needed on DID trunks?

12. What signal is placed on the M lead to signal a request for service?

13. What does a signal on the E lead tell the PBX?

14. How many wire pairs are needed for E&M tie trunks?

15. How are DTMF and rotary addressing handled on an E&M tie trunk?

16. Give two ways a PBX signals over an E&M tie trunk that a register is available.

17. Name three types of E&M signaling arrangements.

18. Is RS–232 used for voice or digital information?

19. Describe the terms handshaking and signaling in the computer and telecommunications fields, and discuss the similarities and differences between them.

20. Over what range of distance is a direct RS–232 hookup used? Is this similar to E&M hookups?

21. What is the purpose of a parity bit?

22. What is the difference between bits per second and characters per second?

23. What pins on the DB–25 plug must be connected to transmit data? What pins are needed to receive data? What pins are needed for full-duplex communication?

24. What is the difference between the DSR (Data Set Ready) and the DTR (Data Terminal Ready) signals? What is a data set, and what is a data terminal?

25. What is a null modem used for in an RS–232 hookup?

26. In which interface standard are balanced pairs used, RS–232 or RS–449? Which requires more pins and wires?

27. What kind of cable is used for a T1 span?

28. What is the speed, in bits per second, of a T1 span?

29. T1 can be used between a PBX and a central office instead of a space-division-multiplexed cable. What kind of multiplexing is used on T1?

30. Give five applications where a T1 span could be used for interconnection.

11

Switching and Networking

Suppose that John, in New York on a business trip, wants to make a phone call to Jenny, in Los Angeles. How does John's station equipment get connected to Jenny's? (Hmmmm . . .)

The simplest way would be to string a wire pair directly from John's phone to Jenny's. (Let's pretend, despite our better judgment, that losses and distortion aren't important on a wire pair 3500 miles long.) This works fine as long as there are only two phones in the United States, John's and Jenny's. Now, suppose that there are 100 million phones in the United States. Even if nobody calls outside the country, if you decide to wire a connection between each phone, and any other phone it *might* call, each phone has 100 million pairs coming out the back (and top, sides, and front, no doubt). Since every phone is connected to all the others, there are 100 million × 100 million, or 10,000,000,000,000,000 wires strewn across the countryside. To give you an idea what this means, the total land area of the United States is about 100,000,000,000,000 square feet, so there are about 100 wires for every square foot of land area in the United States. That would be a bit like living under a spiderweb. It would also be impossible, because there isn't enough copper in the world to make that many wires.

So much for the simple solution. What has to happen to make it possible for 100 million phones to call each other, as well as additional hundreds of millions elsewhere in the world, without creating a runaway webwork of wire? First, we need to recognize the fact that nobody's phone can be connected to 100 million other phones all at the same time. In fact, we can recognize the fact that very few phones are connected to any other phone all day, 24 hours a day, seven days a week. Probably, any average phone is in use, and connected to *one* other phone, less than 10% of the time. That means that we have far fewer than 100 million actual connections between phones at any moment in time.

The public network (which is made up of the telcos, long-distance carriers, and their equipment) makes connections between the callers using a sequence of switching systems. The call from John to Jenny might go through as few as two switching systems or as many as 10. In Figure 11–1, we see the (a) best-case and (b) worst-case situations. The difference between a "best case" and a "worst case" is that, in the best case, the subscriber loop is attached directly to an office with interstate access, such as a dedicated line to MCI, and in the worst case, the call must go through the

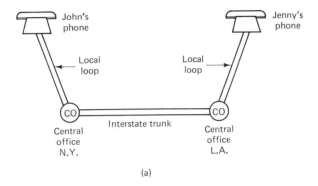

(a)

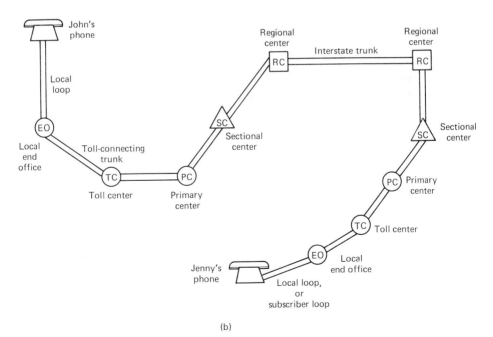

(b)

Figure 11-1   Best- (minimum system) (a) and worst- (maximum system) (b) case call situations.

entire hierarchy of small offices feeding calls into larger ones, and so on, up to the interstate trunk, and back down again at the other end.

In the case of John's call to Jenny, the local office will receive the dialed number. For example, let's suppose that Jenny is at 1-(213)867-5309. This number will be stored in a dial register at the local office, and its switching network will "decide" that Jenny's phone isn't in the local New York area it serves, based on the first digit dialed being a 1. That will force the local switch to attach to a toll office. It is possible that the toll office won't have a trunk to Los Angeles, and it will select a connection to an office that does have a trunk to L.A. We'll assume that this isn't a worst-case routing situation, and that the toll office has got a trunk to area code 213. After the call is placed on the trunk to L.A., it arrives, let's say, at an L.A. toll office serving the exchanges in the 213 area code. The New York toll office sends the phone number dialed to the Los Angeles toll office, and in L.A. the switch decides which local office the 867 exchange is in. This connection brings the call to Jenny's local office, which now connects John's call to subscriber number 5309 out of a possible 9999 subscribers, and Jenny's phone rings.

In this example, the "addressing" that enables John to reach Jenny is a telephone number. It doesn't matter whether the communication is voice or data, or whether a phone number in the North American numbering plan is used, or some other scheme, such as Telex. The important thing is that all information that uses a common carrier, and by extension, information that has to get around in a complex private network as well, must have an addressing scheme to route it through the switching network, and a switching network capable of recognizing the scheme in order to route the message.

In previous chapters we discussed the various links that are used in public and private telecommunications to carry information. The switch is the part of the telecommunication system that decides how to get into and out of the various links, to direct the information to its destination. We will look at methods of completing the connection between two users from the earliest switching schemes to those used at the present time.

11.1 MECHANICAL SWITCHING PRINCIPLES

11.1.1 Switchboard or Cord Board

In the earliest telephone systems, and until quite recently in some private systems, connections were made using a panel of sockets with patchcords, called a **switchboard** or **cord board** (see Figure 11–2). A **patchcord** is simply a cord with a plug on each end. The plugs used here are like the one shown in Figure 1–2. These plug into sockets that complete the connections between lines. Each operator has a number of subscriber loops attached to sockets on the switchboard, sockets going to other operators' switchboards, and in the case of the long-distance operator, a switchboard with sockets attached to trunks going to other cities and other central offices.

Suppose that you are the operator on such a system. When John rings up the

Figure 11-2   Representational drawing of a switchboard (plugboard) in a local office.

local operator in such a system, to call Jenny, he turns a crank on the phone that generates a ring voltage with an electric generator called a **ring generator**. This is a **Request for Service** signal. The Request for Service signal arrives at the switchboard, winking a neon lamp above John's socket on the board. This signaling visually alerts you and you plug your headset into his socket to find out who John wants to talk to. (At first, addressing was done on a last-name basis, but eventually subscriber numbers—phone numbers—had to be invented to handle large numbers of subscribers.) In the event that Jenny is on the same board, you will first ring Jenny's phone by connecting a patchcord from your ring generator to her line and ringing her phone. When she picks up the phone, you plug a patchcord connecting John's phone to Jenny's into the two sockets and the connection is completed. When the conversation is over, you disconnect the patchcord from John and Jenny's phones.

Suppose that Jenny's subscriber loop is connected to the same local office but is on another operator's switchboard. Then you would patch John's call to a socket on your board that connects to the other operator's board. That operator would complete the connection to Jenny's line. What do we call the lines that connect each operator to other operators in the office? They're not subscriber loops, anymore, because any call may be placed on them. By our definition, this makes each of these inter-switchboard lines a **trunk**, because each operator is a "switching system" and the function of a trunk is to connect one switching system to another.

In the event that Jenny's phone is in Los Angeles and John's is in New York, John will ask for the long-distance operator. You will connect John's call to the socket for the long-distance operator on your switchboard. The long-distance operator in your office has a switchboard with a number of trunks going to long-distance operators in other cities and other central offices. That operator will connect John's call to the Los Angeles central office, and it will be routed from there to Jenny.

In this example, all the "switches" are human beings, but what they do is the same as is done by automatic switches in a present-day system. A subscriber loop is connected by the switching system to another subscriber's loop. In the event that two subscribers are served by different switching systems, a trunk (or tie line) is used to complete the connection between switching systems.

The signaling that John used to direct his call to Jenny through your switchboard was manual (ringing you) and verbal (telling you who he wanted to talk to—or giving you Jenny's phone number). With the information provided by this signaling, you knew that John wanted to make a call, and to whom it should be connected. These functions are now carried out by dialing and interpreted automatically by devices rather than people, yet the function of signaling has not changed.

11.1.2 Step-by-Step and Common Control Switching

In the 1920s, the dial-telephone system came into use. Automatic switching devices replaced human operators for some switching operations. Where signaling was manual and verbal in the switchboard system, the dial now became the primary signaling device. We have already discussed how the rotary dial works. Briefly, the combination of the rotary dial with the rotary switch (**Strowger switch**) made it possible for you to use your dial to remotely control the connections made by switching equipment. A Strowger switch, shown in Figure 11–3, is also called a **stepping relay**. It is essentially a rotary switch like the channel selector on a television set, connected to a ratchet-and-pawl arrangement that "steps" the rotary switch to the next contact position every time an electromagnet is activated and deactivated.

When the magnet is activated, it pulls on the armature, causing the pawl to snap down to the next tooth on the ratchet. The armature also compresses a spring, and when the magnet is deactivated, the spring pushes the pawl forward, advancing the ratchet to its next position. A rotary switch (not shown) is fastened to the ratchet, and moves from one contact position to another as the ratchet rotates.

Each pulse from a rotary dial thus causes the stepper switch to move its connection from one point to another. A switching system made of steppers with 10 steps each can permit you to dial decimal numbers.

When the stepper is not in use, there has to be a neutral position where the stepper is when it's not holding a number like 1, 2, 3, 4, 5, 6, 7, 8, 9, or 0. There are various methods of doing this, but all steppers contain some mechanism for resetting the stepper to home position when it doesn't have a number in it.

Jenny, I've got your number

Figure 11–4 shows how an end office switch using steppers would select one subscriber. Suppose that John is calling Jenny at 867-5309 and has dialed the first

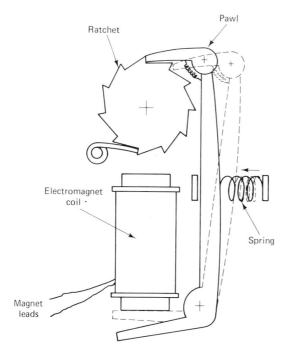

Figure 11–3 Strowger (stepping relay) switch.

three digits. He is now connected to the stepper switch that has the 867-XXXX numbers. This group of steppers is part of the switch for 867 and selects any number from 867-0000 to 867-9999. That's 10,000 phones, and of course, we haven't shown them all. As John dials the first digit, 5, his rotary dial makes and breaks a current loop five times. The thousands' place stepper in the 867 exchange is receiving current pulses from his phone, and steps to position 5. It is now possible for this call to be connected to any phone from 867-5000 to 867-5999. The second number John dials puts his phone's pulses into the hundreds' place stepper. The number he dials is 3, and three current pulses advance the stepper to position 3. It is now possible for this call to be connected to any phone from 867-5300 to 867-5399. With the third number dialed, John puts 10 current pulses into the tens' place stepper (the 0 is really a 10 on the rotary dial), and now his call can only be connected to phone numbers 867-5300 to 867-5309 (some of these are shown on the picture). When he dials the final digit, 9, the last stepper, the ones' place stepper, rotates to position 9. The contacts on this stepper are connected to the end users, and the ninth contact, which is receiving the ring signal, is Jenny's phone.

   Step-by-step switching, as we illustrated for Strowger switches, requires each stepping relay to act immediately on the pulses from the rotary dial. When the last number is dialed, the last stepping relay has to be ready to act. There is no circuit that holds the number dialed if it has to be relayed on to another switching system. For instance, if you are making a long-distance call and your call has to be circuited through a number of intermediate switches, the process of identifying your call as long distance, and routing it to where it is going, might be impossible on a point-to-

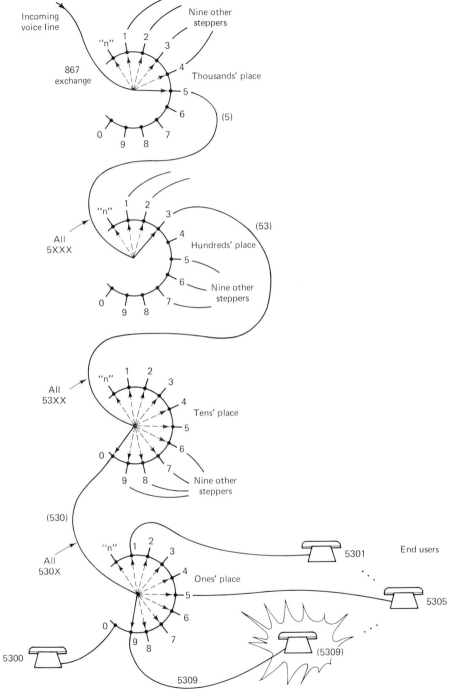

Figure 11-4 How an end office switch selects one subscriber using steppers.

point basis. While the intermediate switches are switching, you may end up dialing a number or two, or possibly the whole rest of the number, before the interstate switching is completed. By the time the switch at the distant end becomes ready to accept your next dialed number, you may have dialed a whole bunch of numbers that the network was too busy to notice. In that case, the network would fail. That type of failure could be eliminated by using common control, which, essentially, adds some smarts to step-by-step switches and is standard on all present-day equipment.

Another problem with step-by-step switching happens because the phone network grew, instead of being built from the ground up in a coordinated fashion. If you are a rational designer, you might start all over again and build the phone exchanges to fit a step-by-step scheme that covers the entire country and will never have to be expanded. You might come up with a way to switch each call to another level of switch banks as each digit is dialed. This would be a way that would always and uniquely route your call to its destination. The only thing that's wrong with this is that once you were done, you would have a worldwide webwork of wire that makes the one we introduced at the beginning of this chapter sound simple by comparison. Each phone would go to a local office, which would connect to nine worldwide offices according to the first digit dialed (1 through 9). (0 is used for Operator, remember?) Each successive digit, in turn, would "tree down" through a network of relays, until the last digit rotated the contacts of the last relay, in a local office connected to the subscriber line.

Sounds like a nice idea. Unfortunately, that's not how the phone system grew up. The way it *did* grow up, a lot of exchanges with similar numbers were in widely separated locations. For instance, within one area code, the 867 exchange might be located 40 miles from the 868 exchange. It is not only impossible for the switch in your local office to decide what trunk to put your call on when the 8 is dialed, it wouldn't even be possible after the 6 was dialed. The switch would only be able to direct your call to the right office after all three digits of the exchange were dialed. This means that somehow your local office will have to "remember" all three digits at once, to be able to figure out on which trunk your call goes out. Once the call gets to the office with 867 in it, you are dialing the fourth digit. Suppose that the office with 867 also has 866, 865, and 864. How will the incoming digits be handled? The call coming in on the trunk doesn't carry with it the information saying for which of the two exchanges it was intended. We would either have to have a separate trunk coming in for every exchange, or come up with a way for the call to "carry" its exchange number along with it somehow. This is where a part of **common control** comes into the picture. Your local office's equipment can remember the digits you dialed, in some sort of a memory device—which we call a **register**—until it has enough digits to decide on which trunk to go out. In practice, the memory device might as well remember all the digits you dialed, so a **dial register** is set aside, which is a device capable of holding all the digits of a number you dial. When the connection to the distant office is completed, the number you dialed can be retransmitted—we call this retransmitter a **sender**—to a register at that office, which will then be able to decide to which of the exchanges attached to the trunk your call should go.

In the case of the long-distance call that had to go through so many interstate switches that we lost the rest of the number, a sender would retransmit the number from one switch to another, all the way to its destination.

The idea that a call can come in on one trunk to an office with, say, four exchanges, is an important one. In a step-by-step system, there would have to be four trunks coming into that office from every other office that contacts it, so the next few digits can be routed to the proper exchange. With common control, only one trunk is needed. It reduces the number of trunk connections by 4 to 1. Multiply that by all the offices with trunks coming in to other switch offices, and you have an idea why common control simplifies the network and the number of trunks and connections that are needed.

11.1.3 Crossbar Switching

Crossbar switches, like Strowger switches, are electromechanical devices that make connections in a telephone system by moving parts. They came into use about 10 years after Strowger switches and were used as part of the telephone switching matrix. Many electromechanical crossbar switches are still in use in this country, and they are quite common in other parts of the world. As the name suggests, a lattice of crossed bars is involved. Each vertical bar carries a set of contacts that are connected to those carried on a horizontal bar when magnets are activated to move the two bars. Contact is made where the two bars cross—hence the name "crossbar." In Figure 11-5(a), a simplified version of the crossbar switch is illustrated. Inputs 1, 2, 3, 4, 5, and 6 can be connected to outputs A, B, C, D, E, and F. Its contacts will latch; that is, they will stay attached after the magnets are deactivated, and contacts can be made elsewhere by the other bars. Of course, when the communication is over, the contacts can be disconnected. Although oversimplified, this picture of a crossbar switch does provide some idea of how this kind of switching matrix is used. One of the vertical lines in the picture goes to another crossbar matrix, and this type of thing can be protracted indefinitely. Through this connection, additional contacts can be made.

Figure 11-5(b) shows another way to connect six inputs to six outputs using three levels of switching. Although the number of inputs and outputs is the same, the number of paths, or ways a call can get through, is much larger. For instance, in Figure 11-5(a) there are 36 crosspoints, and 36 possible connections can be made between an input and an output. That means there is one path possible for each crosspoint in the array. In Figure 11-5(b) there are more crosspoints—63—but there are a great many more potential paths. In that figure there are a possible 8748 ways of getting through from any input to any output. This means that for each crosspoint, there are almost 139 possible paths. By increasing the number of crosspoints by 75%, the three-stage grid arrangement gives 24,200% more possible paths.

It is important to remember that although we have labeled 1, 2, 3, 4, 5, and 6 as "inputs" and A, B, C, D, E, and F as "outputs," real telephones don't talk just one-way (simplex) communication. Each telephone needs to be able to call in or out. Those things marked "inputs" represent telephones that can call out but cannot receive any calls, while those things called "outputs" can receive calls but can-

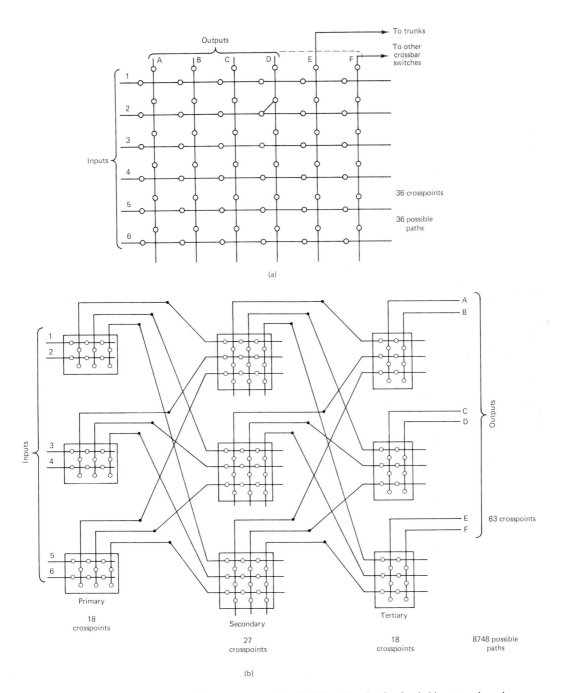

Figure 11-5 (a) Simplified crossbar switch. (b) Using three levels of switching, to reduce the probability of "blocking."

not make any. In reality, phones attached to points such as A, B, C, D, E, and F should be able to call out, and 1, 2, 3, 4, 5, and 6 should be able to receive calls. A should be able to call D, and 1 should be able to call 4, and so on. To do this, we need an arrangement like Figure 11-6. Since there are 12 phones, and two of them are connected together at a time, there are a possible six connections to be made. The six vertical crossbars in the figure are needed to complete all the connections. This means that we need 72 contacts, twice as many as in Figure 11-5(a). In Figure 11-5(b), however, there are over 8000 possible paths, and any of these possible connections could be made without increasing the number of contacts at all. Since the number of contacts in Figure 11-5(b) is only 63, it is a more compact arrangement. As the switching network becomes larger, the advantage of the three-stage grid becomes greater and greater.

Blocking

In the case shown in Figure 11-6, there are enough contacts so that all possible connections could be made at the same time. When this is done in larger, real-world

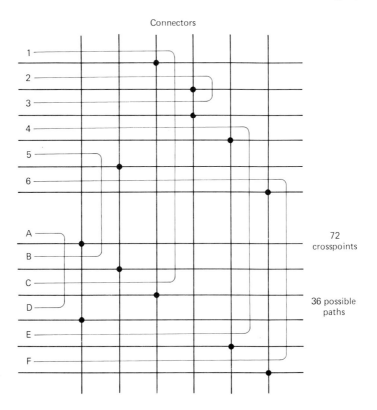

Connections: A-D, B-5, C-1, 2-3, 4-E, 6-F

Figure 11-6 Nonblocking crosspoint grid.

systems, they become too large and cumbersome. Also, it isn't necessary to design the system assuming that every phone will be engaged at the same time. Normally, fewer than 10% of phones are in use at any one time. Designing for "reasonable" use rather than "maximum" use, we might safely design a system with 500 possible connections to handle 5000 stations. If 500 callers are using the switch, and additional subscribers dial numbers in the exchange, the calls will be **blocked**, because there are no more connections to handle them. This **blocking** problem has nothing to do with the telephones being called; they might well be idle, but the switch just doesn't have another connection available to complete any more calls. In designing this system, our goal is to get the largest number of possible paths, to avoid blocking, with the smallest number of contacts. As we saw in Figure 11–6, this is accomplished using a *multistage grid*. Perhaps it is not obvious that these two alternatives yield systems with comparable performance. Figures 11–7 and 11–8 have been provided to illustrate this point.

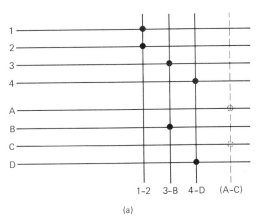

(a)

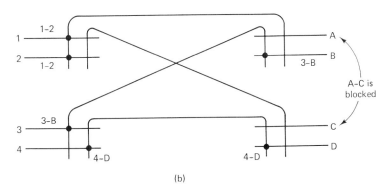

(b)

Figure 11–7 Comparison between single crossbar matrix and two-stage grid. (a) Switch designed for 25% blocking with 24 contacts; first three calls are completed, last is blocked. (b) Two-stage grid designed for 25% blocking with 16 contacts; first three calls are completed, last is blocked.

In Figure 11-7 a single crossbar matrix and a two-stage grid are compared. Since real switches are designed with a certain amount of blocking expected, the circuit of Figure 11-7(a) illustrates how a switch would look if designed for 25% blocking. To handle all possible calls in the system, Figure 11-7(a) would need four vertical crossbars with 32 contacts. Since there are only three crossbars and 24 contacts, the first three calls can go through, but the fourth will always be blocked. In the example shown, 1 calls 2, 3 calls B, 4 calls D, and A tries to call C, but no crossbar exists to make the connection. The fourth call is blocked, and we say that the system has 25% blocking.

In Figure 11-7(b) a two-stage grid connects 1, 2, 3, 4, A, B, C, and D. The same calls are made, with the same results; 1 can call 2, 3 can call B, 4 and call D, but when A tries to call C, it is blocked. This system also has 25% of its calls blocked, but is different in two important respects: first, it has only 16 contacts, compared to 24 in Figure 11-7(a); and second, although there will always be 25% blocking in Figure 11-7(a), there are some cases for which Figure 11-7(b) will have less blocking. We see an example in Figure 11-8. Regardless of which calls are placed, the cir-

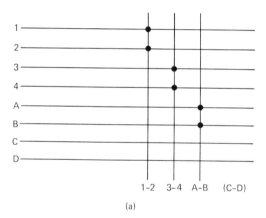

(a)

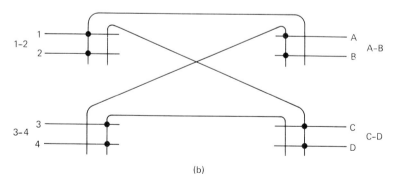

(b)

Figure 11-8   (a) Circuit designed to always block the fourth call, with 25% blocking, 24 contacts; call from C to D cannot be connected. (b) Best case situation, with 16 contacts, 0% blocking; calls 1-2, 3-4, A-B, and C-D can all be completed.

cuit in Figure 11-8(a) always blocks the fourth call (25% blocking). In Figure 11-8(b), the best-case situation, calls from 1 to 2, 3 to 4, A to B, and C to D all go through. In Figure 11-8(a) four calls could never be completed, while in the Figure 11-8(b) four calls are completed, although the switching network has fewer contacts.

In any switching system (other than intercom types), a certain number of trunks must be used to connect it to other switching systems. For example, if there are 50 stations in a PBX, it isn't necessary to have 50 trunks to the public network. For one thing, all the phones aren't going to be used at the same time. For another, they're not all calling outside the PBX into the public network. Therefore, a PBX with 50 stations would probably have no more than 10 trunks. If all 50 stations were trying to call outside at the same time, 40 calls would be blocked (80% blocking). These 40 call attempts from the PBX would be blocked because there are only 10 trunks available. Before, we were talking about calls being blocked within the matrix; now we are looking at blocking because of a limited number of trunks. The same word—blocking—applies to both situations. Usually, blocking is more likely because of unavailability of a trunk than because of unavailable circuits in the switch matrix. Trunks are more expensive than connections in the switching matrix, and in consequence, a minimal number of trunks are used. Switching matrices are usually designed to handle a larger number of connections than the minimum required, so that as more stations are added, there will be enough room in the system to handle expansion.

11.2 ELECTRONIC SWITCHING: SPACE-DIVISION MULTIPLEXED

Electronic switching is distinguished from electromechanical switching by the fact that it does not involve moving parts. In the early years of the twentieth century, the word **electronics** was coined to describe the control of electron flow (electric current) without mechanical or moving parts. An electronic switch is one that involves a vacuum tube or transistor rather than a pair of metal contacts opening and closing mechanically. Electromechanical switches involve mechanically opened and closed metal contacts, usually moved by an electromagnet. Let's look at two of the devices by which electric current may be switched on and off without the use of mechanically moving contacts.

11.2.1 Vacuum Tubes

In 1881, Thomas Edison was experimenting with attempts to prolong the life of filament lamps. He found that before they burned out, his carbon-filament light bulbs deposited a thin layer of carbon on the inside wall of the bulb, and the filament thinned out until it broke. In an attempt to determine how the material was removed from the filament, he inserted various obstacles into the bulb and found that they cast shadows in the carbon deposited on the glass. One day he attached a voltmeter to a metal plate inserted into the bulb and found that it acquired a negative charge when the filament was turned on. The charge went away when the filament was turned off. He was unable to think of a use for this discovery, and it was forgotten, only to be rediscovered by a British researcher named Fleming shortly before the turn of the century. Between Edison's discovery and Fleming's rediscovery, the electron was discovered and identified as the "cathode rays" that illuminated certain tubes when they were electrified. Fleming knew that the plate in the tube was picking up electrons. Since electrons were negative and emitted only from the heated filament, never the cold plate (or even the filament, when it was cold), Fleming knew that he could apply AC voltage to the light bulb-with-a-plate, and current would flow in only one of the two directions an AC voltage pushed. Since current would flow only out of the heated filament toward the plate, and not vice versa, it permitted AC voltage to produce a DC current, which flowed in only one direction. Because it passed current when the voltage was polarized in one direction, and shut off the flow of current when the voltage was polarized in the other direction, he called the light bulb-with-a-plate a *valve*. It wasn't a very sophisticated switch, since the only thing that made it turn on and off was a reversal of its polarity, but it proved useful for radio-wave detection or AC-to-DC power conversion (as a **rectifier**).

The next important step forward, in terms of switching technology, took place about 1908, when Lee De Forest added another element to the light bulb-with-a-plate. He added a fine grid or grille of wire between the filament and the plate [Figure 11-9(c)] and found that if the wires in the grid were tightly spaced and the grid placed close to the filament, a very small voltage on the grid could completely stop current flow between the filament and the plate, even though a much larger voltage were used to attract electrons to the plate. Since this device, too, could be used as a switch, it came into uses where high speed was needed. Since the current flowing between

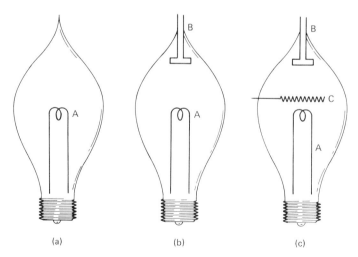

(a)   (b)   (c)

Figure 11-9 Diagrams depicting the development of the three elements of a vacuum-tube triode: Edison, 1880 (a); Edison, 1881, Fleming, 1896, (b); DeForest, 1908 (c). A = filament; B = plate; C = grid.

the filament and the plate responded almost instantly to signals applied to the grid, the light bulb-with-a-plate-and-a-grid could switch thousands of times faster than mechanical switches. The flow of electrons was controlled by the electric field on the grid, which extended itself throughout the tube with the speed of light. De Forest called his device the audion tube, and although the name audion didn't "stick" when other types of tubes were made with more and different electrodes, the name **tube** did. To this day, the British call these things valves, and Americans call them tubes.

In Figure 11–9, the development of the three elements of a vacuum-tube **triode** (as the audion is called today) was shown in three diagrams. Figure 11–10 shows how the elements in a triode finally ended up in commercial products, with the wire **grid** becoming a fine spiral of wire surrounding the **filament**, and surrounded by a tungsten metal cylinder (the **plate**). To use a triode as a switch, we begin with a heated filament and a voltage on the plate that is more positive than the filament. The plate attracts electrons and a current flows as long as the filament is heated. Now, if we apply a negative voltage large enough to push the electrons back before they can squeeze through the spaces between the coils of the grid wire, the current will be stopped. This is called **cutoff** and is equivalent to opening a switch. If we have no voltage on the grid, the electrons pass unhindered from filament to plate and the current is as its maximum value. This is called **saturation** and is equivalent to a closed switch. With electrical signals, the grid of a triode can be used to switch it on and off millions of times a second.

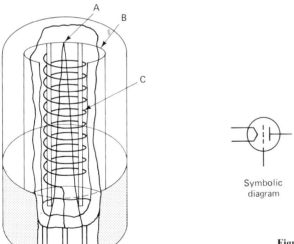

Symbolic
diagram

Figure 11–10   Elements of the triode as they wound up in commercial products. A = filament; B = plate; C = grid.

11.2.2 Solid-State Switching Devices

It has been known for a long time that certain crystals, when attached to a metal contact and with another contact called a "catwhisker" pressing into one surface, tend to act like vacuum-tube diodes. The underlying principles behind this were worked out during the 1930s and 1940s, primarily by teams at Bell Labs using germanium metal. They acquired a detailed understanding of how the "catwhisker" phenomenon worked and were able to achieve diode action with germanium crystals grown in the laboratory.

The semiconductor diode is a device with two layers of semiconductor material, one designed to emit electrons and the other to emit virtual positive charges called **holes**. Wires attached to these two layers can carry current in one direction and not the other. The schematic symbol for a semiconductor diode is shown in Figure 11–11(a). Figure 11–11(b) is a pictorial representation of current flow through the two-layer device. In this case "current" is defined as the direction holes flow through the semiconductor, and in the symbolic diagram, the arrow points in the direction in which positive charges would flow. This may seem a bit backwards to you, since electrons, which are the charge carriers that normally travel through wires, are negative and flow in a direction opposite to the arrows in the symbol.

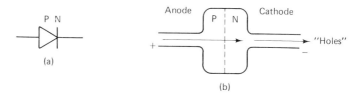

Figure 11–11   (a) Schematic symbol for semiconductor diode. (b) Current flow through two-layer device.

Transistors, technically called **bipolar transistors**, are represented symbolically in Figure 11–12(a). A transistor is made of three layers of semiconductor material. By comparison to a diode, a transistor is like two diodes nose to nose. Since a single diode can conduct in only one direction, it would appear that two connected together this way would conduct in *no* direction. That would be true if not for the third wire attached to the center, or **base** layer, of the semiconductor [see Figure 11–12(b)]. By applying a voltage to the base layer and drawing a small current into that region, we are able to "sweep" far larger amounts of current into the **collector** layer. This current would normally be stopped by the collector-to-base connection, which is like a diode "backward" to the current, but once the base wire's voltage manages to attract some of the holes (virtual positive charge carriers) from the **emitter** into the base layer, most of those holes go straight through the transistor to its collector, and only a small fraction are pulled into the base wire. By turning the **base voltage** on and off, the transistor's **collector current** can be turned on and off, thus making the transistor a switch. When John Bardeen, Walter Brattain, and William Shockley invented the transistor in 1947, they announced it as "a new type of switch." Like the vacuum tube, the transistor can switch currents on and off with no moving parts and at rates of millions of times per second.

A four-layer semiconductor device, called a **silicon-controlled rectifier** (SCR), has a control input called a **gate**, and may be used, like the triode and bipolar transistor, as a switch. It has one important difference over the other electronic switch-

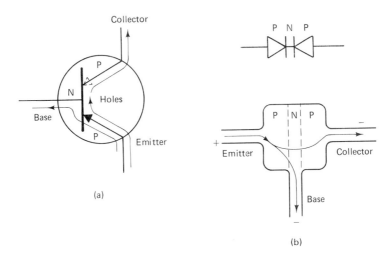

Figure 11-12   Symbolic diagram of bipolar transistors. (b) Pictorial diagram of a "grown junction" bipolar transistor.

ing devices mentioned so far. The triode and bipolar transistor are like momentary-contact switches—pushbuttons—because they stop conducting as soon as you take away the control signal. The SCR does not. It will remain latched in a conducting state after the signal that triggered it is removed. Like a light switch, it stays on until someone turns it off. In Figure 11-13 an SCR is shown in symbolic diagram form (a) and as a four-layer device (b). In Figure 11-13(c), the four-layer device has been taken apart functionally into two transistors, Q1 and Q2, to show how it works. The gate is the base for transistor Q2. It can be used to turn Q2 on so that it conducts a collector current. The collector current from Q2 becomes the base current for Q1, and that turns Q1 on, so that Q1 conducts a collector current. Q1's collector current is the base current for Q2, and that turns Q2 on even if the current coming in the wire marked "gate" goes away. Once this loop has been established, no more gate current is necessary. Q2 will keep Q1 on, and Q1 will keep Q2 on, as long as some external circuit keeps supplying (hole flow) current into the **anode** and taking it out the **cathode**.

Although vacuum tubes are obsolete for switching, they are still used in telecommunications as final-output stages for radio transmitters. Transistors and SCRs are used to switch signals in electronic switching systems, although transistors usually are found in the form of **integrated circuits** (ICs). An integrated circuit, as its name suggests, integrates the functions of many semiconductor devices into a single piece of silicon.

11.2.3 Integrated Circuits

Single transistors in such things as transistor radios are often small metal or plastic canisters between a $\frac{1}{4}$ and $\frac{1}{2}$ inch across (about 1 centimeter across). The actual, ac-

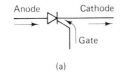

(a)

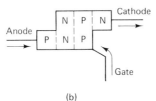

(b)

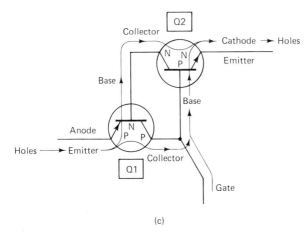

(c)

Figure 11-13   (a) Symbolic diagram of
SCR. (b) SCR as four-layer semiconductor. (c) Four-layer semiconductor
broken into two functional transistors,
Q1 and Q2.

tive part of the transistor (the three-layer device made of silicon or germanium) is
much smaller than the canister. When these individual transistors are *not* cut apart,
fastened to wires, and enclosed in a canister, a great many of them can be fitted together on a single piece of silicon. For instance, in an IC called a "128K RAM *chip*,"
an integrated circuit one centimeter across contains more than 256,000 transistors,
plus an additional 128,000 capacitors or so.

Since the IC silicon chip is the same size as the canister for *one* transistor, you
can see that this is a much more efficient way to put transistors together in circuits
where you want a lot of them.

11.2.4 Transistor Switching

When a transistor or SCR is used as a switch, it is most often employed in digital
switching, where it is always either **saturated** (conducting as much current as it can)
or **cutoff** (conducting almost no current). In these two states, the transistors approximate a closed or open switch. Figure 11-14 illustrates how this switching takes place.
In both parts of the figure, the current in the collector is controlled by the current
in the base. There is no current in the base in Figure 11-14(a), and the current in

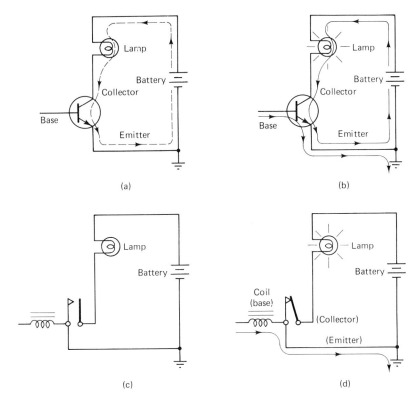

Figure 11-14   Bipolar transistor used as digital switch. (a) No current is in base. (b) Path has been made to conduct current to base. (c) No current in relay base. (d) Relay coil is magnetized by coil current.

the collector is thus very small. Without enough collector current to light the lamp, it remains off. In Figure 11-14(b), a path has been made to conduct current to the base. Since there is a base current, the collector current is now much larger, and the lamp is lit. When the base current is removed, the collector current stops (or at least becomes so small that it won't light the lamp).

Compare the transistor action to the illustrations of a relay with normally open (N.O.) contacts, switching a light on and off in Figure 11-14(c) and (d). In Figure 11-14(c) the light is off because no current has been put into the relay coil, and the relay is not magnetized, so its metal contacts are not pulled together. In Figure 11-14(d) the relay coil is magnetized by coil current. This closes the contacts of the relay and the lamp lights. The base of the transistor is comparable to the coil wire in Figure 11-14(c) and (d), since it controls the conduction of current through the remaining two connections. In Figure 11-14(c) and (d) the connections to the relay circuit are labeled according to which transistor connection they represent. To help you visualize how the transistor switch works, just picture the relay circuit in your mind. As long as the transistor is being used for switching, it will work like the relay.

(*Note:* In reality, there are several differences between transistors and relays.

A transistor is much faster than a relay, with no mechanical parts to wear out. A relay can carry current in either direction whereas a transistor cannot, and the input of the relay may be isolated from its output completely.)

11.2.5 An SCR Switching Matrix

When electronic switching first came into use, the electronic devices were used solely for the control function. It was the job of the electronics to tell the switches which contacts were to open and close. The contacts that were used to complete the voice path were still mechanical or electromechanical contacts. Part of the reason for this was the fact that transistors and other electronic devices have more resistance in the "closed-switch" condition than metal contacts do.

One attempt to eliminate the mechanical switching of contacts was the **SCR switching matrix**. It involves the silicon-controlled rectifier, which, you will recall, is a four-layer semiconductor device that "latches" or remains in a conducting state after the control signal that switched it on is removed. Like other solid-state devices, it can be switched on and off at much faster speeds than relays or mechanical contacts can. The matrix that we constructed with crossbars earlier can be replaced with one that has an SCR at each contact point. This is illustrated in Figure 11-15.

Why replace mechanical switches with solid-state ones? One reason is speed—SCRs are faster than magnet-driven metallic contacts—and another is the "bouncing" of the contacts once two pieces of metal meet. A solid-state connection, once made, is secure. When metallic connections are driven into contact, they bounce for a while before reaching a state of secure contact. This produces a "click", "ping," or "ding" as the contacts hit and adds extra time to the delay in mechanical contact closure. There is no way for dirt to get in between the contacts and prevent closure when a solid-state device is completing the voice path, and the contacts won't erode

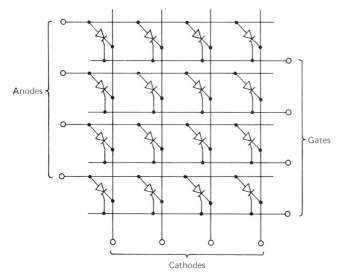

Figure 11-15 SCR matrix.

or burn away from repeated use, as with a relay's. Semiconductor devices cost less than relays, take up less space, use less power, and require less maintenance. It should be clear why a semiconductor replacement for the relay or electromechanical type of switch was desirable.

We paint such a glowing picture of the advantages of semiconductor switching that it would seem that SCR matrices should have been used immediately to replace all crossbars and Strowger switches everywhere. They were not, so you might well ask "What's wrong with them?" First, SCRs are four-layer semiconductor devices, and a "switched on" SCR has, as we mentioned earlier, more resistance than a set of closed metallic contacts. This leads to signal losses everywhere the voice path goes through an SCR. Second, these devices conduct one-way only and must be used with a single DC polarity; current can't go the other way. Third, just as the technology to produce integrated-circuit arrays of SCRs on a chip became possible, better ways of switching signals with time-division multiplexing (TDM) and faster three-layer devices came into use, and since these were more efficient, they supplanted SCR matrices with circuits that used fewer components and worked faster than SCR matrices.

11.3 ELECTRONIC SWITCHING: TIME-DIVISION MULTIPLEXED

11.3.1 Time-Division Multiplex

The crossbar, step-by-step, Reed relay, and SCR matrices we have just seen are all space-division-multiplexed (SDM) switching. To get multiple voice paths, we have a separate set of contacts for every voice path. The place, or *space*, that it occupies determines where it goes and who it reaches. Time-division multiplexing (TDM) uses the same components and contacts at different times to produce multiple voice paths with fewer components. The *time* at which the signal appears determines where it goes and who it reaches, instead of the space it appears in.

In Figure 11–16 we see a TDM system represented as a sort of matrix. The appearance of this matrix is almost exactly like the single-stage crossbar matrix in Figure 11–6, but instead of the vertical lines being crossbars, the lines labeled T1, T2, T3, T4, T5, and T6 are time slots in a repeating cycle. Let's suppose that there are only six time slots in a TDM system, as pictured in this figure, and 12 stations. In the case pictured, station 2 is connected to station C, and station 6 is connected to station 4. At time T1, the connection between 2 and C is completed and held for 20 microseconds. Then it is time T2, and connection between 6 and 4 is completed and held for 20 microseconds. Then at times T3, T4, T5, and T6, each lasting 20 microseconds, nothing is connected. When it would be time T7, we go back to T1 again. Every time T1 rolls around again, we complete the same connections to make the connection between 2 and C, and every time T2 rolls around, 6 is connected to 4. This means that the path between 2 and C is completed over 8000 times a second and disconnected five-sixths of the time, but to 2 and C it must seem that they were connected all along.

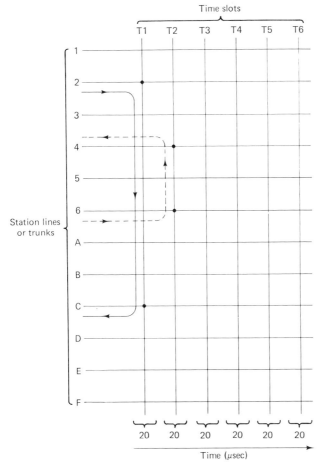

Figure 11-16   TDM matrix.

11.3.2 TDM and PAM

The signal from 2 to C is **pulse-amplitude modulation** (PAM) and looks something like Figure 11-17(a). When the pulses from time slot T1 arrive at station C [in Figure 11-17(b)], they are filtered through an electronic network (a **low-pass filter**) that makes the PAM wave smoother. As it goes through the filter and the amplifier, the wave is restored almost to its original appearance. Even though the connection is there only one-sixth of the time, the signal received at station C is sampled often enough to reproduce the signal sent from 2 quite faithfully. There are many other ways to make a low-pass filter, or one that would eliminate just the 8333-Hertz sampling frequency, keeping information of all other frequencies in the input waveform—a **notch filter**—but since most of those designs involve parts we haven't discussed, we showed a simple circuit involving resistors, capacitors, and a transistor.

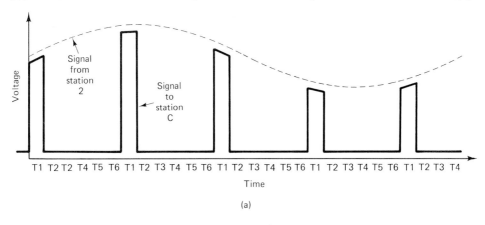

(a)

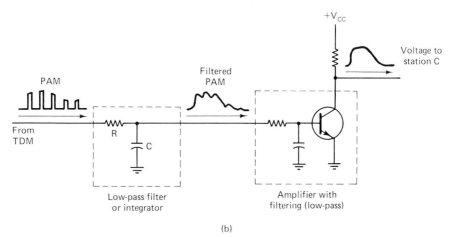

(b)

Figure 11-17   (a) PAM waveform from timeslot T1. (b) Changing pulse sample back to original waveform.

11.3.3 TDM and Blocking

The idea of TDM is that the connection being made depends on what time it is, and the number of possible connections depends on how many time slots are available. In Figure 11–16 there are six time slots and six potential connections that could be completed between different stations. If all six possible connections are being used, there will be a connection made in all six time slots. Is it reasonable to expect all six connections to be needed at once? In fact, it's extremely unlikely that we need six time slots, even though our 12 stations could be connected to six paths. We can usually apply the same rule that we did in the crossbar arrangement, and use a lot fewer time slots than would be required for the maximum number of connections. Instead of six, suppose that we had only two. That would still handle the situation in Figure 11–16, but if a third call were made, we would have *blocking* as we did in the crossbar matrix.

Using a TDM system that permits blocking would result in a reduced number of components, but many PBX manufacturers prefer to increase the number of time slots for a maximum system so that they can advertise that their product is "non-blocking."

11.3.4 Problems with TDM/PAM

For reasons related to analog waveform analysis, large TDM systems with lots of PAM channels offer degradation of the samples due to sampling rate, crosstalk between time slots, and harmonics of the original wave. One solution to this is to convert the amplitude of each sample into a digital code, which is then transmitted and decoded during its time slot. This is PCM (pulse-code modulation) and is generally used instead of PAM in present-day TDM designs.

11.3.5 TDM and PCM

In a TDM system involving PCM, a circuit called an A/D (analog-to-digital) converter converts each channel's analog or voice signal at the transmitter into digital-code numbers. A **digital multiplexer** circuit comes around to each channel's A/D converter in numerical order, sampling the code at the outputs of the A/D converter at some sampling rate. Usually, these codes are sent down a parallel data bus that carries all the binary digits of the code from each A/D converter. At the receiver, the codes are received from the bus, by a circuit called a **digital demultiplexer**, which assigns the code in each time slot to a separate output bus. A circuit called a D/A (digital-to-analog) converter or DAC converts the code number back into a voltage, and the waveform is reconstructed and filtered by a network similar to the one in Figure 11–17.

For four-wire or full-duplex communication, there must be both a multiplexer and a demultiplexer, a DAC and an ADC, at each station. The gizmo that embodies all these parts is called a **codec**—a coder/decoder—and is shown in the block diagram of Figure 11–18. In the figure, a **clock**—a pulse generator—puts "clock pulses" into a **counter**, which counts the pulses in a digital binary code. The number in the counter is the number of the **time slot**. It is used to tell the multiplexer (the MUX) which of its several digital-code inputs it should connect to its output. In this case, the MUX is picking up digital data from station 2 and connecting this input to its output. Station 2 is producing an analog wave which is "digitized" into code by an A/D converter. The output of the MUX—digital code from station 2—is connected to the **TDM bus**, which carries it around the system. The number in the counter also is used to tell a demultiplexer (a demux) on the receiving end of the TDM which of its outputs should receive the digital code on the TDM bus. At the place where the digital code is delivered by the demux, it is converted back into an analog waveform, which arrives at station C.

Although Figure 11–18 shows an analog line arriving at the PBX (switch) and being converted to digital, that is not the only way to do it, or even the best way. Recent designs incorporate A/D and D/A converters directly into the station, to-

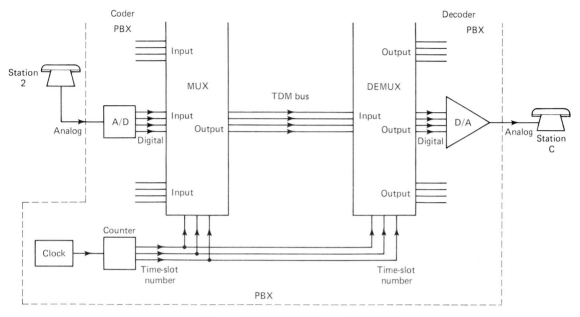

Figure 11-18   Block diagram of CODEC.

gether with a UART—which converts data from parallel to serial, or vice versa—as a package called a codec. These are available as single chips and can be included quite easily in the design of a phone. The advantage of sending serial data on a line from the station to the PBX (where it must be converted back into parallel data for the TDM switching) is that stations that originate digital-only code, such as terminals, can use the same ports on the PBX that the telephones use. This is shown in Figure 11-19.

11.3.6 Bidirectional Digital TDM

In reality, time-division multiplexing of connections isn't as simple as we indicated in Figure 11-16. We are seeing communication in only one direction during time slot T1 in that figure. If station C wants to transmit to station 2, as well as receive from it, especially in a digitally multiplexed PCM system, another time slot will be needed. We need one time slot for transmitted data and one for received data, so stations 2 and C can communicate in both directions. This is sometimes called **four-wire** or **full-duplex** communication.

11.4 COMMON CONTROL

Once upon a time, the **common control** element of a telephone company's local switch was a human operator. The operator of a first-generation telephone switch performed the tasks of registering the telephone number, selecting lines that should be connected, and making the connection. When step-by-step technology was brought in, there was

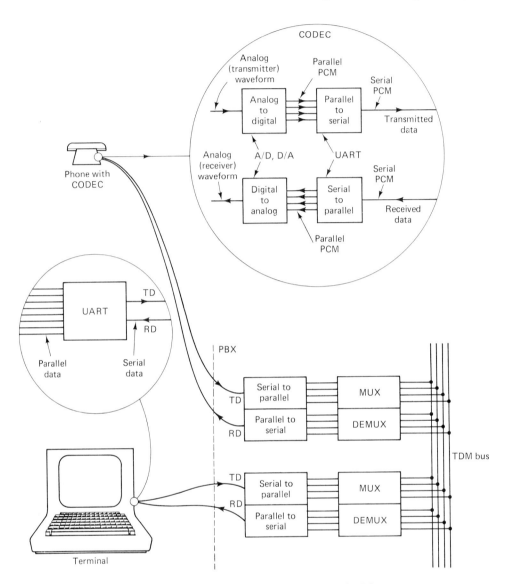

Figure 11-19 Serial transmission using CODEC.

no common control. Step by step was direct control; in other words, it was controlled directly by the person dialing the call, or more precisely, the pulses coming from the rotary dial of their telephone. As we saw, this caused many problems, such as those discribed at the conclusion of Section 11.1.2. In that section we described the significance and function of **registers**, but they are only a part of common control. We look in more depth at registers and examine the other parts of common control in this section.

11.4.1 Line Finders or Scanners

One of the first tasks of common control is **line finding**. What is a **line finder**, what does it do, and why do we need it? A **line finder** is the part of a switching system that identifies when a telephone has gone off-hook and is requesting service. It is also the function of the line finder to attach the telephone to a register when it has gone off-hook (more about registers later).

When your phone goes off-hook, a line finder in your local switch detects the current flowing in your line. When the phone is on-hook, there is no current, you'll recall, so it is the presence of current flowing in the line that alerts the line finder. In early systems, this detection was accomplished by letting the current in the line flow through a relay coil. The relay magnetized, and the current pulled up the contacts in the relay. Present-day phone systems have other ways of sensing current flow, but still identify a phone off-hook (requesting service) by sensing current.

Another method of finding a line that's gone off-hook is a **line scanner**. Instead of activating a line finder when you take your phone off-hook, you wait for the line scanner to find you. It scans all the lines, one after another, in a repeating cycle, and eventually it gets to you. Although a line finder would service only a small group of phones, and there would be other line finders for other groups, a line scanner services all the phones. Typically, a line scanner would look at each line at a rate somewhere between 1 or 2 times a second to perhaps 20 times a second. That means that it would check all the lines in 1/20 of a second or so, checking each to see if it is off-hook.

This line-finding or line-scanning function is comparable to your task as a manual switchboard operator, watching the blinking lights on the board, which light up whenever a phone wants service.

To compare a line finder to a line scanner, think of the difference between calling a taxi company to be picked up, and taking a bus to where you're going. You call the cab, and it deals with you exclusively (and expensively!). In that regard, it is like the line finder, which you engage by picking up your phone. The bus, on the other hand, is making its appointed rounds, and comes around to your stop whether you're there or not. It does not deal with you exclusively but is providing a service to everybody on its route. The bus is like the line scanner, and its bus route is like the line scanner's **polling cycle**; it is the "appointed rounds" the scanner travels in checking every line.

Now we get to the question: Why do we need line scanners and line finders? Line finders and line scanners, like the whole idea of switching itself, provide service with less hardware. By having a common pool of hardware that can be shared by any users who need it, we eliminate the need for redundant equipment that is not in use.

By the way, in step-by-step systems, in which no thought was given to having common control, there were still line finders. Even in these systems, line finders were used as a way to reduce the amount of hardware needed in the step-by-step switching matrix, through sharing. This means that there *was* some common control in a step-by-step system, to reduce the size of the thing, even though the term "common control" was not used.

11.4.2 Registers

A **register** is a memory device capable of holding a number. A **memory device** is just a recording device with some special characteristics. It records numbers instead of music or sounds. Although they were originally electromechanical, registers are now completely electronic and have no moving parts. A register holds only a limited amount of information. In the case of a telephone register, it can hold as many digits as a complete telephone number requires.

11.4.3 Rotary Dial-Pulse Registers

A **dial-pulse register** is a pulse counter. Every time a train of pulses is generated by a rotary dial, the train of pulses goes into a **decade counter**, which is a device that can count from 1 to 10 pulses. There are as many decade counters as digits in the phone number. Whenever another digit is dialed, it is counted into the next decade counter in line, until the number is complete. The entire register is the chain of decade counters, plus the **selector**, which decides which decade counter is receiving pulses from the digit being dialed. The dial pulse register shown in Figure 11–20(a) is receiving a train of seven dial pulses as we dial the number 867-5309, and we are on the third digit, 7. The 8 and 6 have already been stored in the first and second decade counters. They are the first and second digit we dialed. Since the selector is receiving the third digit—the one we are dialing right now, which is a 7—the seven pulses are being directed to the third decade counter. Each counter counts the pulses it receives and holds the number for future use. Connections to take the numbers out of the decade counters and use them are not shown in the diagram.

11.4.4 DTMF Registers

A **DTMF register** is a tone decoder, selector, and a memory device. It has got to have the capability of identifying each of the 12 different dual-tone digits, and converting each digit into a code that can be stored in consecutive binary-coded decimal places in the memory device. As with the dial-pulse register, the DTMF register has as many decimal places as a phone number. Figure 11–20(b) shows a digital latch circuit operated as a DTMF register by a tone decoder that identifies and digitizes the tones, and a selector that enables the appropriate latch to pick up the digital code.

In Figure 11–20(b), the DTMF dial has been used to enter the digits 867 and is in the process of entering the 7. The tone for 7 is decoded at the **tone decoder** into the binary code (0111) for a 7. This code is placed on the **bus**, a group of wires shared by all latches in the register. The **selector** has identified this digit as the third one dialed and is enabling the third latch to pick up the code for 7 from the bus. The previous codes for 8 and 6 have been picked up from the bus and latched into the first and second digit latches at an earlier time. **Latches** themselves are memory de-

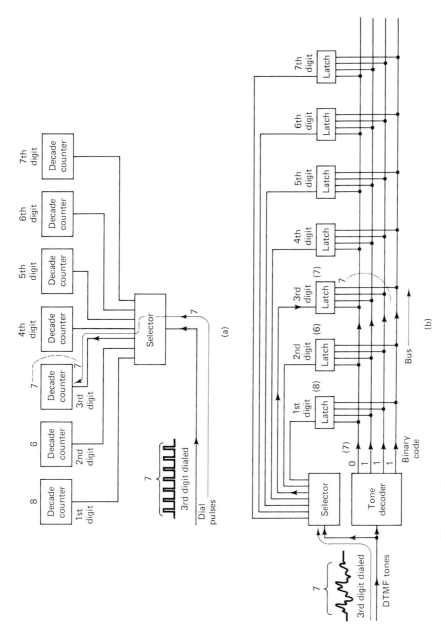

Figure 11-20 (a) Dial-pulse register. (b) DTMF register. *Selector* identifies that the third digit is being dialed and enables Latch 3 to accept the code from the bus. *Tone decoder* converts Tone 7 into binary code for the number 7, and puts that code onto the bus. *Latch* is a memory device for storing one digit in binary code.

vices that can record one digit each, in the binary code being used, whenever they are enabled by the line coming in from the selector. This kind of register would be found in an electronic or computerized switching system.

In Figure 11–20, both the rotary dial pulse and DTMF registers shown represent one way—but not the only way—to register digits dialed from these two types of dials. Some switching systems allow both kinds of registers to be attached, so that it doesn't matter which kind of a phone you connect to the line.

Digression

In fact—we tried this on our local Bell switch—the line finder in our local office connected our phone to both a dial pulse *and* a DTMF register. We—the two authors of this book—were curious and couldn't find anyone who knew the answer. We checked this out ourselves by dialing in, alternately, rotary and DTMF digits using a rotary phone and DTMF phone on two extensions of my telephone line. When we dialed seven digits, two rotary, three DTMF, and two more rotary, nothing happened. When we dialed rotary and DTMF digits, alternately, as soon as there were seven of one kind, or seven of the other, the call was placed, so from this we concluded that each rotary digit went into a rotary register and each DTMF digit went into a DTMF register. As soon as one of these registers was filled, the call was placed, regardless of the other register's contents. I'm not sure how useful that experiment was, but it *did* give us some further insight as to how the registers in the public network work. Private branch exchanges are also capable of accepting either type of dialing. Can you do both at the same time on a PBX? The proof is left to the reader.

11.4.5 Senders

Another task of common control—in fact, the main thing that makes common control different from step-by-step—is handled by a device called the **sender**. The sender is the part that repeats the number you dialed to the next switch when you dial a number that can't be connected locally. For example, if you are in Chicago, dialing New York, the number you dial will have to be repeated to the switch in New York by a sender. At the beginning of this chapter, we pointed out the need for common control. In a step-by-step network, there was no way to relay a dialed number to another switch. Long-distance calls had to be connected by a long-distance (human) operator, and the connection could not be dialed directly or connected automatically. With senders, this can be done automatically, and calls can be relayed from switch to switch through any available path from one station to the other.

11.4.6 Call Setup

The interaction between the register and the matrix of a switching system may be accomplished in many ways, depending on what type of switch it is. This interaction takes the information stored in the register and translates it into instructions that can be used by the matrix in setting up the path. Regardless of the type of matrix that's used, the **control logic** needs to set up a path through that matrix. The line finder and register have already determined that a call is in progress and know the number of the line to be connected to. The next step is to look at the state of the line being called; that is, we have to determine if it's busy or idle. Once we determine that the line is idle, we send it a ring voltage and ring the called party. The calling party is given the familiar ringing tone to verify that the called party is being rung. When the called party goes off-hook, the path through the matrix is completed and a voice path is established. The voice path, that is, the path through the matrix, is not complete at the time you hear the ringing. The connection is not really established—"cut through"—until the called party answers.

While all this takes place, many other things are happening, or *might* happen, if conditions are different. A simplified picture showing when things happen in a call setup is given in Figure 11–21. This diagram looks suspiciously like a flowchart for a computer program, doesn't it? Do you suppose that this control logic could be done by a computer program rather than by a complex set of switches activated by a set of variable conditions? We'll give you a hint: A computer *is* a complex set of switches activated by a set of variable conditions. The conditions are called a **computer program**.

11.5 COMPUTER-CONTROLLED SWITCHING

We just said that a computer is a complex set of switches activated by a set of variable conditions. The telephone switching network is also a complex set of switches activated by a set of variable conditions. Does that make a telecommunications system a computer?

Actually, no. A telecommunications system is a special-purpose switching network, which performs a limited number of tasks. A computer, on the other hand, is a general-purpose switching network, which can perform a very wide variety of tasks. Any switching function you can imagine may be performed by a computer.

In that case, it sounds like a computer could be programmed with instructions that would make it "imitate" any specialized switch network you want, including a telecommunications system. This is exactly what computer-controlled switching is. By making a computer imitate the switching functions of a telecommunications system, we make it possible to replace the telecommunications system with a computer and a suitable program.

In precomputer designs, the switching system consisted of custom-designed circuits, or **hardware**, constructed to fit the tasks done by a telecommunications system, and nothing else. Now we use a general-purpose machine—a computer—and a program made of **software** that imitates the hardware of the earlier systems. To some people, including computer programmers, it might seem wasteful to use a computer to run only one program, and spend all its time **emulating** (imitating) a single hardware system. This is *not true,* for two reasons.

First, software is easier to change than hardware. When the time comes for alteration, expansion, or improvement in an all-hardware switching system, the system must be torn down and rebuilt to some degree. Software can be changed while the

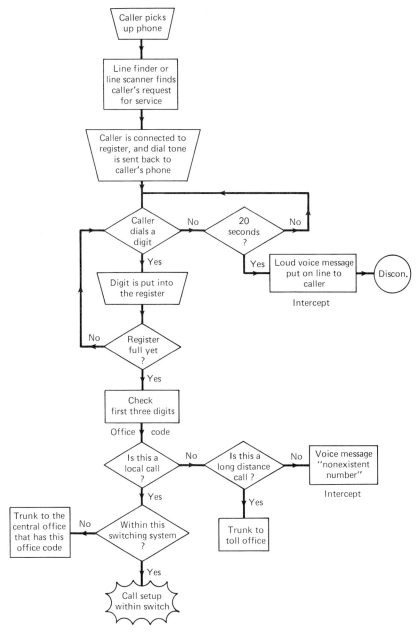

Figure 11-21 Simplified flowchart for call setup.

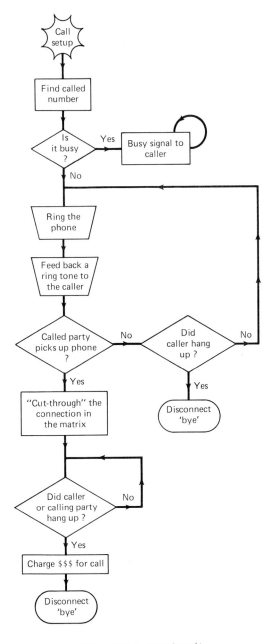

Figure 11-21 (*continued.*)

old system is still running, and the old system is replaced with new software in a very short time. With hardware modifications, you have more "downtime." The more extensive a set of hardware modifications are, the larger the part of the system that is "down" (not usable) and the longer the time it takes before it is "up" again.

Second, the computer is not "wasted" simply because it is running the same program all the time. We call a computer that is used this way a **dedicated machine**, and it is quite common to use microprocessors and minicomputers this way. Dedicated machines spend *all their time*—100%—of it—running the program that is held in their memory storage. A common control computer is working 100% of the time, performing the tasks that were performed by hardware in earlier switching systems. The fact that it cannot also be used to play Space Invaders at the same time is no big loss.

11.5.1 A Computer Is a Computer Is a Computer

. . . to paraphrase Gertrude Stein. What we mean by this is that it doesn't matter whether you are looking at a "toy machine" that only plays "Super Deluxe Star Trek" when it is turned on, or the microprocessor in an Amana Radarange microwave oven, or the computer in a Mitel PBX (see Figure 11–22). The computers in each case contain the same basic parts and work in the same fundamental way. The only thing different between each of the **dedicated machines** we just mentioned is the **software** they contain.

It is possible, for instance, that the three machines we mentioned contain the same **processor**. If they did, we could swap programs by interchanging the ROM integrated circuits that contain the programs. In that case, the PBX might be valiantly trying to preheat a turkey, while the Star Trek box attempts to set up least-cost call routing through a WATS line, and the Radarange attempts to zap the next Klingon battlecruiser we find in our quadrant of the galaxy.

How will the Radarange survive the next Klingon attack? Will that turkey in the PBX ever get basted? And what part of the galaxy did the WATS line in the Star Trek box go to? The use of the peripherals in each box would be ridiculously inappropriate (to put it mildly!). While the processors may be the same, and it's even possible that the program ROMS (we'll tell you what this means later) can be swapped from one board to another, we have to remember that the peripheral devices are the computer's way of interacting with its environment. Each machine has its own peripheral "environment."

In the example above (which I know is ridiculous) we see the three major parts of any computer system. These are the **processor**, the **memory**, and the **peripherals**, as shown in Figure 11–23.

11.5.2 The Processor

The processor is the part of a computer that responds to the commands of a program. In the processor, circuits may be switched on and off automatically by numbers that arrive from outside. Each circuit has an identification number, called its **machine**

Figure 11–22 No matter what the purpose, computers contain the same basic parts. Processors and programs may be interchangeable, but their peripherals are not, necessarily.

code or **opcode**, which switches the circuit on. These machine codes are stored in **memory** devices that permit them to be recalled into the processor and used to activate the circuits in the processor. Some processors use the same identification numbers (machine codes) as others. For instance, both the Intel 8085 processor and the Zilog Z-80 processor use the number 70 as a code for the adder circuit. If you want to add a number stored in processor register A to one in an external memory device, both the 8085 and Z-80 will add them when a 70 is "fetched" by the processor from its memory. Not all processors, however, can use a 70 to activate the adder. In the Motorola 6809 processor, for example, the same process is done by circuit 155, and *that* is its opcode for "ADD A."

All processors aren't interchangeable, but they all do the same thing, generally using different codes. Computer programs that operate these processors arrive as strings of numbers that say which circuits to switch on, and in what order they should be activated.

Some of the circuits that the processor can switch on are outside the processor

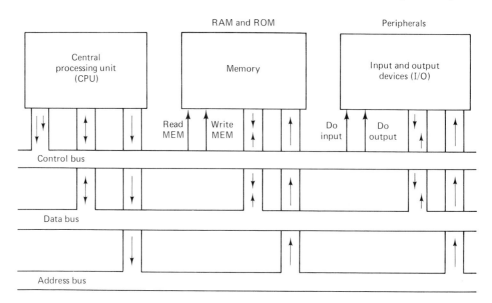

Figure 11-23 Block diagram shows three major parts of a bus-oriented computer system: processor, memory, and peripherals.

itself. For instance, the processor must fetch each instruction from a memory device, so it needs the capability to tell the memory device when to turn on and where the next instruction the computer wants is located. Similarly, the processor needs to be able to scan its input devices and activate its output devices. These peripherals do things like reading which button you pushed on the microwave oven's control panel (input), displaying the time remaining before the turkey is done (output), and shutting off the magnetron when the timer reaches zero (output). Since the processor of any computer is the control center for all the actions of its peripherals, it is often called the central processing unit (CPU) of the computer. Ninety percent of the logic, or more, that does "actual computing" in a computer is located within its CPU. In microcomputers, the CPU is called a **microprocessor**, and is often a single integrated-circuit device.

11.5.3 Memory Devices

The external memory device of a computer works like a magnetic tape recorder with an *awfully fast* rewind and fast-forward. An ordinary tape recorder has a counter that gives a rough idea where each part of a recording may be found on the tape. In the memory devices used with computers, the data is recorded in locations called **memory cells**, which each have a number like the number on the tape counter. Each number is called the **address** of the memory cell. Like the number on the tape counter, it tells you *where* to find a piece of information, and allows you to store information in a place where you can find it again. There are two answers to *what* is stored in each place in a memory device. Information stored in a computer's memory may be instructions or data.

The **instructions** are the opcodes we mentioned before. They are the numbers that control the computer. A list of instructions make the computer activate its circuits in a particular order we call a computer program.

The **data** are the numbers that are controlled by the computer. They are the victims of its action rather than the agents that cause it to act.

Most computer programs are stored in addresses that line up like the numbers on the tape counter. That is, when you run a program, you do the instructions in the order the addresses indicate. The instruction stored in address 0 is done before the instruction in address 1, the instruction at address 1 is completed before the computer does the instruction at address 2, and so on. Like the tape recorder, the instructions in the memory device—the computer program—are usually recorded in the same order in which you want to play them back. Unlike the tape recorder, the instructions can easily be done in another order, or out of order, as easily as in 1–2–3 sequence.

We may store computer programs in two basic types of electronic memory. **(ROM)** is **read-only memory**. Information stored in a ROM is recorded permanently at the factory. Like a phonograph record, a ROM, once recorded, is only used for playback. Information stored in ROM, once recorded, is not easy (and in some types, not even possible) to alter. The normal place to store a program, which must remain constant and unchanging, is in ROM.

RAM is **random access memory**. The name "random access" refers to the statement made earlier that computer instructions can be done in 1–2–3 order, or out of order, equally well. This is not the characteristic by which RAM differs from ROM, however. RAM always refers to memory devices that have inputs for recording as well as outputs for playback. Information recorded in RAM can be altered at any time simply by writing new information in the place of old information. The RAM forgets the old information that "lived" in a memory address and stores new information in its place. This is the type of memory that would be used for variable quantities such as data. When power is turned off, most RAM devices lose their ability to hold information, and the information is forgotten. Memory with this characteristic is called **volatile**.

The main working program—or **operating system**—of a computer is intended to be available when the machine is turned on. It is supposed to be the same every time you turn it on, and shouldn't be "forgotten" simply because the power is turned off. It seems like a good idea to store this operating system program's instructions in ROM memory. ROM memory contains memory cells that do not "forget" what has been stored in them merely because the power is turned off.

If it is certain that changes will be made in the program at reasonable intervals, a type of memory called PROM or EPROM makes it possible to field-program changes into the operating system without making an entirely new ROM at the factory. Unlike RAM, ROM memory is **nonvolatile** and will retain every bit of information it contains, even if power is turned off.

PROM—**programmable read-only memory**—is programmable. That means that it has inputs like a RAM memory, but once the information is "burned in" to the PROM, it will not be forgotten if power is turned off. If you make a mistake programming a PROM, or want to change some of its data, the mistake, or earlier data, is just as permanent as the correct data. You will have to throw out the PROM and start again.

EPROM—**erasable programmable read-only memory**—is not only programmable, but erasable. If mistakes are made in EPROM, or changes are desired in the existing data and instructions, they may be erased with ultraviolet light and the corrected data and instructions programmed into the PROM in their place. There is a small quartz window over the silicon chip, into which we shine ultraviolet light, and it "sunburns" the data out of the chip, leaving it "clean" to be rewritten. Like erasing a blackboard, the erased EPROM must be entirely rewritten with data and instructions.

EAROM—**electrically alterable ROM**—is the answer to the question: "Do we have to erase it all, and rewrite it all, to correct one lousy mistake?" It is even more tolerant than EPROM of errors and corrections. Each of these developments—from ROM, which is hardwired at the factory with the data it must contain, to PROM, EPROM, and EAROM—is better and more flexible, but also more expensive.

Warning note

There are two terms, **dynamic memory** and **static memory**, which refer to the hardware of a RAM, and there are also two terms **dynamic data** and **static data**, which refer to software inside memory. These terms are easily confused.

In hardware terms, a **dynamic RAM** is one whose information must be rewritten periodically, or it will fade away. **Memory refresh**, or **refresh cycle**, are names given to the restoration of "fading" data in dynamic memory. This is accomplished by reading (playback) of the weak data, which is then written (recorded) into the same memory cells. These memory cells are often made of capacitors, which hold information as an electric charge and which gradually lose their charge unless recharged.

A **static RAM** keeps its data intact as long as the chip has power, without being rewritten periodically. Its memory cells are made with active devices such as transistors, and they are arranged in units that can hold a high- or low-voltage condition while power is maintained. Such units are called **flip-flops**, since they can be flipped (high voltage) or flopped (low voltage).

Dynamic data is variable information stored in a memory, information that is subject to change. RAM memory is the ideal type of device to store dynamic data.

Static data is constant data which do not change throughout the course of the program, the kind of data that could be stored in a ROM. The term "data" in this case also includes program instructions, which should not change or be destroyed while the program is running.

11.5.4 Input and Output Devices (Peripherals)

Computers must have the ability to communicate with the world beyond the CPU and memory. All external devices that serve to communicate information *inward to the CPU* (**input devices**) and *outward from the CPU* (**output devices**) are collectively called **peripherals**. The computer controls its input and output devices through interfaces called **input ports** and **output ports**. They are not the same as the peripheral devices themselves. Instead, the ports are general-purpose logical connectors that enable the CPU to present data to slower output devices until they can use it (output ports) and select, or admit, data that is waiting to come in from input devices (input ports).

11.5.5 How a Computer Uses Its Memory to Run a Program

In Section 4.2, we used, as an example, a computer made from a simple four-function calculator, and examined how it might be made to automatically compute the sum of any number plus 48. This is the method used to convert decimal digits into ASCII code, when done in binary. To refresh your memory on the action of the CPU, the program counter, the instruction decoder, and the program memory, it would be a good idea to review Section 4.2 at this time.

Did you just go back to Section 4.2 and then return here? If you did, you just completed a **subroutine call** (see Figure 11-24). "I did?", you ask. "You betcha!", I reply. A subroutine is simply a procedure where you (or your computer) go to another place—besides the place that would normally be next—and do what you find there. At the end of a subroutine, you return to the stuff you were doing at the moment you went to the **subroutine**, and pick up where you left off, reading the next line down. Getting back from a subroutine is called a **return**.

"Fine," you say, "but what does a computer do when it does a subroutine call and return?"

Well, you might begin analyzing the action of a computer's subroutine, by looking at what you, yourself, did. You started on this page, then jumped back to page 79 where Section 4.2 begins. Before you did that, though, you stashed the number of *this* page, *right here*, into your short-term memory. Maybe, if you don't have a

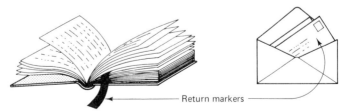
Return markers

Figure 11-24   "Subroutining" in everyday life.

very good short-term memory, you put your thumb in between the pages of the book right here so that you could find your way back. Computers have a very good short-term memory where they can stash things like the return location. (Since computers don't have thumbs, the other method wouldn't work!) This short-term memory is called the **stack**, and it is a part of the RAM memory of a computer. The computer saves its **program counter** in the stack just before it jumps to the subroutine. The combination of "save to the stack" and "jump to the subroutine" is called a subroutine **call**. How does the computer "jump"? The computer puts the address of the subroutine's first line into the program counter, erasing the address of the "normal" next line. Thereafter, the computer fetches all instructions after that from the subroutine. At the end of reading all the lines in the subroutine, the computer "pops out" the number it put into the stack, getting that number back into its program counter, and picks up where it left off. Getting that "return address" into the program counter is called a **return**.

You understand how a self-addressed stamped envelope (a SASE) works? The return address on the envelope saves the mail-order operator the trouble of having a mailing list containing the address of each customer and the trouble of addressing every envelope. A mail-order business can be conducted this way without the operator bothering to learn the addresses of his customers. A change of address is no trouble at all, and new customers make no extra work either. The return address stacked up during a subroutine call works the same way. The programmer who writes a subroutine does not have to know the place it is called from in order to get back. The return instruction uses the stack to find out where to go back to, and this is just as well. Subroutines are often reused by calls at many different points in a program, and each time the subroutine is called, it must return to a different place. Since this can't be planned in advance, the call must carry its return address with it each time the subroutine is called.

11.5.6 Entry Points and the Load Map

The **entry points** into subroutines in memory are vital information to the programmer who is sharing those subroutines with other programmers writing other parts of the operating system program. In a common control program, for instance, the coding may be the product of many programmers' work, and some routines may be used in common by almost all the programmers. They will all need a listing of entry points, either in address form or symbolic form, to commonly shared routines. A listing of these subroutines, or **tasks**, available for "global" use by all the programmers, may be called a **load map**.

11.5.7 Addressing and Gating

In mini and microcomputers, there are many peripheral devices attached to the computer, but only one path to carry information to all those devices. The path is called the **data bus**, and the information, of course, is called data. How do we get the data to the "right" device when all the peripherals share the data bus? We are asking, in this case, *where* the data should go. The **address bus** is another path that carries the "where" information. It is also shared by all the peripheral devices. Each device must be able to distinguish its own **address**, through the use of a digital circuit called a **decoder** (an address decoder). The address identifies one, and only one, peripheral, and is not the same for any other device. The peripheral device is switched on, or, as we say in digital electronics, **enabled**, when its unique code appears on the address bus. Since no other peripheral device has the same code, only one device is switched on at a time.

Imagine, for example, a system in which a computer controls the routing of data between 20 peripherals (see Figure 11–25). To make it simple for us human beings, we will assign each peripheral a **device code** from 1 to 20, although this might not be the most efficient method of coding for a computer. Some of these peripherals are input devices, and they will be attached to input ports, which only place data onto the data bus. Others are output devices, and they will be attached to output ports, which only pick data up from the data bus. (Again, this is a bit of an oversimplification, as most real peripherals or stations perform both input and output functions.) Now suppose that device 5—an input device—has data for device 7—an output device—and the computer has to control the transfer of data between the peripherals. Usually, a bus-oriented computer does it this way: First, the computer places the input device's code—5—onto the address bus and issues a **control code** identifying an input device (this is optional). The port connecting device 5 to the data bus switches into the conducting state and places device 5's data onto the data bus. The computer then picks up the data and stores them in its **accumulator**, an internal memory register inside the CPU (the central processing unit). This is shown in Figure 11–25. You can see that the 5 on the address bus is not specifically sent to device 5—all the devices receive the 5 on their address-bus inputs—but device 5 is the only one that *uses* the address 5 to do something; the others ignore it. When device 5 puts its data onto the data bus, it is not specifically sent to the computer, either—all the devices, including the computer, receive that data on their data bus inputs—but only the computer, or rather, the computer's CPU, actually picks up the data that comes in on its data input lines and does something with it (stores it in the accumulator register). Then, with the data from device 5 safely stored inside its own innards, the computer turns off the port for device 5 and then places the code for the output device—7—onto the address bus, issuing a **control code** identifying an output device (also optional). The port connecting the data bus to device 7 is activated. Any data placed onto the data bus will now be **latched** into port 7 and held there after the transfer operation is complete. Almost immediately, the data held in the accumulator is placed onto the data bus to be picked up by the latches in port 7. The address and control signals are kept on long enough for the latch to "catch" the data placed on the data bus, then shut off. The transfer of data between port 5 and port 7 is now complete. If this seems like a complicated process to do such a simple thing; perhaps it is, but even in a slow computer, it could be done in less than 10 millionths of a second.

This process is called **gating**. The ports attached to the input and output devices are like gates that are opened and closed to let the data in and out.

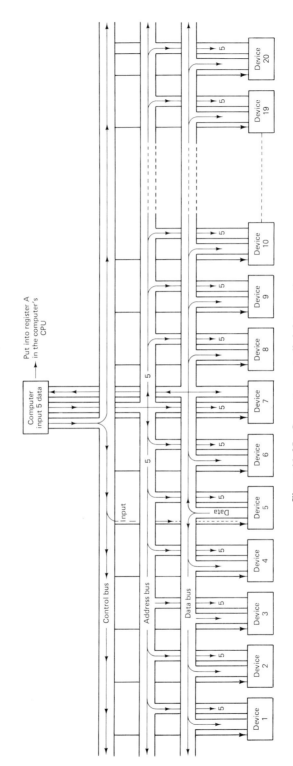

Figure 11-25 Computer-controlled data routing.

There is another way to accomplish the same task (see Figure 11–26). It is faster and involves fewer steps. This method is not the standard way that data-processing computers transfer data among their input and output devices, but it *is* the method used for telecommunications switching.

First, our new system now has two data buses instead of one. The internal data bus connects the computer with those peripherals that must communicate directly with it. The local data bus connects the peripherals that need to talk to one another. In a telecommunications system, these local peripherals would be the stations connected to the PBX's common control computer. The address bus has also been split, into an input address bus and an output address bus, each capable of carrying a separate address at the same time. With this arrangement, the data does not have to go into the computer and then out, to be transferred from an input port to an output port, and the computer does not have to "know" what's being transferred to it. (And why should it? It's none of the computer's business, anyway!)

To transfer data between port 5 and port 7, as before, our computer now places the input port's code—5—onto the input address bus; that activates port 5 to place its data onto the local data bus. This determines the source of the data on the data bus. Meanwhile, the output port—7—code is placed on the output address bus, and that port is activated to "pick up" any data defined on the local data bus. This design assumes the existence of separate input and output system activators, and that the input (data source) and output (data destination) systems are separately controlled to allow only one source and one destination at any time.

Another method uses only one address bus, carrying numbers that identify the source and destination of data. When the number on this address bus changes from 5 to 7, the input port—5—remains active while the output port—7—picks up the data from the bus. In this case, when the address on the address bus changes from 5 to 7, the input port—5— remains active, and "remembers" to keep putting its data onto the bus, even when the new address—7—is activating the output port. This requires some complicated hardware and might make this kind of a system more complicated to *build* compared to the standard input/output system. It does, however, simplify the operations carried out by the system, and speeds up those operations.

11.5.8 Enabling and TDM

In Section 11.5.7 we said that the common control computer was responsible for enabling the connections between the source of data and its destination. What is **enabling**?

When data was enabled in the way we just discussed, data was transferred to the data bus from an input device when certain digital circuits switched from a nonconducting state to a conducting state. Although this is *one* way to enable signals to transfer from one place to another, it is not the *only* way. In the past, switching of voice signals took place when metal switch contacts were physically closed and opened. Now the same functions take place without moving parts by enabling and disabling the connections from one bus to another in a computer system in which stations are the peripherals. Unlike the operation of reed relays and Strowger switches,

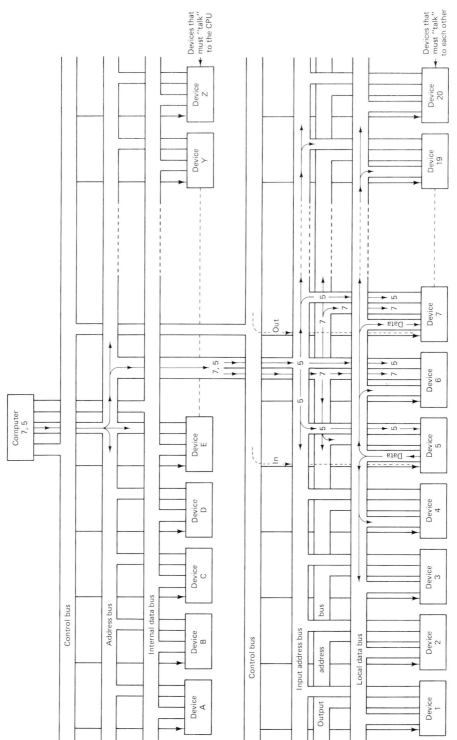

Figure 11-26 Method used for telecommunications switching.

294

this enabling is digital switching. To make this possible when voice stations want to talk to each other, the voice signal must be converted into digital information. A device called a codec, as you will recall, converts the voice signal into a serial bit stream. This means that voltages which appear from moment to moment on the voice line are converted into digital codes, often using eight or more binary numbers. Since station equipment is generally not connected to the switch by a bus of eight or more wires, the binary numbers are transmitted serially, a scheme by which a single wire can carry multibit binary codes, transmitted one after another at separate times (time-division multiplexing).

In some cases, the codec is built right into the phone, and in those cases, you do, indeed, have a serial bit stream coming directly out of your phone on the pair of wires going to the switch.

It is also possible for your voice to be digitized at the PBX. In that case, your voice signal arrives at the PBX in analog form, and is converted there, by a codec. Your phone is one of the many stations the PBX is handling, and since the PBX is not limited to a single pair of wires, internally, it puts the codes for voice signals on a **TDM bus**, a path in which a group of wires carries each digitized number for the voltages on the voice line. This has the advantage that other numbers, from other stations, can be placed on the bus and routed to other destinations, in between the times when the PBX "samples" the signal on your line. This is also called time-division multiplexing, but in this case, instead of permitting multiple-digit binary codes to be transmitted on one wire, the multiplexing permits multiple **channels** (signals from different stations) to be switched and controlled by the same processor. This system depends, for its efficiency, on the fact that the processor is able to switch at rates much faster than the rate at which signals change on any of the channels.

REVIEW QUESTIONS

1. How many digits can follow any individual office code?
2. Why are area codes needed? Are they similar to access codes in a PBX?
3. What is the common control element of a *cord board*?
4. What types of switching elements were used in the first types of mechanical switches?
5. What were some of the problems with using step-by-step switching on long-distance, or "tandem" calls?
6. What is the purpose of a register?
7. What is the purpose of a sender?
8. In a crossbar switch, what is the purpose of the horizontal and vertical crossbars?
9. What causes blocking in a crossbar switching matrix?
10. What type of matrix would be used on an electronic space-division-multiplexed system?
11. What was the first nonmechanical electronic device that could be used for switching and repeating a voice signal?
12. Name three advantages solid-state (transistor) devices have over vacuum tubes and relays.

13. How can a transistor be used as a switch? What turns it on and off?

14. What are the major differences between using an electromechanical switch and a transistor switch?

15. What does an SCR do that is different from a transistor?

16. Why are mechanical (Strowger or crossbar) switches still in use?

17. What can be done to prevent blocking in a TDM matrix switch?

18. Where is an analog-to-digital converter used in a codec? (In the coder or the decoder?)

19. What are the advantages of putting the codec at the station instead of at the PBX?

20. How do line finders and line scanners reduce hardware in a switch, and what is their function?

21. Why do switching systems use both DTMF and rotary registers?

22. What are the primary functions of the common control element in a switching system?

23. What are the advantages of using a computer for common control instead of hardwired logic?

24. Can the same processor be used in different dedicated machines?

25. Explain the difference between RAM and ROM memory.

26. What is a refresh cycle? Why is it needed for DRAM (dynamic memory)?

27. What is a subroutine? What is an entry point? What is the load map?

28. Describe what kinds of information are carried on the address bus, data bus, and control bus of a computer.

29. How is data switched by being enabled and disabled through logic gates? How does this resemble the actions of an electromechanical switch?

30. How is the destination of information selected in a computer system by the device's device code?

12

Private Networks

and Routing

When you place a telephone call through you local telco, your call is routed to its destination through a network that is already in place. The telephone company has taken care of connecting everything, and you don't have to think about it.

When you need to set up a private network, however, the way things are connected together is *your* concern. In this section we explore some of the ways in which switches and stations are connected together that you may find in use within a private network.

The simplest of private systems has just a single switch that connects calls from one station to another within an organization. A single switch—a PBX (private branch exchange)—originally was just a branch of the public network. It was a way to extend the public network inside the organization's headquarters, plant, or whatever. There was no concern for least-cost routing, alternate route switching, or call-detail recording. All a PBX was, was a switch that would allow all the telephones within the organization to call one another and to reach the public network outside the organization. Therefore, a private branch exchange was nothing more than an extension of the public network (a branch).

Today's PBX is much more than just an extension of the public network, although it has kept the same name. Perhaps a better name for a PBX would be a private switching center. Some PBXs are no longer even tied to the public network, but function as privately owned local offices, and are connected to other private switching centers across the country, to form a network that may seem to parallel the public network. Private communications within the organization would always travel along its own trunks and would not require the use of the public network's facilities and services. However, organizations often do have to communicate to organizations other than themselves, and when they do, their switching centers must somehow gain access to the public network.

12.1 A SINGLE-SWITCH CONFIGURATION

In a **single-switch configuration,** there is one PBX that switches voice and data between all the telephones and/or data terminals within an organization. This configuration is called a **star network,** because conceptually, the PBX is at the center, with the stations attached to it like the points of a star. Whenever one station wants to communicate with another station, the connection is made through the PBX. A star network is a centrally controlled system, with every station acting as a satellite to the PBX at the center. Signals from one station to another station go through the PBX, and can't go from station to station directly.

12.1.1 Routing within a Single-Switch Configuration

In Figure 12-1 we see a **star network.** In this configuration the PBX is in a position like the central processor of a computer, with each station acting as a peripheral to the PBX. This configuration works fine if every station in the organization wants to do nothing but talk to other phones within the organization. All the stations are assigned an extension number, and any station can call another station by simply dialing the digits of its extension. If you want to establish a communication path between yourself and the stations elsewhere in an organization, you need a **numbering scheme.** You would assign each station its own unique number within this scheme. If there were less than 100 stations, for instance, a two-digit decimal number would suffice for the extension number. Similarly, if you had a larger system, with up to 9999 stations, you could use four-digit numbers to identify all the stations. In the public network, the North American Numbering Plan identifies telephones using 10-digit numbers, three for the area code, three for the office code, and four for the subscriber code of the telephone within that exchange. In addition, an eleventh

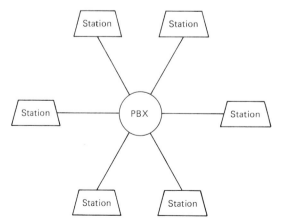

Figure 12-1 Local star network.

digit (a 1) is added to identify the fact that the next three numbers are an area code when the same digits may be used to begin area codes and office codes.

When you dial another extension within a private system, the digits you dial are like a subscriber code. In a single-switch system, you don't need to dial the exchange, just the subscriber code. (In some places, if you're calling someone in the same exchange of the public network, you don't need to dial the office code number either—for instance, if you're 867–5309 and you want to call 867–5300, you would just dial 5300, to place the call, in certain communities. This is not, generally, true for most communities. For instance, skipping the area code digits is always permitted for calls placed within the same area code.)

Within multiple-switch systems, where an extension connected to one switch is calling an extension connected to another switch, the first digit of the extension number is often a code to identify that switch. That's most often the case where tie trunks are used—the PBX uses the first one or two digits of the extension number to identify which switch is being called. This first digit of the extension number would be equivalent to the office code in the public network.

12.1.2 A Single-Switch Configuration

We've already discussed OCC (other common carriers) and SCC (specialized common carriers) and described the advantages of using their services for long-distance calls. The major advantage of using specialized carriers is cost reduction on long-distance calling. The specialized carriers can provide limited services for a lower cost than AT&T long lines, which provides a wider variety of more general services, but costs more because of this flexibility. The public network is changing very rapidly. One of the big changes taking place is the choice of long-distance carriers. On the public network, you will no longer automatically be connected via AT&T long lines when you dial a long-distance number directly. If, for instance, you have specified MCI as your long-distance carrier, you won't have to dial MCI first if you want them to carry your long-distance telephone traffic. You can "specify" that MCI is your long-distance carrier, and calls you direct-dial to long-distance locations will automatically be transmitted by MCI's equipment. Of course, AT&T, the long-distance carrier that you chose in the past via monopoly, you can today choose voluntarily, if you wish.

A single-switch configuration can make use of this, and other options, by selecting the **trunk group** through which a call would be routed. At present, your home telephone has only one choice once you've specified your long-distance carrier. PBXs, unlike your home phone, have more than one pair of wires coming in from outside. These pairs (trunks) could be connected to different common carriers; for example, some could be connected to AT&T and others to MCI. All the MCI trunks would make up a **trunk group**—the MCI trunk group—for instance. Since the PBX has several trunk groups to select from, it can switch to select any trunk group that it needs.

If you're the user of a PBX with access to AT&T long lines and a number of other specialized long-distance carriers, you could personally route the calls through

AT&T, MCI, or whatever you choose. You would dial the access code of the carrier's trunk group and then dial the call.

Let's work out some examples here, using a table for some hypothetical SCCs, and see how you would select the trunk group on which each call should go out:

TABLE 12-1 XYZ COMPANY ANALYSIS OF SPECIALIZED COMMON CARRIERS

	Carrier	Destination	Cost	Quality of service (A, B, C, D, F excellent poor)
1.	Pigeon Drop Communications	Podunk Junction	$0.10/min	A
		Grover's Corners	$0.12/min	A
		Peoria	$0.28/min	B
		Disneyland	$0.40/min	B
		Los Angeles	$0.45/min	B
2.	Swiss Yodeler Audio Telecom GmBH	Podunk Junction	$0.12/min	B
		Grover's Corners	$0.15/min	B
		Peoria	$0.30/min	C
		Disneyland	$0.45/min	C
		Los Angeles	$0.50/min	C
3.	Red Eyeball Express Optical Datacomm. Inc.	Podunk Junction	$0.08/min	B
		Grover's Corners	$0.10/min	B
		Peoria	$0.25/min	B
		Disneyland	$0.38/min	D
		Los Angeles	$0.40/min	D−
4.	Tin Can Telecom Ltd.	Podunk Junction	$0.25/min	B
		Grover's Corners	$0.30/min	C
		Peoria	$0.35/min	C
		Disneyland	$0.50/min	D
		Los Angeles	$0.55/min	D

In Table 12-1 we can see XYZ Company's analysis of their long-distance carriers, for five frequently called localities, compared in terms of cost and quality of service. Cost is rated in terms of how much money is charged for the first minute of the call, and quality is a highly subjective parameter computed by combining such factors as quality of transmission, attention to customer needs, frequency of billing errors, and accuracy of addressing, using a complex and secret formula. Although there is no uncertainty about billing rates, the "quality of service" factor is one the customers must decide for themselves, according to their needs. Subjective evaluations such as "quality" also cannot be transferrable from one subscriber to another. If you were making the trunk-selection decisions manually, and you were working for XYZ Company, you could use this table as a guide to which carrier to choose for each individual call. You would use access codes like those used to reach another PBX, to identify which trunk group the call would be connected to, and make the call on that carrier. Let's say you want to make a connection to your L.A. office, and you go according to cost only. The cheapest call goes to Los Angeles through Red Eyeball Express Optical Datacomm, Inc. However, that disregards the fact that smog gives Red Eyeball's L.A. heliograph operator eyestrain, and that means that his reliability is very poor. To select the Red Eyeball company, anyhow, you would dial the access code for Red Eyeball, 3, and then dial the number. The 3 would tell the system on which trunk group to place the call. In a real world, this approach isn't very practical because, even looking at price alone, with so many long-distance carriers to use, you'd spend so much time doing the search through the table that your company would lose more in labor expenses while you looked up the best option than they would gain in cost savings on the call. What if, after all this searching, you find that all the trunks in the Red Eyeball group are busy? You'd have to go back to the table and look for a second choice. What if the second choice's lines to Los Angeles are down? There are a couple of things you could do. One way would be just to select one common carrier on its overall merits, and make it your long-distance carrier all the time. This is the same option that you choose for your home phone when you specify your long-distance common carrier. The second option is to make the entire operation automatic; put it into the hands of the PBX's common control computer, so to speak.

The common control program of a PBX should contain an automatic route-selection scheme; that is a system feature available on most computerized PBXs. A greatly expanded version of the table in the example above could become part of a computer's **database. A DBMS (data base management system)** is a program that could look through a table such as this and find all the references to Los Angeles. Once those items in the database are found, the references can be sorted in order of cost, from the cheapest to the most expensive, allowing the PBX to access the trunks in order of cost until an available trunk is found. With this kind of sorting program, the criterion used by the database could be changed from cost to quality of service, and the higher-quality carriers sorted in order from the least expensive A-quality carrier to the most expensive. It's unlikely that the database would include estimates of quality for infrequently called places, so the "quality factor" would probably not be included as a factor under consideration.

12.1.3 Automatic Route Selection

Figure 12–2 shows a flowchart for **automatic route selection** (ARS), assuming that the cost factor is the only important determining factor in selecting the trunk group. It is a typical ARS scheme, similar to that used by the more-sophisticated PBX. It is "transparent to the user"—when you use the telephone, you don't have any idea any of this is going on—you dial 9, to tell the system you're making an outside call, and get a dial tone; then you dial the number. The system will automatically select a trunk in the most economical trunk group if a trunk is available in that group. If all the trunks are busy, the next-least expensive carrier's trunk group will be selected,

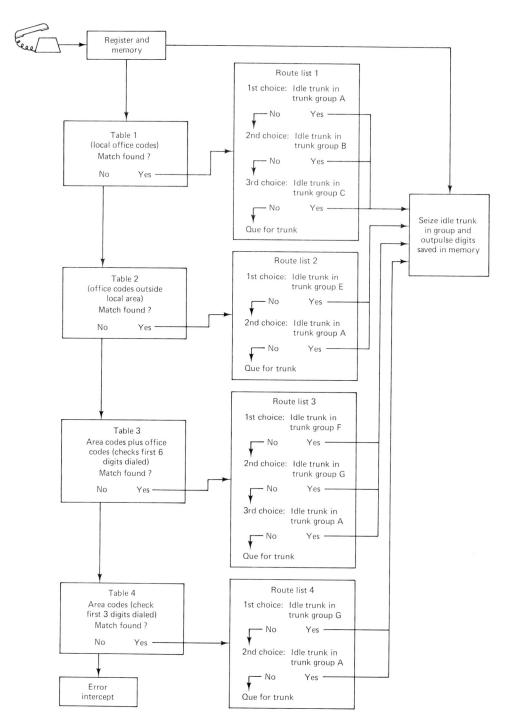

Figure 12-2 ARS flowchart.

and so on. If you follow the flowchart, you will be following the steps that the computer is doing. You can see from the flowchart that several tables are stored in the computer's memory, containing lists of area codes and/or office codes. For each table entry, the choice of which trunk group will be selected first, which one second, and so on—the *priority* of each carrier—is matched up with the area code and/or office code entry. It is the telecommunications manager's responsibility to see that these tables are updated to reflect the current cost of each carrier.

Outline of ARS System Operation in a Single-Switch System
Access Code
You dial 9, signaling the PBX that you want to make a call on an outside line. The PBX returns a dial tone to let you know that it has accepted your request to go ''off-system.''

  I. *Dialed number is stored in memory:*
      You dial a seven-digit or 10-digit telephone number, which goes into a register in the PBX and is stored in a portion of its memory while you are dialing.
 II. *Compare digits in memory to entires in tables:*
      Your PBX scans through tables of numbers in its memory, comparing your digits to those it finds in the tables.
  A. **If** first three digits match three-digit entry in Table 1:
      This could be a table of local office codes for the telephone exchanges in the immediate area. The PBX may find that this is not a toll call, and send the dialed digits directly out to the local telco.
    1. **Go to** Route List 1, first choice trunk group A, for idle trunk in group
       If idle trunk available in A, send dialed digits in memory to A trunk.
          A is probably the local telco, the first carrier of choice for a local call.
    2. **If** no idle trunk, check Route List 1, second choice, trunk group B for idle trunk and send digits.
          B is an OCC carrier that can handle the local call on an alternate route if no trunk is available from the local telco. It is more expensive than the local telco.
    3. **If** no idle trunk in B, check C for idle trunk and send digits, or issue a busy signal.
          In this case, C is an even more expensive carrier than B for the local telco.
  Else. . .
  B. **If** the first three digits match the three-digit entry in Table 2: This table could include office codes outside your local area, but within the same area code. These would be dialed as seven-digit numbers but would be charged as a toll call.
    1. **Go to** Route List 2, first choice, trunk group E, for idle trunk in group.
       If idle trunk available, send dialed digits on selected trunk.

Let's suppose that the local telco offers MABS (Metropolitan Area Business Service) to your greater metropolitan area at a reduced cost. This could be trunk group E.

2. **If** no idle trunk in E, check A for idle trunk and send digits, or issue a busy signal.

   If the MABS trunks are all busy, the next choice is the local telco, again.

Else. . . (This is a long-distance)

C. **If** the first six digits match the entry in Table 3:

This is an area code and an office code. The area code may be served by a number of carriers, but some carriers may not be able to provide service to certain office codes (exchanges) within the area.

1. **Go to** Route List 3, first choice, trunk group F, for idle trunk in group. If idle trunk available, send dialed digits on selected trunk.

   This is the most economical carrier that serves both area and office codes in Table 3. It is your "specified" long-distance carrier, but unlike your home phone, your PBX does have some alternatives.

2. **If** no idle trunk, check Route List 3, second choice, trunk group G for idle trunk and send digits.

   This could be a WATS (wide-area telephone service) line trunk group. It is more economical than direct distance dialing over local central office trunks.

3. **If** no idle trunk in G, check A for idle trunk and send digits, or issue a busy signal.

   This is direct distance dialing over the central office trunks that are supplied by your local telco. They can connect you to anywhere, but they are going to charge you plenty for it.

Else. . . (All other long distance)

D. **If** the first three digits match the entry in Table 4:

This handles any long-distance calls that "fell through the net" and weren't serviced by your first choice of long-distance carriers.

1. **Go to** Route List 4, first choice, trunk group H, for idle trunk in group. If idle trunk available, send dialed digits on selected trunk.

   This is probably WATS. You would immediately choose a WATS line trunk group if the OCC of your choice does not provide service to the area you want to call.

2. **If** no idle trunk, check Route List 4, second choice, trunk group A for idle trunk and send digits, or issue a busy signal.

   This is a direct-distance-dialed call over the central office trunks of your local telco. They can be using any long-distance carrier they want to, but they'll charge you the same, regardless.

Finally, there's the possibility that what you dialed is impossible to find in a list of area codes and office codes, and is not found in Tables 1, 2, 3, or 4. You probably goofed in dialing, and will be connected to some error tone, voice message, or the operator, since you dialed something "illegal" or nonexistent.

12.1.4 Queueing the Class of Service

We described what a class of service is in the section on PBX features and functions. How does it apply to the ARS aspect of PBX operation?

Some of the more expensive routes that a call may take are denied to certain users. You cannot place some calls at all from a station with a restricted class of service. Some businesses and hotels have phones that are completely restricted from making outside calls, such as courtesy phones in the lobby. The courtesy phone, restricted to internal calls within the PBX system, is in the most restricted class of service. A less restricted class of service permits outside calls to local numbers only. Perhaps some phones may be restricted by the PBX to long-distance calls only on the most-economical trunk group (the "specified" common carrier). If the carrier is not available, a busy tone will be sent to the station instead of the call being re-routed to a more expensive trunk group. The most privileged class of service is that which allows unlimited access to any available carrier. Even the most privileged class of service may encounter a busy signal if all possible trunks are busy.

What are the options when you encounter this type of busy condition? If your class of service allows **queueing,** you could "take a number" and wait for the system to get back to you when a trunk is available. You might push a "camp-on" button on your keyphone, or key-in a special code on your terminal, or enter some special access code through your DTMF keypad. The system may allow you to go out a more expensive route if you have been blocked from a cheaper trunk group beyond a certain time limit. For instance, if you are dialing long distance on a station whose PBX would prefer you to use MCI, but that PBX finds the MCI trunk group is busy, the PBX might let you use WATS after 30 seconds if no MCI line becomes idle (call completed). You might hear a "fast-busy" for 30 seconds, but if you hang on, you may be rewarded by a ringing tone as your call "overflows" to the more expensive line. If you are on one of the "underprivileged" stations, you will never hear the fast-busy go away, and might as well give up anyhow if more then 30 seconds go by. The same program in the common control computer that searches tables for ARS must also decide if your station is entitled to use the route you are trying to use. At the point in the route-selection process where the program finds the first choice (a trunk group) in the route list, it checks back to see if your station is permitted to use that trunk, before connecting the trunk. It is at this point that you are blocked, permitted to access the trunk, or queue until a trunk becomes available.

12.2 MULTIPLE-SWITCH CONFIGURATIONS

A private telecom system may have more than one PBX. It may have any number of PBXs at different locations that are all linked together to form a private network. A large corporation like General Motors is a good example. A person at GM headquarters in Detroit calling someone at a local GM office in Los Angeles could just dial their extension number. The PBX could automatically route the call over a tie trunk that links the two PBXs. The call would never have entered the public network and there would not be a charge for the call. The cost of leasing a tie trunk from a common carrier (such as AT&T or MCI) would, of course, have to be considered. A private network that can carry data as well as voice communications is usually cost-beneficial to a large organization. PBXs used at each location may be of different manufacture, but all must be able to support network abilities. Another benefit of a private network is the ability to go "off-net" at some distant switch on the network. In other words, we use the private network as only part of the link, then connect into the public network to complete the path, thereby making a cheap call, locally, through the distant PBX.

The links that connect parts of a private network are most commonly tie trunks leased from the public telco. They may also be T1 spans, coaxial cable, fiber optics, or microwave. Whenever we refer to tie trunks, the medium carrying the signal could be anything. It will interface to the ports of the PBX through an E&M interface, and once connected, the medium carrying the trunk group to the PBX will be unimportant. When these multiple PBXs are connected together, what kinds of connections are possible, and how are they arranged?

We now look at three important ways to arrange the connections between switches in a private network: the **star, mesh** and **ring** arrangements. A fourth possibility, which we will also consider, combines the attributes of two or more of the others.

12.2.1 Star Arrangement Network

A **star arrangement** was shown in the beginning of Section 12.1.1. It is the natural way to connect one central point of control to a number of controlled devices in peripheral locations. In a network, these devices, instead of being stations within a single switch environment, would be complete PBX systems (these would, in turn, form little stars with stations at the points of the star). In Figure 12-3 we see a diagram of a system in which a centrally located PBX—or the PBX at the organization's headquarters—is the **hub** PBX. The subordinate PBXs—**satellites**—route all their "on-net" calls—traffic—to, or through, the hub. Why would an organization need a hub and satellites in a network? One answer might be "concentration." By having a hub PBX, we can concentrate all our trunks to the public network at one location. Say, for instance, that the four satellite PBXs in Figure 12-3 were independent. If, at times of heavy use, each one needed 50 trunks, there would be 50 trunks times 4, or 200 trunks in all, going to these independent PBXs. At any one PBX, the 50 trunks would seldom be in use at one time. What are the chances that *all four* PBXs would have every trunk busy? That would be unlikely. With all the PBXs connected to a common hub, a lot fewer trunks could be used, because 200 trunks would not be needed unless all four PBXs became busy at the same time. Now, 100 to 125 trunks concentrated at the hub could replace 200 trunks parceled out (50 each) to the four PBXs. In a smaller system, suppose that each of the satellites needs only *one* trunk, and

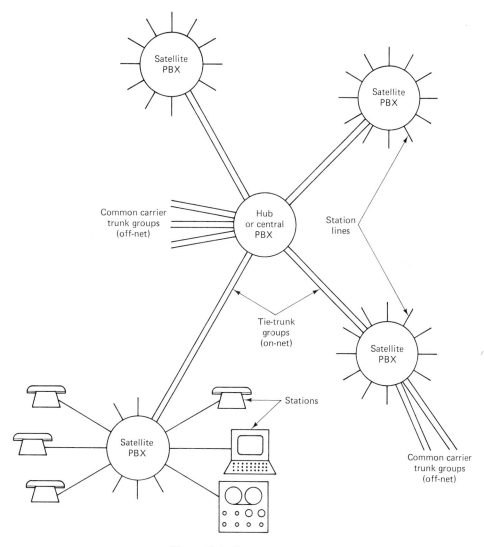

Figure 12-3 Star arrangement.

uses that one for long-distance calls only 10% of the time. The hub PBX of a **star** could get by with only one trunk, sharing it with four satellites, since the combined satellites need the trunk only 40% of the time. This saves money, because it is cheaper to use one trunk 40% of the time than it is to use four trunks 10% of the time. In fact, the closer you get to 100% usage, the cheaper each call becomes. Also, some of the more expensive optional equipment and features can be concentrated in a large hub PBX as well, and the smaller satellite PBXs would not have to duplicate this equipment.

Another reason for using the **star arrangement** might be call detail recording. If headquarters wants to keep statistics on telecom use at every one of its locations, routing the calls through the headquarters PBX, or at least passing information to it about every call made in the system would be a way to keep accurate statistics on all traffic throughout the network. Since most organizations would want this information available centrally, the headquarters office's PBX would serve as the clearinghouse for this information. Another reason might be that the headquarters gets the calls for the organization and distributes them to the satellites. This makes it possible to reach any branch of the organization, however far-flung, by dialing the same number. We mentioned before that the star network is similar to a computer with peripherals. Like the computer system, if the central processor goes down, the system ceases to function as a whole. Without the hub PBX, the star network falls apart as a network. The individual satellite PBXs may continue to function at their own locations, but communications with headquarters, and perhaps with the public network, are lost. As far as the satellite PBXs go, they are still alive; their little hearts continue to beat, but they no longer beat as one.

12.2.2 Ring Network

The **ring network** of Figure 12-4 is connected in a loop. Traffic comes into the ring at some source PBX and "gets off" at the destination PBX, after being passed around the loop from one PBX to another. The PBXs along the way have to pass information from one to another bucket-brigade fashion. Addressing information is passed along or repeated along the ring until the destination PBX is reached. For instance, PBX 1 is calling PBX 3 in Figure 12-4, when an extension in PBX 3's stations is dialed from one of PBX 1's stations. The extension number arrives at PBX 2, which identifies (through its routing tables) that the incoming trunk should be connected to an outgoing trunk, and sends (repeats) the extension number to PBX 3.

Since there is no hub in a ring network, there are no satellites, either. What do we call the PBXs? In this case, and in the star network as well, we can call any switch connected to other switches a **node.** Unlike the star network, loss of any node in the ring will not "down" the whole system, but it will slow down the passing of information considerably. A lot of the traffic will have to take the "long way" around, and on the average, half of the traffic will have to travel farther. Also unlike the star, there isn't any one node (like the hub) that is more important than the others.

12.2.3 Distributed Network

The **distributed network** of Figure 12-5 is more "democratic" than the star, faster and more efficient than the ring, and more complex—hence more expensive—than both. In this arrangement, all PBXs cooperate in the control of information transfer, and are cross-connected in ways that ensure direct connection between any source node and any destination node in the network. A completely interconnected network like this is called a **mesh,** although a distributed network does not always contain complete interconnection.

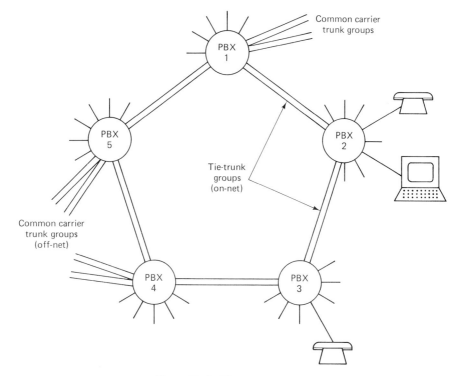

Figure 12-4   Ring arrangement.

The distributed network also has the advantages of a ring, in that if one of the direct paths is not available, any node can reach another by using a ring approach. In Figure 12–5, if tie trunks between PBX 1 and PBX 3 were unavailable (busy or down), traffic might reach PBX 3 via PBX 2. This is called **alternate routing,** and this ability requires each node to be equipped with sophisticated routing programs. Of course, alternate routing is done all the time by the common carriers, having originated with AT&T long ago as a necessary adjunct to long-lines operation.

12.2.4 Combination Network

The **combination network** of Figure 12–6 is a compound of ring, star, and mesh networks. Chicago-based Orbital Resorts, Incorporated, has just completed another addition to their orbiting vacation resorts. Located in geosynchronous earth orbit over the Indian Ocean, this third facility required the addition of a Canberra, Australia, ground station to the existing facilities of OR, Inc. The other two orbital resorts, over the Atlantic and Eastern Pacific Oceans, have been in place for several years and are accessible from the Quito, Ecuador, ground station.

PBXs 4, 6, 7, and 8, the Chicago-area facilities, form a **mesh,** while 1, 4, and 5 form a **ring.** PBXs 1 and 5 are the ground stations at Quito and Canberra, used to maintain communications with the three OR, Inc. resorts. PBXs 2 and 3 are the

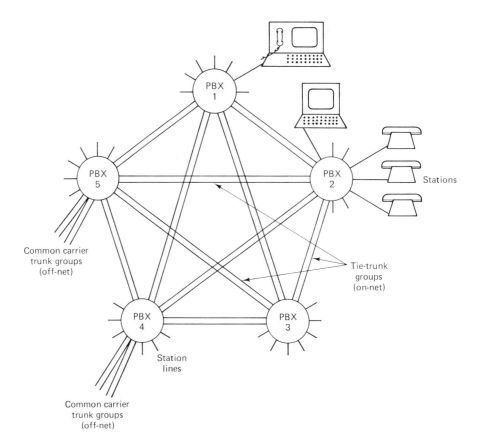

Figure 12-5 Distributed (mesh) arrangement.

Atlantic and Pacific orbital resorts, but to the telecommunications system, they are the satellites of a star network, with PBX 1 at the hub. PBX 9 is an isolated satellite, in more ways than one. It is, in the telecommunications sense, a satellite of a star with PBX 5 at its hub, and in fact, being the Indian Ocean Orbital Resort, it is really a satellite, with its only link to the system being the Canberra ground station.

Let's suppose that the maintenance man at PBX 1—the ground station—wants to call PBX 8 to report a problem with a new transmitter they are manufacturing. The most direct route would be through PBX 4 to PBX 8. There are nine alternate routes in the event that the direct route is blocked. Some alternate routes are through PBX 5 to 4 to 8—PBX 5 to 4 to 6 to 8—through PBX 5 to 4 to 7 to 8—PBX 4 to 6 to 8—and PBX 4 to 7 to 8. Can you find the others?

In some cases, there are no alternatives. PBX 2 or 3 may want to call PBX 1, or PBX 9 may want to call PBX 5. If the trunk group connecting them is busy, there's no other path for the call to take.

Most large networks end up being some kind of a combination network because of the amounts of traffic going between various points and the economics of using

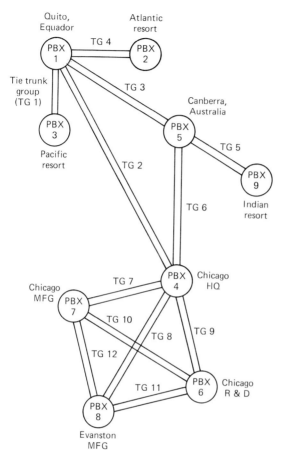

Figure 12-6   Combination network arrangement.

various links. Critical paths should have an alternate route, but the lower cost of a star network may be a deciding factor in selecting the hub and satellite arrangement.

12.3 ROUTING IN MULTISWITCH NETWORKS

In a multiswitch network, the characteristics of a single-switch network are all still in place, and additional features are added. Some of the operating characteristics are different as well. As already mentioned, the extension numbers dialed to reach other stations that are on-net in a multiswitch network include information about which switch is desired, as well as the extension number where the desired station is located. The first digit often specifies which switch is being selected, then the remaining digits which station in that switch's facility will be addressed.

One problem that arises in a multiple-switch configuration is how to access the public network through a distant switch. We may want to call our home office in New York from the Chicago branch, but not to talk to anybody in the New York office. Instead, we want to use the New York office switch as a "conduit" to reach another New York number. We want the New York office's switch to place a local call for us, and then connect that call to our telephone. How is this done?

An **FX trunk**—a foreign exchange trunk—is a link supplied by the public network. It can be used, among other things, to make a local telephone number available to customers in Chicago for a company in New York that hasn't got a Chicago office. This provides the appearance that the company has a business office in Chicago, without the company actually having to maintain one. The links that connect the switches of a private network could be used in the same way. We are turning around the problem we just discussed in the last paragraph, and instead of using the system to place calls off-net, we are using it to receive calls from off-net, and deliver the calls to one of the other switches in the net. The 800 number is not a free lunch. Someone has to pay for it, and it isn't the customer; the company supplying the number to its customers pays the telco. Now, if that company could use a private network, and let its customers dial in to the network as a local call, the expense to the customer wouldn't be much more than the 800 call (compared to a long-distance call), and the company wouldn't have to pay the telco a cent. The customer pays a small local call, which is a little more than the 800 call would cost—nothing—but using their own private network, the company can now provide a service like the 800 number which they might not have been able to include in their budget at all before. If there is a number (a local call) to call, where the options otherwise would have been a long-distance call or nothing at all, everyone comes out ahead (except maybe the telco).

12.3.1 Manual Selection in a Private Network

A network with **manual selection** is one in which you must dial an access code, wait for a dial tone indicating that you have reached the other switch, then dial further digits to access extensions or trunk groups in the other switch, waiting for a dial tone at each step. When you dial the extension, you are waiting for a ring, not a dial tone, but until the process terminates at a station, each access code leads to another dial tone.

For instance, you might dial 81 to reach the New York switch we've been talking about. When you get a dial tone indicating that you're in the New York PBX, you would dial 9 for an outside line. Another dial tone, this one from the New York telco, would be put on your line, and you would dial the New York number you wanted to call. The New York division of your company would now be billed for a local New York call instead of the Chicago division being billed for a long-distance call to New York.

The distinguishing feature we see when using the manual approach is that we must supply all the switching information at each step through the process, and wait for the switch to act on it. This is a little like the step-by-step switching we described in the public network, before we discovered what common control, and the repeating of dialed digits, could do.

12.3.2 Automatic Selection

The designers of the public network figured out a way to improve on step-by-step switching, and private switching equipment designers did, too. **Automatic selection** is a little more sophisticated than manual selection, and one aspect of this is **digit translation.** Suppose that the access code to the New York office in our example—81—will be followed by an extension number—454—that exists only in the New York office. In that case, if all extensions that begin with 4 are in New York, when we dial the first digit—4—the PBX will identify (by software) that the extension is in the New York switch, and the PBX will act as though you had dialed 81–454. Digit translation has made the 4 into 81–4 and the trunk group going to New York has been selected within the PBX.

On calls that have to go off-net, instead of the trunk selection being limited to trunks from common carriers, a switch in a private network may also select tie trunks between the switches of the network. We may be calling a city where our network goes, via tie trunks. These trunks would be the ARS routing program's first choice before any common carrier trunks.

In placing that off-net call to New York that we have been talking about, we might dial 9-(area code)-(office code)-(subscriber code). The PBX would identify a call in the New York area and say (figuratively), "Aha! I've got a tie line going there, so I will connect to the New York branch via the tie line, and then dial the number." The PBX connects to the New York switch, and then sends the 9 that we dialed, along with all the other digits, because our PBX has to tell the other PBX to seize an outside line. After the other PBX has accessed a trunk from the local telco, our PBX sends the (area code)-(office code)-(subscriber code) digits. (*Note:* The New York call may need an area code, although it's going to a phone in the same city, since some New York boroughs have a different area code than others.) In some cases, the numbers repeated to the second PBX will require an area code; in some cases, not. If the off-net call we want to place is in the same area code with the second PBX, our PBX should strip off the area code before the digits are sent. Deciding when to do this, and when to add digits—like the 81 access code we added to reach the New York PBX—are functions of the digit-translation and digit-manipulation software. Digit manipulations in the PBX might include adding digits, or transforming one code to another (digit translation), or removing digits, as in the case of a call placed to the 212 area code when the PBX is already in the 212 area code.

12.3.3 Tandem Switching

Tandem switching is just the ability to go through a switch in a network, when a call must go more than one step along its route from caller to destination. Tandeming is relaying the connection from switch to switch. The multiple switches between the caller and destination are used "in tandem" as one big switch. Tandeming is what was required in Figure 12–6 to route calls from one place to another in the combination network.

In the public network, some end offices (local offices) are connected to one another through a **tandem switch.** The tandem switch relays calls from subscribers of one end office to subscribers attached to the second office, yet has no subscribers itself. It is like an end office whose subscribers are other end offices.

In a private network, calls may be tandemed through a switch, but a private system wouldn't generally place a switch where there was no organization (or station loops) to be serviced. Since the private network is an affiliation of PBXs that serve organizations already, there won't be tandem switches—in the telco sense of the word—that have no stations attached. Instead, tandem switching is done through PBXs that are "end offices" but also have *other switches* as "subscribers."

The tandem switch must also do digit manipulation to tandem a call from one of its "subscribers" (a tie-trunk group) to another. This manipulation involves routing the call to the second trunk group by identifying the destination of the incoming call, and deciding on what outgoing group of trunks that call must be relayed. The digits for the call are translated into information that will access the outgoing trunk group.

12.4 DIGITAL SWITCHING IN PRIVATE NETWORKS

We have spent most of our time in describing networks as though voice were the only thing being switched. Data, too, may be switched through a private network. In this section we look at how this is done.

In a digital network, data is switched from terminal to terminal, host computer, or whatever, in binary form. In most cases, the PBX uses binary information inside itself, whether voice or data is being conveyed. When the TDM bus is used to switch through a matrix and complete a call, if it is a voice call, the voice has been digitized before being switched by the multiplexers in the PBX. Data or voice, as binary signals, take the same kind of paths, and are switched by the same logic gates, once inside the PBX. It is only after the multiplexing and switching is done that the binary information is separated again, rather artificially, into data and voice. The data is formatted into some digital standard, perhaps RS–232, and the voice reconverted to analog form, so that a voice signal goes to the telephone and RS–232 data goes to the terminal. There is no reason why information couldn't be kept in binary form—once outside the PBX—and switched externally through the links and switches of the private network, or, for that matter, through the public network, entirely in binary form. In that case, it wouldn't matter whether voice or data was being switched.

T1 spans are being used in some designs instead of analog tie trunks. The 24 channels in each T1 span are used as though they were 24 tie trunks in a tie-trunk group. The 24 voice channels digitized into binary format (T1–D3) can be further subdivided into many data channels for each voice channel if data is being transmitted rather than voice.

Fiber-optics links have been used to carry binary information for distances of more than 75 miles at data rates of 1 gigabaud (one billion bits per second), the equivalent of 14,000 simultaneous voice conversations on a single fiber without repeaters. In terms of data, that's equivalent to more than 3 million 300-baud terminals. (300 baud is the most common data rate used for sending modemized data over the voice network.) A small group of fibers like this could supply data enough to keep every teletypewriter in the world occupied, full time.

12.4.1 Packet Switching

In a digital network, where data is not converted by a modem to analog and transmitted on analog lines, it may be transmitted in the form of **packets.** A packet is about 2 kilobits of binary information, which includes data, address, and error control information. The packet transmission concept is not used for voice, but for data only, because the transmission is not carried out continuously as a voice signal would have to be. What is this packet transmission like?

Imagine that you have won a lifetime supply of free, postage-included envelopes from the Ace Envelope Company, and you'll never need to send a letter on your own again, but can only send stuff that fits into the envelopes. You've just written the great American novel—or maybe, the great American telecommunications book—and you want to send the manuscript to your publisher. You hate the thought of sending the 1000 pages you've typed parcel post in one big package, when you have all these free envelopes, so. . . .

What you do is put a few pages here, a few pages there, into a whole lot of envelopes, stuffing each one with as much as it can hold, and address each one to your publisher, putting them into your local mailbox until it's stuffed full, then starting on the one down the block, and so on. Eventually, the envelopes all arrive at the publisher, one by one. They may not all arrive in the order you sent them, or at the same time, because they've gone through different post offices and been carried over different routes.

That's what **packet transmission** is like; the whole data block you want to transmit is parceled into **packets,** each holding a certain amount, and each separately able to convey its contents to the same destination. If you were clever with the envelopes, you might mark them "Chapter 1," "Chapter 2," and so on, so that the publisher would know what he's got, what hasn't arrived yet, and in what order to put it when it's all together. (P.S. I don't know of any publisher who'd like it if you did what I just suggested, so when you write your next book, use parcel post!) That's what happens with packets. Each unit of data carries an address, to tell where it's going, and perhaps some identification as to the sequence it belongs in when it arrives.

How are packets sent? There are two methods. Packets may be sent by the **datagram** technique or the **virtual circuit** technique. In both, a group of packets is sent out in some order from a computer-based switching center. Each packet is routed over a path chosen according to a scheme called **adaptive routing**—in which the path is chosen that gives the best performance. A message consisting of several packets may arrive in a different order than it was sent, because each packet may have traveled along a different route. Some packets may have been dropped to ease system congestion and will need to be transmitted again by the transmitting terminal. In the **datagram** technique, the terminal receiving the packets has the responsibility of putting the packets back into order and inquiring about the whereabouts of "lost" packets. That would be like your publisher telling you that he never got Chapter 6, and could you send another copy of those pages? (This happens to envelopes put into mailboxes sometimes—but we hope it's not because somebody at the post offices decides he's got too many letters and throws some out—and it doesn't mean you didn't mail Chapter 6. It just didn't get there.)

In the **virtual circuit** approach, the packet-switching network takes responsibility for correctly reordering the packets, and delivers the packets in the order in which they were sent. Since the message's packets all arrive as though they had been sent down a single circuit and arrived in the order they were sent, there appears to be a single circuit—a "virtual" circuit. This would be like leaving instructions for the postmaster in your publisher's city, asking him to save up all the envelopes and stack them up in order before giving them to the publisher.

We need a standard interface protocol to exchange packets between the network and its terminals, computers, and host processor. In the datagram technique, the receiving station equipment is responsible for reassembling packets into sequence. A comprehensive end-to-end protocol will be required in this case, operating in conjunction with the transmitting station. The network, in the datagram scheme, relies on the terminal to reassemble the packets, and recover missing packets, based on its protocol.

12.4.2 Message Switching

Message switching is different from packet switching, in that the whole transmission goes out as a single block, but may not be transmitted right away, as packets are. This is like sending our manuscript parcel post in a big box. In message switching, we may have to wait for service, since the message can't be sent a little bit at a time on various routes, like packet switching. There is no virtual circuit concept in message switching, since an actual circuit is used to transmit the whole message, and we don't have parts of it taking one route and other parts taking another. This means that the circuit must be connected for the duration of the transmission, and disconnected when the message is completed, which is similar to an ordinary telephone call. Messages still do have to carry address information and error-detecting information, but only once, for the entire message. Another attribute of message switching is **store-and-forward.** What you want to transmit is stored in a memory device, and transmitted (forwarded) whenever the system gets the time. Higher-priority messages are sent before low-priority messages, and if your message is of low-priority, it might not even be sent until the middle of the night. In a store-and-forward message system, a backup copy of the message is kept in memory after the transmission for a while, perhaps several days or weeks. The message is stored on some mass-storage device, perhaps a floppy disk or hard disk, magnetic tape, or even punched-paper tape. Eventually, the disk or tape is "purged" and reused, but there is a backup of your message for retransmission, if needed, for a reasonable time.

While packet switching is interactive, involving the exchange of more data than just one-way transmission, message transmission is one way at a time. (Computer people refer to this type of operation as **batch processing.**)

12.4.3 Local-Area Networks

A **local-area network (LAN)** links data processors with a type of private network. It is possible that our LAN uses a computer-based switching center to route the data from one peripheral to another within a single business office or organization, in an arrangement similar to the single-switch arrangement we discussed earlier. More often, the LAN carries data around the organization in a "bus" arrangement, one of the forms we didn't discuss with voice networks, because it's almost never used for voice. In either case, the links between processors and terminals carry data in binary digital form, but are otherwise like links between stations and the PBX in the business office. Although nodes in this LAN network typically aren't PBXs, they could be. PBXs contain common control computers with all the capabilities of the

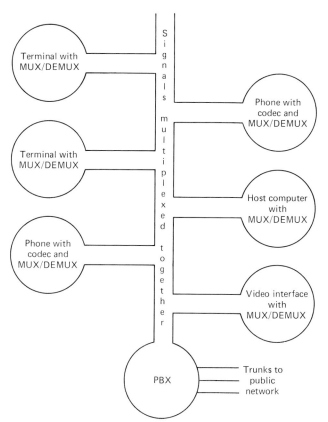

Figure 12-7 Bus-oriented local area network.

packet-switching (computer-based) switching centers. If used to perform switching tasks for the LAN, as well, PBXs could also give the LAN access to a larger private telecommunications network if the PBX is part of one. Since modern PBXs combine voice and data switching capability, the nodes of an LAN probably should be the PBXs.

The most common type of architecture used in LANs is the **bus-oriented** arrangement. In Figure 12-7 we see a bus with several kinds of digital stations attached. Unlike the bus in a computer, the bus used in this LAN is one shielded line rather than a group of parallel wires. Because of this, data must be transmitted serially, a bit at a time, rather than in bytes or words. Star, ring, and mesh networks are possible in LANs as in voice switching networks, but usually, all the peripherals and processors are tied to a **common bus,** and all share the same path for data on a TDM (time-division-multiplexed) basis. The information is sent along the bus in packets. Typically, coaxial cable is used in this kind of network rather than the twisted-pair used in telephony or T1 connections. One advantage to this is that you can tap into a coaxial cable with a T connector anywhere along the line with little trouble, whereas the twisted-pair arrangement requires another pair for each additional station. Dozens, hundreds, or even thousands of digital stations could share the same bus in an LAN.

There's no reason why a PBX couldn't be equipped with some kind of coaxial, serial data interface, and attached to a LAN, or even several LANs. The PBX could then be used to tie various LANs together, creating a network of LANs.

REVIEW QUESTIONS

1. Define routing.
2. What kind of trunk groups may be used in a single-switch private network? In a multiple-switch private network?
3. What factors does a switching system use to determine which trunk group is selected to place an outside call on?
4. Why is it sometimes desirable to wait in a queue for an available trunk group?
5. Does an ARS system always place calls on the least costly trunk group if programmed to do so? What are the advantages and disadvantages of this strategy?
6. How many PBXs may be linked to form a private network?
7. What is meant by on-net and off-net when referring to a private network?
8. Describe the network arrangements known as ring, star, mesh, and bus. Is a network that is a combination of these possible?
9. How is addressing information given from PBX to PBX in a ring network?
10. What is a node and what is a hub in a private network? What is a satellite?
11. Explain the difference between a distributed network and a mesh network.
12. Are access codes always necessary to call a station at a distant PBX in a private network?

13. What is digit translation? How is it used to simplify dialing in a private network?
14. What is a tandem switch used for?
15. Waht information is contained in a packet used in packet switching transmission?
16. What are the two methods for sending packets?
17. What is the store-and-forward form of message switching?
18. What is the difference between the datagram and virtual circuit systems in packet switching?
19. Which is more efficient in getting digital information from the source to the destination *quickly, message* or packet switching? Which is likely to involve the longest delays?
20. Are LANs (local-area networks) used for voice or data networking?
21. In local-area networks, is the arrangement of stations and switching equipment usually a ring network, a bus network, or a mesh network?
22. What advantage, if any, would accrue to using a data-switching PBX to form a digital local-area network?

13

Emplacement of a Private Telecommunications System

In this chapter we discuss the basic concepts used in planning and installing a telecommunications system. A telecommunications system installed in any organization could be similar to one on an Air Force base, a university, a hotel, or a corporation. Principles embodied in the layout of one system would be used in the design of any of the others. The needs of each customer will vary, but the needs of total system planning won't change much. If the system will replace an existing system, the job will be easier because you already know the structure of the existing system and will be able to repopulate the system with new equipment that "lives" where the old equipment had been before. For example, if Mr. Jones had a telephone and a data terminal, most likely in the new system, he will have a telephone and a data terminal again, although (we hope!) with much more powerful capabilities.

13.1 DESIGN COORDINATION

A business complex that has more than one building on the same grounds will have to be connected by underground or overhead outdoor cables. If the buildings are connected by walkways or utility tunnels, inside cable can be installed through them. An architect will usually plan for telephone cable installation when a new building is designed. In one case that we remember, the telecom person didn't touch bases with the architect, and each planned a separate trench with conduit for telephone cables. When the trench planned by the telecommunications person was dug, it crossed the existing trench placed there by the architect. This broke the conduit, so it couldn't even be used later for expansion. This type of duplication of efforts often arises from lack of proper communication. In cases like this, it's just as likely that there will be nobody planning the cable connection, if each one thinks it's the "other guy's job" and they never talk to each other.

When you begin to plan a telecommunications system for installation, it's a bit like designing a building. There are a lot of interdependent parts which perform different functions, and you want to know which, and how much of each, you need.

If growth is anticipated, system expansion with the least amount of trouble and cost will be a factor. Can the switching equipment be expanded, or must it be replaced and a larger switch put in? When the cabling is put in, are there enough extra conductors in the feeder cables to handle additional stations so that new cable won't have to be pulled in?

As additional equipment is put in, it will require space and power. If we anticipate future needs to a degree, we can provide room to expand and additional electrical power to grow on. We can put equipment in a room that we don't fill with equipment, leaving space for adding future equipment. When the electricians bring in power lines, we could have them install 100-Ampere servce even though we might only be using 50 Amperes right now.

Since telecom systems offer such a wide variety of features and auxiliary equipment, your planning should include investigating what different vendors can offer. You should look for the features you want at a reasonable cost, good servicing and maintenance, flexibility, and utilization of state-of-the-art technology.

It's a good idea to stick with one vendor for the total installation so that he or she will have the responsibility for the equipment and the installation. If you have multiple vendors, you are in the middle when you have a problem. Vendor A blames vendor B for all the problems, and vice versa, and you are left holding the bag.

13.1.1 Steps in System Emplacement

There are three main steps in emplacing a private telecommunications system:

1. Planning
2. Preparation
3. Installation

Planning

Planning the system—you have to start somewhere—is the first step in emplacement. You have to plan the installation by deciding what is required for each of four components in the final system:

1. The switching system
2. The station equipment
3. The links
4. Peripheral and optional equipment

Each of these components has four requirements:

1. Type, quantity, or size
2. Interfacing requirements
3. Power and environment
4. Location

Preparation

The next step is preparation. What preparing do we have to do? We ask ourselves what work has to be done before the components of a system can be installed.

For each of the four system components, the site has to be prepared for installation of that type of equipment. These preparations require:

1. Space for the components
2. Access routes for links
3. Proper power and environment
4. Interfacing for nonstandard or previously installed equipment

Installation

When the planning is complete and the preparation is in progress, the installation of all these components can begin.

Although delivery and siting of all the components must take place before the installation can be completed, the majority of the work lies in connecting things together. Pulling the cables, installing sockets and jacks, connecting the power, and checking that everything is working as it is put in constitutes the bulk of the work. Finally, the system is ready for "cutover"—the moment when the whole system goes on-line—and the users can begin using the system.

13.2 PLANNING

The first step in planning is the selection of components. One component that must be selected is the switching system. How do you select a system? The first question you ask yourself is "What do I want it to do?", and the first thing you must decide is whether to do voice/data or just voice communications. Then you must ask yourself what size the system must be to handle existing needs and how much expansion capability it must have to handle anticipated growth. This naturally brings up the question of what station equipment you need—how many phones, terminals (if any), and other peripheral equipment will you need. You then need to decide what features and functions the users will need, and what system features the manager will require to administer the system responsibly.

From whom will you obtain the system? You will need to talk to suppliers of switching and station equipment. Most electronic station equipment is not compatible with the public network, and in fact, works with only one supplier's PBX. Unless all you want is standard "2500" sets (translation; ordinary telephones), you will find it necessary to obtain the station equipment from the same supplier as the switching system. Some vendors supply data terminals that will interface only with their switch, but in general, terminals that use RS–232 and similar standard interfaces can be connected to any data switch. The general rule that you should buy station equipment from the same supplier who provides the PBX is not quite as important in the case of data terminals. However, the PBX suppliers will add "bells and whistles" to their terminals, so that they work directly with their PBX, without code conversion. These features will enhance the user-friendliness of the terminal.

A good place to find a vendor is in the Yellow Pages, under "telephone." Really. A quick browse through the telecommunications trade magazines in the technology department of your public library would give you a good picture of who the major manufacturers are, through their advertising. Reputable vendors will supply you with a list of their customers on request, and you can talk to the customers about how they like the product.

Station and system features, and optional equipment

The next stage in planning an installation, after beginning to explore the selection of a vendor, is to decide what features the stations and system should have, and what optional equipment you will add to the system. The details of station features are listed in Chapter 6, and you will want to familiarize yourself with what's in that chapter to make an informed decision on the features your stations should have to satisfy your needs. System features are described in Section 6.3. You will want to review the features available on PBXs, to decide which will fulfill your organization's requirements. In addition to those we listed, new features are entering the market every day. In the process of researching possible vendors, you will probably find several new—and hitherto unmentioned—features that the vendors will be promoting, which did not exist at the time we wrote about PBX features and functions.

Now, you've been researching what's available, and have decided what you do need and what you don't need. What you need, now, is a systematic way to organize the information that you've gathered. This information is best organized in the form of charts and tables. When you complete them, these charts and tables document what you've decided to include in the system. Documentation is, first, for yourself, to make sure that you have everything down in black and white and won't forget anything. Once completed, this documentation helps you in organizing a Request for Proposal for prospective vendors. This organized information can then be passed along to potential vendors, to see if they can meet your needs. These charts and tables can be used by prospective vendors to estimate the cost of the installation, and to submit proposals in response to your request.

13.2.1 Typical System Planning Chart

Table 13–1 shows a chart of the type we just described. It contains some specifics about the components of a "minimum" private telecommunications system. Of

TABLE 13-1 TYPICAL SYSTEM PLANNING CHART

The switch (PBX)

Type/Attributes of Switch	(Yes/No)
Analog matrix	_____
Digital matrix	_____
Voice switching	_____
Data switching	_____
Redundant common control	_____
Redundant critical components	_____
Battery backup (memory)	_____
Battery backup (system)	_____
Operating system (ROM based)	_____
Operating system (disk based)	_____
Nonblocking system	_____
Other (specify) _____	
_____	

Size of switch	Number equipped	Wired for expansion
Ports		
Single-line stations	_____	_____
Multi-line stations	_____	_____
Data stations	_____	_____
Off-premises extensions (OPX)	_____	_____
Trunks	_____	_____
Tie trunks	_____	_____
Digital trunks	_____	_____
Operator console(s)	_____	_____
Paging port(s)	_____	_____
Music-on-hold	_____	_____
System administrator port	_____	_____
Maintenance port	_____	_____
Other (specify) _____	_____	_____
_____	_____	_____
_____	_____	_____

Miscellaneous		
	_____	_____
DTMF registers	_____	_____
Rotary registers	_____	_____
DTMF senders	_____	_____
Rotary senders	_____	_____
Tone generators	_____	_____
Other (specify) _____	_____	_____
_____	_____	_____
_____	_____	_____

Simultaneous conversations	Number _____
Calls attempted per hour	Number _____
Call traffic density	CCS/hour _____

System Features	Strong Need	5	4	3	2	1	No Need
Automatic route selection		—	—	—	—	—	
Call detail recording		—	—	—	—	—	
Electronic mail		—	—	—	—	—	
Voice message recording		—	—	—	—	—	
Automatic call distribution		—	—	—	—	—	
Direct inward dialing		—	—	—	—	—	
User-group partitioning		—	—	—	—	—	
User-programmable MAC		—	—	—	—	—	
Remote access for above		—	—	—	—	—	
Other (specify) _____		—	—	—	—	—	
_____		—	—	—	—	—	
_____		—	—	—	—	—	

Power and Environment Requirements

Power type	Ac _____	Battery _____
Amount required	Volts _____	Amps _____
Heat dissipation	Btu _____	
Air conditioning	How much _____	(Btu)
Floor space	Square feet _____	
Weight	Pounds _____	

Station Equipment

Type	How Many
Standard 2500 set single-line phones	_____
Multiline phones	_____
Featurephones	_____
Display phones	_____
Speaker phones (hands-free)	_____
Attendant consoles	_____
Data terminals	_____
Video display terminals	_____
Printing terminals	_____
Voice/data terminals	_____
Answering machines	

Station requirements

Cabling	
Single-line cable	_____
Multi-line cable	_____
Data cable	_____
Station power supplies (if needed)	_____

Station Features	Strong Need	5	4	3	2	1	No Need
Call transfer		—	—	—	—	—	
Call pickup		—	—	—	—	—	
Call forwarding		—	—	—	—	—	
Conferencing		—	—	—	—	—	
Automatic dialing		—	—	—	—	—	

TABLE 13-1 TYPICAL SYSTEM PLANNING CHART (Continued)

Station Features	Strong Need	5	4	3	2	1	No Need
Abbreviated dialing		—	—	—	—	—	
Class of service		—	—	—	—	—	
Flexible numbering plans		—	—	—	—	—	
Forwarding on ring-no-answer		—	—	—	—	—	
Forwarding on busy		—	—	—	—	—	
Call restriction		—	—	—	—	—	
Ring-down phones		—	—	—	—	—	
Camp-on callback		—	—	—	—	—	
Trunk queueing		—	—	—	—	—	
Message waiting		—	—	—	—	—	
Other (specify) _____		—	—	—	—	—	
_____		—	—	—	—	—	
_____		—	—	—	—	—	

Links to the Outside	Type (2-wire, 4-wire, etc.)	How Many
CO trunks	_____	_____
MABS trunks	_____	_____
In-WATS trunks	_____	_____
Out-WATS trunks	_____	_____
MCI trunks	_____	_____
SCC trunks	_____	_____
OCC trunks	_____	_____
Tie trunks	_____	_____
FX trunks	_____	_____
DID trunks	_____	_____
OPX lines	_____	_____
Data lines	_____	_____
T1 spans	_____	_____
Other (specify) _____	_____	_____
_____	_____	_____

Special interfacing requirements for any of the above:

Special power requirements for any of the above:

Special signal conditions for any of the above:

Special terminations, plugs, or connectors for:

Peripheral and Optional Equipment	(Yes/No)
Dictation tanks	_____
Modem pools	_____
Teleconferencing units	_____
Call cost accounting device	_____
Paging system	_____
Security system	_____

Peripheral and Optional Equipment	(Yes/No)
Environment monitoring and control	_____
Automatic recorded message dialer	_____
Special space and/or power requirements for above:	

course, this is just an example, and you will need to modify and expand it for any real system. The chart is broken down into the four component parts of a private telecomm. system—the switching system, station equipment, links, and peripheral and optional equipment—and in each of these areas, the requirements of that component type are listed. The first part of this chart—the switch (PBX)—outlines what types of PBXs are possible, and we have checked some of these options. The same approach is used for all the other features of the chart, as you can see in the example.

13.2.2 System Size

The size of the system is determined by the number of people who need stations. This is called the **line size** of the system. The amount of traffic the system will carry to the outside world is a factor in determining the number of outside trunks to which the system should be connected.

Support S/L (single-line), electronic sets, data sets—how many of each?

You have to find out what you are going to need. In an existing system, everybody already has telephones, so you know that this is one type of equipment you need, and you know pretty much how many you need from what's in existence. This would be determined by examining the existing system, and on a survey basis, by finding out if existing telephone service is adequate, excessive, or insufficient in each division of the business. An electronic phone can generally replace a keyphone, even though it may not have as many lines. An ASCII keyboard may soon replace the dial on your home telephone, giving it full terminal capabilities. In a private data system, you don't use a dial to access a computer; you use an ASCII terminal. It's just a matter of time before the ASCII keyboard merges with your home telephone keypad, as well.

For data terminals, data sets, and data communications interfaces, the system requirements again may be determined from a survey of the existing system. Added features and enhanced data rates may be added to the new system.

Plant expansion or restructuring by upper management may be important considerations in deciding the size of the new system. For systems that are not replacing existing equipment, more of the decision making will rest in the hands of corporate planning and management. The same survey approaches will be used with upper management and planners, but will be done at a more general level, and more of the details of the design will be filled in by the telecommunications expert. For example, you might have the problem of deciding who's going to need a terminal. If the office is the newsroom of a major metropolitan daily paper, every reporter is going to need a terminal. Typically, during the day, they're going to be using their terminals for word processing, saving and calling up former articles, and getting information from data retrieval libraries such as Dow Jones Information network (financial), ERIC (educational research), DIALOG scientific journal publications, and periodicals.

The production supervisor in a plant might need a terminal for rapid access to a central database with schedules and quantities of orders, what parts are needed to fill each order, and so on. The foreman on the production floor and the workers at the machines would *not* need terminals—in fact, they would be a distraction, as would telephones. The foreman would need a telephone but not, necessarily, a terminal.

13.2.3 Optional Equipment

Services that are optional require extra space and power in the overall plan. The site at which this optional equipment is located varies. In an earlier chapter we introduced peripheral technology, which included equipment for these optional services. Here are six optional services you may find in any installation:

1. Call detail recorder
2. Recorded announcements
3. Paging/music
4. Dictation tanks
5. Interface port to system/terminal
6. Electronic mail or voice message

In dealing with any of these pieces of equipment, you will have to ask yourself the following questions:

1. Will it be integral or external equipment?
2. Where should this equipment be installed?
3. Where will we get power for it?
4. Is special interfacing needed?
5. What coordination is needed with suppliers of communication equipment?
6. Are special environmental conditions needed for this equipment?

Integral or external?

Integral equipment is built into the main cabinet in the equipment room. This is the same cabinet where the rest of the switching equipment is found. It would be ordered along with the order components of the switching system, from the same vendor. **External equipment** is found outside the main cabinet, perhaps outside the equipment room itself. It may be supplied by vendors other than those who supplied the switching equipment.

13.2.4 Location

Will it take up space in the equipment room, and how much? If not, where is it? When you're making up floor plans for the equipment room, you have to decide how much space, and where, to allocate to special peripherals. All equipment should be easily accessible for servicing, but some must be more accessible than others. An amplifier that must have frequent adjustments of volume levels, or a cassette unit where the cassettes must be frequently changed, are going to require special accessibility. For instance, if a recorder has cartridges that must be inserted and removed from a drive frequently, the drive must face a corridor big enough for someone to walk in. Most of this equipment is rack-mounted (see Figure 13-1). These racks are frequently 19 inches wide (48 centimeters) and 6 to 7 feet tall (2 meters). It would be unwise to place the cartridge drive just mentioned in the top slot of a 7-foot rack.

Dictation tanks may be located on a table in the area where word processing is done. Stenographer/typists would walk over to the tank, pick up a cassette, and return with it to their desks to transcribe whatever it contains. An ideal location for the dictation tank should be centrally located within easy reach of all typists.

13.2.5 Power

There are several questions to consider about providing adequate power for special equipment.

> When added, will this equipment's electricity requirement force us to change fuse/breaker ratings or bring in additional circuits?
> If battery backup is used, does the inverter that powers AC equipment have the capacity to handle this optional equipment?

A lot of telecommunications equipment can be supplied with DC power directly from the batteries and does not require an inverter.

> Will isolation transformers and AC line regulators/noise suppressors be needed to "condition" the power to this equipment?

13.2.6 Interfacing

When we are trying to interface optional equipment to the switching system, we should consider the following:

> What sort of port do we have to attach the equipment to?
> Do "exotic" signal levels require level-shifting circuits to make the optional equipment compatible with the telecom system?
> Does the optional equipment have unusual power requirements?

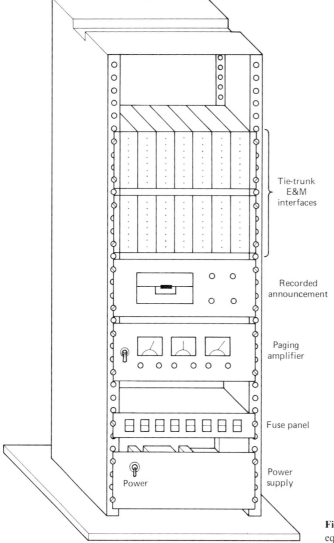

Tie-trunk
E&M
interfaces

Recorded
announcement

Paging
amplifier

Fuse panel

Power
supply

Power

Figure 13-1 Rack-mounted optional
equipment.

13.2.7 Main Equipment Room and Location

The area where the switching equipment is located must be considered with space
and environment in mind. Computerized systems require air conditioning. Some additional, specialized equipment may be located outside the main equipment room, so
smaller remote equipment rooms may be needed in some areas. These smaller rooms
will also be used for interconnecting the multiconductor cables to individual station
lines.

Switch room layout

To lay out the floor plan of the switch room, you should choose an existing room, or partition off an area in an existing open space, large enough to accommodate the equipment, and allow access corridors between equipment so that it may be easily serviced. One method is to lay out a floor plan on graph paper, or obtain a blueprint of the area where the switch room will be located. Scale-model paper cutouts of the chassis outlines of switch equipment can be moved around on the graph paper or blueprint to evaluate various arrangements of equipment in the room. In a big installation, there should be room for a desk or work area, and a place for storage of spare parts or extra parts (a storage cabinet). A place to keep system records—like a file cabinet—should be provided, although a file drawer in the desk is sometimes sufficient. Figure 13–2 shows an example of a typical layout for a switch room. Chassis and racks that would be found in this room include the PBX cabinet and the MDF (main distribution frame). The MDF is the place where the "demarc" connections to the public network are made, and also where connections are made between connecting blocks attached to the station equipment and the connecting block that goes to the PBX. Typically, the main distribution frame is made up of connecting blocks mounted against the wall, attached to a plywood board. Also associated with the switch room, but not inside it, is a battery room, for DC systems with a battery backup. The batteries are generally in an adjoining room, which must be well ventilated to prevent the buildup of noxious and/or explosive gases.

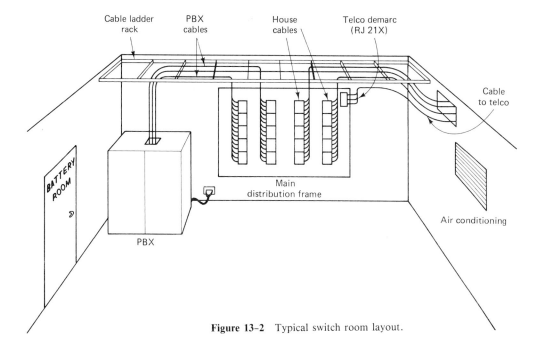

Figure 13–2   Typical switch room layout.

The floor plan has to include electrical outlets, lighting fixtures, cable raceways, and the electrical distribution panel. Another thing that has to be considered is the ceiling and floor material. Raised flooring is sometimes used, as in computer rooms, so that cables connecting equipment do not have to be run overhead on "ladder racks." Another advantage of raised flooring is that elevating the equipment chassis and racks provides some protection against water damage if the switch room is in a basement or other area that may be subjected to flooding. In the event that a raised floor is not used, cables will be run across the ceiling of the room using ladder racks. Even in rooms with lowered ceilings (the sort used to conceal overhead pipes and wiring for lighting fixtures), the cables are usually exposed for easy service access.

Although some municipalities require sprinkler systems for fire protection, these are not desirable in a switch room (or a computer room). The water would ruin the switch equipment. A Halon-gas fire extinguisher system would be the best choice for fire prevention. For this system—which floods the room with Halon gas rather than water—to be effective, the room must be relatively gastight. When the room is filled with Halon gas, which does not support combustion, the air, which is lighter than the Halon gas, is displaced toward the ceiling, and the fire, which relies on a return draft of oxygen-bearing air from underneath, is snuffed out. A Halon-gas fire-suppression system is not a simple thing like a fire extinguisher canister on the wall; it requires ducts to bring the gas from its canister to all parts of the room, is quite as extensive as the electrical outlet wiring in the room, and must be included on the floor plan just as the electrical outlets are.

13.2.8 Proper Environment

We mentioned that the switch room must be air conditioned to handle the electrical heat produced by the switching equipment. The manufacturer or vendor can provide the information needed to determine what amount of air conditioning must be added to handle each piece of switch equipment. Watts of electrical power consumed by the cards in the switching cabinets are converted directly into heat energy once they've done their job. Those watts are more generally expressed by air-conditioning engineers as Btu or tons of cooling. A watt of electrical power would produce 3.415 British thermal units (Btu) of heat energy every hour. The amount of air conditioning in Btu per hour (carried away by the air conditioner), is therefore about triple the number of watts dissipated by the electrical equipment.

We also mentioned that the lighting would be a part of the floor plan. It is also a part of the source of heat that is generated in the switch room and must be taken care of by the air conditioning. We say that the lights' wattage is part of the air conditioner's "thermal load." A 40-watt fluorescent lamp, with associated ballast coils and starters, might dissipate more than 200 Btu of heat per hour. You could make a good estimate by adding up the wattages marked on the lamps, then tossing in an extra 50% for ballast coils, and multiplying the result by 3.4 to estimate the total Btu load developed by the lighting fixtures.

Air conditioning is required to cool the switch room and keep its temperature within safe operating limits. In addition to air conditioning for cooling, the battery room needs special ventilation, because storage batteries produce hydrogen gas. To prevent the buildup of explosive amounts of hydrogen, the air in the battery room must be "turned over"—the air must be completely replaced—every two to three hours. An air-conditioning system that simply recirculates air after cooling it will not be enough for this situation. The battery room is, therefore, a separate room adjoining the switch room, with its own ventilation and air-conditioning requirements. Like air conditioning, the ventilation requirements of the switch room will be determined by the vendor, who will supply charts and tables that specify what amount of ventilation or air conditioning is needed for a specific installation.

13.2.9 Power Requirements

To assist the vendor in planning for the system, you will need to make a list of electrical devices that will be located in the switch room, and how much power each one requires. These data will be used by the contractor who installs the electrical power lines. We are going to make sure that the contractor, who may not know anything about switching equipment, knows exactly where every outlet and appliance has to go. The vendor who supplies the system will help with this. They will have installation charts and planning guides and can assist you in organizing your information for the electrical contractor. In addition to switching equipment that the vendor supplies, power will be needed for lighting, interface equipment supplied by the telco (such as E&M interfaces), convenience outlets for test equipment, and a service teleprinter. You may need a radio tuner for music-on-hold, and a paging amplifier. Some devices might require "exotic" power, such as 220-volt AC, although most equipment runs on 110-volt AC. If this system is DC powered, 48-volt battery power will not handle everything, and an inverter to transform the 48-volt DC into 110-volt AC will be necessary for emergency lights and other 110-volt equipment. How much power do we need from inverters? How long should the battery backup system "hold up" the switch? How much of the switching system's functions should we support in a blackout situation? We can't tell you the answers—these are decisions on which you will have to use your judgment. Battery backup systems keep the switch going anywhere from a half-hour to eight hours in most systems. In many instances, generators supply power for the institution after a short while, and long-term battery backup is not needed.

13.2.10 Premise Cabling

There will have to be access to the switch room, to get all the premise wiring to the main distribution frame (MDF). All the cable that runs through the building, and all the trunks from the common carriers, will terminate on the main distribution frame. (That's what we mean by "premise wiring.") In an installation where the room is being designed specifically as a switch room, access for these cables can specifically be built into the room. Otherwise, holes will have to be drilled through floor, walls, or ceilings to allow access for the cables. This would, of course, have to be included on the floor plan. We know of a specific installation where the air-conditioning ducts were put through the openings the telecommunications people thought were for cabling—and additional holes had to be drilled through the walls to admit the cabling to the MDF.

13.2.11 Local Equipment Rooms

Another name for the local equipment room is an equipment/wire closet. This is the place where large feeder cables are terminated on a local distribution frame (LDF). On a big installation (in a large organization), large feeder cables will connect the main switch room to these local equipment rooms, and from there, station wires will fan out to the individual station jacks in the rooms and offices nearby.

Other equipment that might be found in a local equipment room could include power for low-voltage station equipment such as electronic sets and keyphones. A low-voltage supply provides power to the lamps on the keyphone buttons, and to the electronics in the electronic-phone circuit boards. Rather than distributing this power from the main switch room, it would be provided from transformers plugged in at the local equipment rooms. Power from these transformers is supplied on a pair of wires contained in the station line, which usually contains two or three pairs. Since these equipment rooms or cabinets are mainly used for termination between feeder and station cables, and for transformers that don't require any special environment, no special ventilation or air-conditioning requirements have to be met for these rooms.

13.3 SUMMARY OF PREPARATION AND INSTALLATION

Once the planning has been completed for the system emplacement, the system vendor and the other contractors involved in preparation of the system will be engaged in carrying out your plans. The preparation and installation steps we mentioned earlier will be their responsibility.

The preparation of the system would involve carpenters, electricians, and other tradespeople who provide the physical framework on which the system is installed. Their work enables the private telecommunications system to be fitted into its environment.

Installation is primarily the responsibility of the vendor. It is the vendor who supplies the tradespeople that pull the cables, install the jacks, and connect the station equipment. They terminate the cables onto the main distribution frame, the local distribution frames, and connect PBX and peripherals to the MDF. An installation engineer is responsible for putting the system on line, loading the operating software, and checking the system for faults.

REVIEW QUESTIONS

1. Why is it necessary to anticipate growth when planning a system installation?
2. What are the three main steps in emplacing a system, and who is responsible for each step?
3. What kind of preparation is needed before a system can be installed?
4. What criteria would you use in selecting the components of a private telecommunications system?
5. What requirements are necessary for the emplacement of optional equipment?
6. If you were planning to put in an electronic mail or voice message system, what kind of requirements would you need to incorporate into your plans?
7. What is the difference between integral and external equipment?
8. What special requirements would a DC battery-backup system have?
9. What information must be included on the switch-room floor plans?
10. What precautions against fire should be made in the selection of switch-room equipment? What provisions are made for other natural disasters, such as floods and earthquakes?
11. What environmental considerations should we take into account when laying out a switch room?
12. Why is proper ventilation of the battery room needed?

Glossary

AC *See alternating current.*

AC signaling Frequencies that carry address information or system status information. An example is the Touch-Tone® signal you produce when you dial a number on a DTMF keypad.

Access code Special character-sequences that you dial-in to connect to a trunk or trunk group, feature, or special equipment in the system.

ACD *See automatic call distribution.*

Acoustic coupler A system that uses sound waves transmitted through the air, to send and receive data from a *modem* to a telephone handset.

Adaptive routing A system in which the network's "smarts" select the best path (route) for each packet or message in a data transmission.

Address An identification number that a communications system uses to enable a signal to reach a specific destination. A telephone number is an address for the station you want to reach. A device code is an address that the common control computer in a PBX would use to select a particular port.

Addressing Using an identification number to select a connection in a switching system. *See address.*

Advisory tones Signals like the dial-tone, busy-tone, ringing-tone and call-waiting tone, which are provided by the system for the person using the telephone.

All trunks busy signal A fast-busy tone, indicating that no trunks are available. This identifies a form of *blocking*.

Alphanumeric   Containing both alphabetic and numeric characters—usually, the blank-space character and some punctuation marks are included among *alphanumeric* characters.

Alternate routing   In switching, alternate routing is the choice of the next-best-path when the best path is blocked.

Alternating current   A system in which electric power is distributed by moving electric charges back and forth in the wire, first in one direction, then the other, by an equal amount in each *alternation. See direct current.*

Analog signal   A signal in which the information can take on an infinite number of values between its maximum and minimum levels. An analog signal varies continuously, and can take on *any* value between its maximum and minimum. Frequency, amplitude, phase shift and pulse-width can all be varied in analog fashion.

AM   *See amplitude modulation.*

Amperes   Units that measure how quickly electric charge moves in an electric circuit. The units are basic charge-elements (Coulombs) per second.

Amplifier   A device that produces an output signal which has more power than its input signal, but which contains the same information. This is the basic principle of the *repeater.*

Amplitude   In a waveform, this is the "height" of the wave—the maximum value of the signal, above or below the center of the waveform. Also called the *peak* of the wave.

Amplitude modulation   Varying the strength of a signal so that its "wave height" varies according to some controlling factor. Usually, an audio-frequency signal varies the amplitude of a much-higher-frequency signal called the *carrier wave.*

Area code   A three-digit portion of the digits dialed for addressing in the North American Numbering Plan. These digits define the specific geographic region in which you will find the station being addressed.

ARS   *See automatic route selection.*

ASCII   American Standard Code for Information Interchange; a binary code that represents decimal numbers, letters of the alphabet, and punctuation marks with a seven bit or eight bit code.

Asynchronous   (NOT synchronous) digital information transmitted without a separate clock—or synchronization—pulse. Instead of a clock-pulse transmitted with each bit of information, synchronizing information is transmitted at the end of each byte or word.

Attendant console   An operator-station "switchboard" at the PBX.

Attenuation   Reduction in the strength of a signal due to a loss of power (usually in the resistive portion of a circuit).

Auto-answer   A device that contains a circuit to detect when it is being called; this circuit "picks up" the call when it is activated. This is used on *auto-answer modems* and telephone answering machines.

Automatic call distribution A method for distributing calls equally to all members of a group having equal priority (for instance, airline ticket agents answering incoming calls).

Automatic data collection unit A storage device used to record information or statistics from one or more sources. Traffic statistics regarding all the calls that pass through a PBX are examples of the kind of data we are talking about here.

Automatic dialer A device that can produce either rotary dial pulses or DTMF tones automatically, to dial a previously-stored sequence of digits. The PBX *speed dialing* feature is an example of this.

Automatic route selection A method used to select the best trunk group for a call going out of a PBX (usually the most inexpensive). This usually incorporates alternate routing as part of the method.

Balanced line A circuit in which neither side of the line is grounded. This minimizes crosstalk or noise pickup between pairs in the same cable.

Bandwidth A range of frequencies, between a minimum and a maximum, in which a circuit receives at least 50% of the power directed to it.

Battery backup A system in which batteries (usually rechargeable) are included in the system, to supply power in case power from the public utilities fails to be delivered.

Baud The number of signaling-elements transmitted per second in a data communication. These signal elements are often, but not always, equal to one bit (binary digit) of information.

Bell operating company A local Bell telephone company that provides service in a limited geographic region (example: Illinois Bell telephone).

Bidirectional Able to communicate in either direction on the same signal-conductors.

Binary Having only two states, or conditions, by which it can communicate information. A number system that only contains the numbers 0 and 1.

Bipolar transistor A solid-state active switching device that controls the flow of electric current without moving parts. It is distinguished from a *field-effect-transistor* by having two junctions instead of one (hence the name *bipolar*). It is used for signal amplification, as well as switching.

Bit A single binary digit—a 0 or a 1. The state (such as a voltage) that is used to represent that digit in a circuit is sometimes also called a *bit*.

Bit stream A series of bits transmitted, usually on a single conductor.

Bits per second The rate at which bits (binary digits) are transmitted. This is the same as the *baud rate* for systems in which signaling-elements are binary digits.

Block A "chunk" of digital information that may make up a *packet* used in data communications.

Blocking The inability to complete connection of a signal-path because no channels are available.

BOC *See Bell operating company.*

Bootstrap   In programming, a group of instructions (program) used to get an outside peripheral device to "load in" a larger program. When a computer is first turned on, the *bootstrap* directs the transfer of the main (operating system) program into the computer's memory from (for example) a floppy disk.

BPS   *See bits per second.*

Buffer   In logic gating, an amplifier-element that will boost the power of a signal without changing its logical value. (Sometimes, this is called a *bus buffer* or *bus driver*.) Often, this device has the capacity to be enabled or disabled (and may be called a *tri-state buffer* or *open-collector buffer*.) In computer systems, a block of memory cells used to temporarily hold a block of information for later output. (Example: a *print buffer* used to hold a page of print that can be completed while the computer goes on to do other processing.)

Bus network   A network in which all the stations and switches share a common path. Only one source and one destination use the path at any one time. *See time-division multiplex.*

Busy   A circuit that is already in-use.

Busy hour   A period of time in which traffic is sampled to get a representation of the maximum amount of traffic. The peak-use hour of the day.

Byte   A unit of binary information now standardized as eight bits of data. In documentation for older systems (pre-1977), this term was not standardized, and may have been used to represent larger or smaller units of binary data.

CCITT   Committee Consultative International Telegraphique et Telephonique (International Consultative Committee on Telegraphy and Telephony). International committee set up under the supervision of the United Nations to establish standards for international telecommunications.

Call cost accounting   An aspect of call detail recording that keeps track of what expenses are accumulated for a certain extension, department, or division of an organization.

Call detail recording   A device, and method, used to record such statistics about telephone calls as the number dialed, cost of the call, extension from which the call was made, duration of the call, and trunk or trunk group used to place the call.

Capacitance   A property of circuits that allows them to store electric charge. Charges will flow into a circuit with *capacitance*, even if it is not a complete circuit loop electrically. The amount of charge depends on the amount of voltage (potential difference) developed across the capacitive portion of the circuit.

Capacitor   An electric circuit component specially designed to have a large capacitance in a small space. *See capacitance.*

Carbon microphone   *See transmitter (handset).*

Carrier wave   A high-frequency wave usually modulated by some lower-frequency form of information. *See modulation.*

Carterphone decision   A 1968 FCC decision which ruled that telephone company tariffs preventing attachment of customer-owned or customer-provided equipment to the telephone system were discriminatory, unlawful and unreasonable.

Cathode ray tube   "Picture tube" or imaging device found on television receivers or data-terminal monitors.

CDR   *See call detail recording.*

Cellular radio   A system in which a geographic area is divided up into smaller "cells," each serviced by a separate transceiver. This allows a small number of frequencies to handle a large number of channels.

Central office   Telephone company switching center, also called an "end office." This is where you would find the local telco switch that connects to your telephone.

Central office line   The link connecting a station to a central office.

Central office trunk   The link connecting a central office to a PBX or another switch.

Central processing unit   The part of a computer that has arithmetic, control and switching capability. This is the "brain" of the computer.

Centrex   A type of direct inward dialing (DID) offered by local telcos. *See direct inward dialing.*

Channel   A communication that contains a single conversation or data transmission is carried on one *channel*.

Check bit   A parity bit. This is a binary digit added to a character to make its parity odd or even. To make a character with an even number of "1s" in it into odd parity, the check bit is given a value of "1." If the character already has an odd number of "1s," its check bit is given a value of "0."

Circuit   A complete path around which electric charges can circulate. Another name for a circuit is a *loop*.

Circuit switching   A method of connecting a communication path so that the same path is used for the entire duration of the call (ordinary telephone calls are always circuit-switched). In other methods, the call may be rerouted and reconnected through different paths during its course (for example, cellular radio).

Class of service   The abilities allowed to a station (such as the ability to dial outside calls) are defined by its class of service. Each station's class of service is defined in a table stored in the PBX, and may be changed by altering the table.

Clock   A source of timing pulses used to synchronize action in digital circuits.

Coaxial cable   A conductor having a single, central wire surrounded by a shield conductor, separated from each other by an insulator.

Codec   A contraction of coder/decoder. It is used to couple analog station equipment to a digital line. *See coding.*

Coding   Conversion of an analog signal into digital format for transmission on a single circuit. The analog level of the signal is sampled, the sample converted into a binary code by a A/D analog-to-digital converter, and the digital code transmitted in serial form, using PCM or delta modulation. Compression and expansion of the analog signal's dynamic range are sometimes done to reduce the number of bits needed for encoding into binery.

Coil   *See inductor.*

Color code   A system of representing numbers with colors. Color codes are used to identify wire-pairs in telephone cable, and to represent the resistances of electrical components like resistors and capacitors.

COM group   *See intercom group.*

Common carrier   A telecommunication resource available to the general public.

Common control   A method of switching in which a central logic system (usually a computer) is responsible for routing the call through the network.

Communications Act of 1934   The law that established the United States Federal Communications Commission. It regulates all interstate communications within the country.

Computer   A programmable calculating machine. Since the most common type of computer uses binary digital logic, and has the same switching and control capabilities as a digital switching circuit, the term *computer* is usually assumed to mean *digital computer*. There are *analog computers*, as well, but they are of no practical importance to telecommunications.

Concentration   A technique in which communications from a large number of incoming lines are sent out on a smaller number of outgoing lines or trunks. For example, suppose that no more than five percent of the station lines coming in to a concentration stage are in use at any one time. The number of trunks needed would be only five percent of the number of station lines.

Conductor   An electrical conductor is made of a material that passes the flow of electric charge freely, with little resistance. Other forms of energy, such as light, may be carried by other conductors, such as a fiber-optics "light pipe."

Cord board   The earliest form of manual switch. Patchcords that terminated in phone plugs were used to connect calls by being manually plugged into phone jacks by the operator. The tip and ring connections on the plugs gave the names *tip* and *ring* to the two lines of a *pair. See switchboard.*

CPS (cycles per second)   *See Hertz.*

CPU   *See central processing unit.*

Crossbar switch   An electromechanical switch that uses a two-dimensional matrix to connect calls. Many crossbar switches are still in use, although the technology has been made obsolete by ESS (electronic switching systems).

Crosstalk   An undesirable condition that happens when a communication from one line can be heard on another independent line. This is usually caused by inductive or capacitive coupling, or by an electrical short circuit between the lines.

CRT   *See cathode ray tube.*

Current   A quantity that measures how much electrical charge (measured in Coulombs) passes a point on a conductor, per second. Current is measured in units called *Amperes.*

Cursor   A symbol on a CRT screen that appears where the location where the next character will be typed (from the keyboard).

Cutover   When a new system is brought "on-line," the transfer of control to the new system is called *cutover*.

DAA   *See data access arrangement.*

Data   Information, usually digital in nature.

Data bank   A memory device for storing data. This term is also sometimes used to describe the data held in this memory device.

Data access arrangement   An interface device used to connect a terminal or other digital equipment directly to telephone lines.

Database   The information stored in a *data bank*. This information usually contains parameters that allow a special access program (called a DBMS or database management system) to rearrange or sort the data into any order, and according to any criterion, that the user wants.

Data communication   Transmission of information by digital means.

Data rate   *See baud.*

Data processing   Altering, rearranging or responding to digital information in a way that is useful to the end-user.

Data set   A device for coupling digital information to an analog communications system. The dataset is usually a **modem**, but may include an *acoustic coupler*, as well.

Data terminal   A workstation that communicates digital information. It may be a small computer, a teleprinter, or a CRT-and-keyboard combination. A data terminal is considered station equipment.

Datagram packet transmission   A technique used in *packet switching*, in which the receiving terminal is responsible for placing the packets it receives in the correct order.

DB   *See decibel.*

DB-25 plug   The standard 25-pin connector used for RS-232 serial data communication. This term is often used for both the plug (male) and socket (female) RS-232 connectors.

dBm   A logarithmic scale for measurement of power. On the dBm scale, one milliwatt is 0 dBm. *See decibel.*

DC   *See direct current.*

DC signaling   A technique in which the status of a line is represented by a voltage level on that line. E&M and DX signaling are examples of *dc signaling*.

Decibel   A unit used to measure relative increase or decrease in power, voltage or current, using a logarithmic scale. A *bel* (Bell) indicates an increase or decrease of ×10. Ten *decibels* make up a *bel*.

Decoding   Conversion of "coded" information into "clear" form. What is "code" and what is not, is in the eye of the beholder. The "decoder" part of a *codec* converts digital code into voice, for example.

Dedicated machine   A computer or microprocessor used to run only one program. This machine cannot be easily reprogrammed to run other programs, like a *general-purpose-machine*. The microprocessor in an arcade video game is an example of a dedicated machine.

Delta modulation   A technique which is used to digitize an analog signal by indicating only whether the level had increased, decreased or remained the same, since the last sample was taken. By contract, PCM (pulse-code modulation) encodes every sample. Errors in delta modulation affect the accuracy of the waveform transmitted in a cumulative way, while an error in PCM affects only the sample containing the error.

Demarc   The "line of demarcation" between the telco's lines and equipment, and the subscriber's lines and equipment. The point where the connection is made between the two may be a terminal block or other connector.

Demodulation   Recovering information from a modulated signal. *See modulation.*

Demultiplex   Separating channels of information that were transmitted together on a multiplexed link.

Deregulation   Removal of certain tariffs and restrictions placed on AT&T and the Bell companies, following the breakup of the Bell System.

Dial pulse   A signal produced by a rotary dial when its contacts are used to interrupt the flow of loop current. Many other circuits have been designed to imitate the action of a rotary dial for signaling purposes.

Dialtone   An audio-frequency signal placed on your line to indicate that the switch is responding to a request-for-service and is waiting for you to dial a number.

Dictation tank   A recording device for voice-documents, which are then available for transcription.

DID   Direct inward dialing.

Digit   The basic signaling-element of a digital system. In a *binary* system, the basic signalling-element is a *bit*, but in an *ASCII-code* system, decimal digits may be seven or eight bits each, and in *packed BCD*, digits are four bits apiece.

Digital   Information that is composed of a limited number of discrete levels or states, as opposed to *analog*, in which a waveform can take on an infinite number of values between its minimum and maximum limits.

Digital transmission   A technique in which all information is converted into digits before being transmitted. Digital repeaters reconstruct the signal as it is amplified on long-distance transmissions, with much more reliability, and at a lower cost, than analog repeaters.

Diode   A semiconductor (or vacuum-tube) device that allows electric charge to pass through it in one direction, but not the other. In real semiconductor diodes, current in the "reverse" direction is not zero, but is many thousands of times smaller than current in the "forward" direction.

Direct current   A system in which electric power is distributed by electric charges that always move in one direction. The polarity of voltage in a DC system is constant, although the amount of voltage may vary, from time to time.

Direct inward dialing   A technique which allows an outside party to call an extension in a PBX directly, by dialing an ordinary telco telephone number. The telco assigns a sequence (block) of DID telephone numbers to the PBX. When any of these numbers is called, the telco connects the call on any idle DID trunk, to the PBX. The telco then redials the last digits—which *are* the extension-number—of the telephone number called, so that the PBX knows which extension to connect the trunk to.

Direct inward system access   The ability for an outside caller to dial into a PBX by dialing an ordinary telco number. The PBX answers with internal dial-tone, and the caller can then dial an extension, or make an outside call through the PBX, just like any other extension inside the system. DISA can permit you to call an extension without going through an operator, but is more commonly used to bill long-distance calls to the PBX account.

DISA   *See direct inward system access.*

Disable   In digital gating, this is the condition of blocking or inhibiting a signal from passing through a certain level of logic gates.

Display   A device that indicates numeric, text or graphics information visually. A CRT, LED array, or LCD screen are examples of displays, also called *readouts.*

Distortion   An undesirable alteration of the waveshape of a digital pulse or analog signal. If it is bad enough, this can cause garbled voice or loss of data on a transmission.

Distribution frame   A framework of connecting blocks where connections are made between components and system cabling. Distribution frames generally use ''66''-type connecting blocks, which allow quick connection of cables and jumper wires. *See main and local distribution frames.*

Distributed network   An arrangement for connecting switches to one another that connects every switch directly to every other switch. It is also called a mesh arrangement.

Distribution group   A group of stations to which the PBX routes calls on some basis that gives every station an equal amount of traffic.

Divestiture   The breakup of the Bell System. This resulted in various companies being formed to handle various aspects of Bell/AT&T business, according to the U.S. government's orders.

DTMF   *See dual tone, multifrequency.*

Dual tone, multi frequency   The type of signaling done by Touch Tone® keypads. Each key generates a pair of audio frequencies, one indicating the row the key is in, and the other indicating the column it is in.

Dumb peripheral   An I/O device whose encoding and decoding functions are performed by software in the computer, not by hardware on-board the device itself.

Duplex   Carrying information in both directions on the same link. If the transmitted and received signal alternate on the same line, the term is *half-duplex*. If transmitted and received signals can travel in both directions at once, the term *full-duplex* is used.

DX signaling   A form of DC signaling in which voltages are superimposed on the potential levels of the two pairs of a four-wire trunk. The difference between the transmit and receive pairs carries "supervision" information.

Dynamic RAM   Random access memory device in which the information is stored on capacitances that are periodically recharged or *refreshed* through MOS driver transistors. Power loss will wipe out data stored in DRAM; also, if refresh cycles do not occur frequently enough, data will be lost.

E&M   A form of DC signaling in which the "battery" voltage is used to signal between various items of switchroom equipment (and tie trunk equipment).

Electromagnet   A coil of wire, generally wrapped around an iron core. Its magnetic field is usually used to activate mechanical devices, or switch contacts. When used in electrical systems with no moving parts, the *electromagnetic induction* of the coil is its important feature, and it is called an *inductor*.

Electromagnetism   Force field produced when current passes through a wire. Energy is put into this field, which may either be used to attract or repel other magnets.

Electronic mail   A system for delivery of messages in data-form to a central data-retrieval facility. People with access to this facility can open the files, and "read the mail." Electronic mail does not have to be left by one person and picked up by another. If there are active terminals or printers that are "in the net" all the time, you can send a message directly to a printer or terminal.

Electronic workstation   *See terminal.*

Enable   In digital gating, this is the condition of passing or permitting a signal to travel through a certain level of logic gates.

End office   *See central office.*

Entry point   In a software system containing a series of program modules, the address of any module's first instruction is its *entry point*.

Erase head   On a magnetic tape recorder, this is the "head" that erases tape by demagnetizing it, before a new recording is put onto the tape.

Extension   One of the stations in a PBX. It can be reached from another extension inside the system, by dialing its "extension number." Outside calls are transferred to an extension by the operator. Using DID or IESA, extensions may be dialed without the help of an operator.

Facsimile   A machine that reproduces alphanumeric or graphics information transmitted by telephone lines. It works similarly to slow-scan TV, and used to be called "wirephoto."

Faraday cage   *See shielding.*

Featurephone   A 'smart' telephone capable of providing some of the features available as options on a PBX.

Fiber optics   A conductor for light-pulse transmission that confines and propagates light-pulses using the principle of *total internal reflection*.

Flash   A signaling technique often used on PBX station equipment that doesn't have special-feature buttons. It involves interrupting the loop current for a time longer than a rotary-dial pulse, but not as long as a "disconnect." Sometimes, this has to be done manually, and works only as well as the user's innate sense of timing.

Floppy disk   A magnetic data-storage medium that records information on a two-dimensional flexible surface. It is faster than magnetic tape, because information can be retrieved from a two-dimensional surface using random-access techniques, but it is slower than random-access semiconductor memory or "hard" disks (which are designed to operate with a higher degree of mechanical precision, and at a higher speed).

FM   *See frequency modulation.*

Foreign exchange trunk   A trunk that connects a distant central office to an organization's main office. Callers at the distant location think they are dialing a local call, although the trunk may be connecting them to a telephone or system in another city. *FX trunks* can be used for outgoing calls, as well.

Four-wire circuit   A circuit having two pairs. These are the *receive pair* and the *transmit pair*, and they may be used for full-duplex data or voice communication. Four-wire circuits provide a better quality signal than a two-wire circuit.

Frequency   Measured in cycles-per-second or Hertz, this is the rate at which an electrical event repeats itself per second.

Frequency division multiplex   Transmission of many channels of information on a common link, but modulated onto different *carrier frequencies*.

Frequency modulation   Information is superimposed (modulated) onto the *carrier wave* by varying its frequency in proportion to the momentary DC level of the input information.

Frequency response   In a circuit whose components are sensitive to changes in frequency, this describes how the circuit reacts electrically throughout a whole range of frequencies. It is usually shown in the form of a graph.

Frequency shift keying   A form of binary frequency modulation. It gets its name from the Morse-code key used in radiotelegraphy which varies the frequency of the carrier wave to two frequencies, one for open, and one for closed contacts.

FSK   *See frequency shift keying.*

Full duplex   Two way communication at the same time on the same link. This is usually done by having the transmitted and received signals modulated onto two different carrier frequencies (called the *answer* and *originate* frequencies).

FX   *See foreign exchange.*

Gain   The amount of amplification of a signal passed through a circuit. It is calculated by taking the output signal strength and dividing it by the input signal strength. If the number is larger than one, the signal is *amplified*; if it is less than one, the signal is *attenuated*. Gain can be calculated for voltage, current or power.

Gating   Enabling or disabling a signal through a certain level of logic. If it is *enabled*, the signal gets through; if it is *disabled*, the signal does not get through.

General purpose machine   A computer that can run a wide variety of programs and is equipped with a large array of peripheral devices for different uses.

Ground start   A form of signaling in which one side of the trunk loop (ring conductor) is momentarily grounded to get a dialtone.

Half-duplex   Two-way transmission on a single link, with the users taking turns using the link for transmit and receive.

Handset   The part of a telephone containing the receiver and transmitter (speaker and microphone).

Handshake   Two-way transfer of signaling between a controller and a peripheral device requiring each device to inform the other if it is ready or busy when information-transfer is desired.

Hard disk   Also called a "Winchester," this is a magnetic two-dimensional recording medium which records data on a rigid surface (unlike the flexible surface of a "floppy"). Its faster speed and more-precision tolerances make it a faster, more high-volume storage medium than floppy disks; although the capacity of magnetic tape can be larger, its speed is *much* slower.

Hardcopy   A printout or graphic diagram output from a digital system to paper.

Hardware   Anything that works the way it does because it was *wired* to do things that way. The "nuts and bolts" of a digital system; the physical circuits it is made of.

Hertz   The basic unit of frequency. It measures the number of times an electrical event is repeated per second.

Hookswitch   The switch on a telephone that connects the voice circuits when the handset is picked up. Also called a "switchhook."

Host computer   The computer that co-ordinates all the digital traffic in a digital communications network. This is the computer that answers the calls when somebody wants to "log-on" to a computer through a "dial-up-line."

House cable   The main multi-pair cable that goes through a building from its main distribution frame. This cable is connected to the various local distribution frames in the building. Pairs of this cable are then connected (jumpered) to the wires going to the various stations in the building.

House pair   An individual wire-pair in the *house cable*.

Hunt group   A group of lines or trunks all addressed by the same lead number. The switch would "hunt" through the stations or trunks in that group, routing the call to the first idle line it finds in the group.

Hybrid circuit   A transformer arrangement that permits the transmitted and received signals to be separated, and then put back together. One place where they are used is to connect a two-wire line to a four-wire line.

I/O device   An input or output device (or both). The word *peripheral* is often used as a synonym in computer technology. *Input* is defined as transfer of information from the I/O device into the CPU of a computer, and *output* as transfer of information the CPU to the I/O device.

Idle Not in use. A telephone circuit which is not busy or on-hook is *idle*.

Impedance The opposition to the passage of AC current when an AC voltage is applied. *Impedance* is measured in Ohms, just like resistance, but varies according to the frequency used in the circuit.

Inductance Opposition to a change in current passing through a coil or electromagnet. Increasing the current causes a *back EMF* or voltage drop, opposing the additional current. Decreasing the current causes a *counter EMF*, or voltage gain, aiding the current flow and opposing the reduction of current. It is measured in units called Henries; every volt induced by a current-drop of one Ampere per second indicates an *inductance* of one Henry.

Inductor A coil or electromagnet, having inductance. *See electromagnet.*

Input Transfer of data from an outside device *into* a system.

Input port An interface that places data from an *input device* onto a computer's data bus, when the computer is ready to accept that data.

Instruction decoder Part of a computer that switches on its internal circuits in response to coded numbers from outside memory (called *opcodes*) that identify which circuit is needed.

Instruction *See opcode.*

Insulator A nonconductor. A material through which electrical charges cannot pass freely, and which has a large amount of resistance.

Interconnect companies Companies that sell, lease, and install private systems, which they interconnect to the public network.

Integrated circuit A semiconductor device that combines the functions of many transistors and other components in a single package.

Interface A device that is used between two types of lines or equipment. An interface contains circuits to change the information to a form that can be "understood" by the receiving equipment. An interface may also supply power and a signaling path for equipment. A PBX is an interface for each station, trunk or peripheral device connected to it.

Jack A connector or socket.

Jumper A wire pair that is used to connect the conductor's o one cable pair to another. The jumper is used so that the connections can be easily changed.

Key system A telephone system that uses multiline phones (keyphones). Key systems are used in smaller organizations, with, usually, less than 50 phones. Each of the lines of a multiline phone in a key system can be connected to the local telco switch or a PBX. Most key systems have an intercom line used to call between phones in the system. *See key service unit.*

Keyboard Usually a typewriter-like array of switches or pushbuttons used for terminals, printers or electronic workstations. The keyboard is used to enter information or data.

Keyphone   *See multiline phone.*

Key service unit   Controlling equipment for a multiline phone system. It contains the interfaces, common circuits and connections for the lines, stations and power supplies of the system.

KSU   *See key service unit.*

LAN   *See local area network.*

LCD (liquid crystal display)   A readout device that controls reflected light, and is used to display alphanumeric or graphics information. These displays are common on digital wristwatches, pocket calculators and some types of portable "briefcase" computers, as well as electronic phones.

LED light emitting diode)   A solid-state device that converts electrical energy into light. A solid-state lamp or illuminator, these devices are available in many colors of visible light, and in infrared wavelengths. Arrays of LED segments are often used in self-illuminated displays, for the same purpose as LCD's mentioned above.

Line   A channel or path for communications between a station and a switch. Line is a general term often used to describe any type of link or port in a system.

Link   A communication path that uses any media or technique. A link does not have to consist of a single channel, and can be multiplexed.

Lite pipe   *See fiber optics.*

Load (program load)   To enter information into a memory device that contains the instructions which operate a computer during a computer program.

Load map   A listing of the entry points into routines and subroutines (programs and program-segments) in a software system. Entry points are the addresses of the first instruction in each program or segment.

Loading coil   An inductor used to help counteract the effects of capacitance between the conductors of wire-pairs. These are used only on long transmission lines; one is required for every few thousand feet of line.

Local area network   A data network used to tie together digital equipment in a limited geographic area, such as an office building. The current trend is to use the organization's PBX as the 'hub', or controlling element of this network.

Local distribution frame   A framework where house-cable pairs and station-wire pairs are terminated. Jumper wires are used to make connections between the two. Local distribution frames are found in a "local equipment room," or a wall panel that serves a limited physical area within a larger building or complex. *See distribution frame.*

Local equipment room   A room or closet in an area of a building where telecommunication gear and a distribution frame are found. Equipment found in this room is used locally, like the key-service unit used to service nearby multiline phones.

Location   The place where a station jack is found. A location is assigned an identifying number (on the jack) that corresponds to the number on the distribution-frame where the wire that goes to that particular jack is connected. These numbers are needed for installation and reference.

Logon   When you want to use a computer through a terminal or dial-up line, the logon routine asks you to provide the computer with passwords and other information before it will give you access to "the system."

Loop   A complete electrical circuit. In telephone jargon, the loop is the pair of wires that goes to a station, or the pair of wires used for a trunk.

Loop current   Current that flows through a line or trunk loop when it is in-use (for instance, when you pick up a telephone handset.) (About 20 milliamperes, depending on wire length).

Loop start   A line or trunk that is seized by placing a load (telephone or equivalent device) across the tip and ring conductors. This causes a current that is interpreted as a request-for-service. *See loop current.*

Loss   Attenuation or reduction of the signal's strength.

Main distribution frame   A framework of connecting blocks where connections are made between the components of the system. These components are the cables to the switching system, cables to the telco demarc, house cables, location wires, and peripheral equipment. *See distribution frame.*

Maintenance port   A port where an input/output device is connected in a system, to allow programming changes, and to look at system status information. Maintaining system operation, testing, and listing configuration-information are features that are often available from this port.

MDF   *See main distribution frame.*

Medium   Any form of energy suitable for conveying information from one place to another, or a carrier for that form of energy. (plural: Media).

Memory   A device that uses some form of energy (usually electrical or magnetic) to retain information after the signal conveying that information is gone.

Message switching   A data communications technique that uses a switch in a network to receive a message, then store it and retransmit it when a link becomes available.

MF signaling   *See multifrequency signaling.*

Microprocessor   A computer central processing unit whose circuitry is contained entirely in a single integrated circuit (the term is sometimes extended to include "chip sets" of a few chips or packages).

Microwave   A radio-frequency band whose electromagnetic waves are less than a meter in length.

Modem   (short for modulator/demodulator) an interfacing device that modulates by converting digital logic signals into frequency shift keying (FSK) modulated information. It demodulates by converting FSK frequencies back into digital signals.

Multifrequency signaling   An AC signaling method used internally by telcos and other common carriers, which is similar to DTMF.

Multiline phone (keyphone)   A telephone that has a number of buttons used to select different lines or trunks. *See key system.*

Multiplex   A method used to combine many different channels of information onto one link. *See time division multiplex, space division multiplex,* and *frequency division multiplex.*

Network   An arrangement of links and switching systems used to permit communications between stations. Public networks are connected to one another to allow calls around the world. Private networks are usually developed to allow communications with distant branches of an organization, but can be tied into the public network as well.

Noise   Electrical interference that contributes unwanted energy to the signal. Any unwanted energy of the same form as the signal, but not carrying any information, is noise.

North American dialing plan   The method for identifying stations in the public network of the North American continent. It consists of three parts, a three-digit "area code," a three-digit "office" code, and a four-digit "subscriber code."

Null modem   A coupler for RS-232 communications that is used for direct hookups between digital equipment, where a telephone and modems are not needed.

OCC   *See other common carriers.*

Off-hook   A term that designates a piece of equipment that is in-use. This term originates with the handset or earpiece of early telephones that was suspended from a hook until picked up (producing a loop current).

Office code (exchange code)   The three-digit number used in the North American Dialing Plan to identify the local central office to which the call is being placed.

Ohm's law   The law which relates the behavior of current (I), voltage (E), and resistance (R), in a resistive circuit. The relationship is usually expressed as an equation giving you a third quantity based on the fact that the other two are known, namely, $E = I \times R$, $I = E/R$ and $R = E/I$.

Ohms   Units of resistance. A resistance of one Ohm permits a current of one Ampere to flow when a potential difference (electromotive force) of one volt is applied to the ends of the resistance.

On-hook   A condition that indicates a piece of equipment is not in-use. No loop current flows when a telephone is on-hook, and the switch "recognizes" that this telephone is available for incoming calls.

Opcode (short for operation code)   A binary number that activates some operation or sequence of operations inside the central processing unit of a computer.

Open circuit   A circuit that does not have a complete path for current flow. If a component becomes open, its resistance becomes infinitely large, and current stops flowing.

Operator console   *See attendant console.*

Optional equipment   Generally, anything in a telecommunications system that isn't a station, switch, or link.

Other common carriers   Common carriers that weren't part of the Bell System were called *other common carriers,* or OCC's. They still are. Bell isn't. Examples are MCI (Microwave Communications, Inc.), SBS (Satellite Business Systems), and Sprint.

Outpulse   Sequence of addressing information, like DTMF outpulsing and rotary outpulsing. This may be called "dialing," but it is usually used to refer to automatic, rather than manual, production of these signals.

Output   Transfer of data in a digital system outwards to an outside device.

Output port   An interface that picks up data placed on a computer's data bus, and transfers it to an output device. It will hold that data until the output device has completed its cycle, or until new data is sent to it by the computer.

Packet switching   A technique used in data communication for transmitting information in blocks (packets). Since each packet contains its own addressing information, the packets that make up a total transmission can be sent on different paths, and at different times, and still be reassembled into the original message at the receiving system.

Pager   A little "beeper" you would wear on your belt, that is part of a simplex communications system. It is usually used to inform you when someone wants you to call them back.

Paging system   A public-address system, for announcements, messages, to call someone to the phone, etc.

Pair   Two conductors. Often, one conductor carries current outwards, and the other to return it. In other cases, pairs may carry independent signals. Pairs are often combined to make multipair cables, for example 25-pair, 100-pair or 1300-pair cables.

PAM   *See pulse amplitude modulation.*

Parallel circuit   A circuit in which electrical current can take more than one electrical path to go from one point to another. The most common example of a parallel circuit is house-wiring. Every outlet is in parallel with every other outlet on a specific circuit-breaker. If one light burns out, current continues to flow through other lights or appliances in the circuit.

Parallel transmission   Transmission of data in which several bits are transmitted on separate lines at the same time. *See space division multiplex.*

Parity bit   A method of error-checking in the transmission of data. In "odd parity," a bit is added whose value makes all the codes used for alphanumeric characters contain an odd number of 1's. In "even parity," the parity bit makes every character contain an even number of one's. This is sometimes called a "check bit," and doesn't have any information value, but is used only for error checking.

Party line   A common line shared by several stations, which may be signalled separately, but have the same voice path, and can only take turns using the line.

PBX   *See private branch exchange.*

PCM   *See pulse code modulation.*

Peripheral   A device that is connected to a system to enhance its abilities. In digital computer technology, this term refers to any input or output device.

Phase shift modulation   Information is superimposed (modulated) onto the *carrier wave* by varying the *phase angle* of the oscillator in comparison to a fixed-frequency source. The phase is shifted in proportion to the momentary DC level of the input information. Digital information is often broadcast via this type of modulation.

Playback head   On a magnetic storage device, this is the part that converts magnetic information on the storage medium into an electrical signal. Moving the magnetic fields on the medium past the playback head generates a voltage, which is picked up in a conductor (coil) in the head and sent on to electronic equipment.

Port   Any input or output channel on a voice or data switching system. The same term is used in digital computer terminology to indicate the interface between the CPU and its peripherals. In a PBX, a port is required for each line, trunk, or peripheral device connected to that PBX. *See interface, I/O device.*

Power   The electronic term describing the amount of work an electric current can do in a unit of time. The units—watts—measure one *joule* of work done per second. A *joule* is the amount of work done in lifting a quarter-pound weight a distance of one yard. (*Metric:* one Newton force used to move an object one meter)

Power supply   A circuit that converts AC line voltage to DC, usually at a lower voltage. Another use of this term refers to a DC source of any kind.

Printed circuit board (card)   A plastic or fiberglass sheet with copper-foil conductors bonded to its surface(s). These conductors are used to make connections between components soldered onto the sheet. Usually called a PC board.

Private branch exchange   A telecommunications switching system used by an organization to connect its stations to each other and to the public network.

Program   A list of instructions for a digital computer. These instructions are carried out in a particular sequence that accomplishes a task.

Program counter   A counter inside a digital computer's CPU that keeps track of which instruction is next, in a program.

Protective coupling arrangement   A device used to isolate the telco's lines from station or switching equipment. Not required if the equipment has passed FCC approval.

Protocol   A standardized set of give-and-take signaling between both ends of a communication path, which must be done to set up a transfer of information.

PSM   *See phase shift modulation.*

Pulse amplitude modulation   A method of pulse-modulation in which the amplitude of the information being modulated controls the amplitude of the pulses. This is used for analog multiplexing. Samples of each input channel voltage are interposed between voltage samples from other channels. The cycle is repeated often enough for the sampling rate at any one channel to be greater than 2 × the highest frequency transmitted.

Pulse code modulation   The voltage level of an analog input signal is sampled and converted into a digital code. This code is transmitted serially (as a series of digital pulses), and converted back into an analog wave at the receiver.

Pulse width modulation A technique in which an analog input signal's DC level controls the pulse width of "digital" output pulses.

Queuing Waiting for a circuit to become available. When the PBX puts you in a "waiting line" waiting for a trunk to become available, for instance, it is putting you "in the queue" for the next idle trunk.

RAM *See random access memory.*

Random access memory Memory that can be over-written with new information (read/write memory). The "random-access" part of the name comes from the face that the next "bit" can be reached at any location in an equal amount of time, regardless of the location presently being accessed. By contrast, access time to reach a recorded "bit" in a *serial access* arrangement, like magnetic tape, depends on how far away the "bit" is from the tape's present location.

Reactance Ohms (opposition to AC current flow) in a circuit that is inductive or capacitive.

Read only memory Memory that cannot be over-written, and contains information that is not affected by power shut-down. This information is also random-access, rather than serial access, but the name "RAM" is *only* applied to read/write memory.

Readout *See display.*

Receiver (handset) The thing you hold in your hand when you answer a telephone call.

Record head An electromagnet that magnetizes the surface of a magnetic recording medium in proportion to an electrical signal.

Regional holding company The central hub of a group of Bell Operating Companies formed after the breakup of Bell and AT&T. Ameritech, for example, is the regional holding company for Illinois Bell, Indiana Bell, Wisconsin Bell, Michigan Bell, and Ohio Bell.

Register A storage device capable of receiving and holding a number of digits. DTMF registers hold a number dialed in Touch-Tones® , rotary registers hold numbers dialed in rotary pulses, computer registers hold numbers that arrive in binary.

Repeater A power-amplification device used to restore lost energy from a signal that has traveled down a long transmission line. The earliest repeaters were relays used to repeat telegraph signals.

Reset A control line on a computer that forces its address counter back to zero, or some other "beginning" point in its memory. Often used to get the computer out of a hopeless jam.

Resistance Opposition to current flow in any circuit.

Resistor Device used to limit the flow of current in a circuit.

Restriction dialing unit A dial that prevents you from dialing certain numbers—usually long-distance.

Ring One of the two conductors (tip and ring) of a line or trunk circuit. (The name originates from the fact that the "ring" of a phone plug was attached to this conductor.)

Ring detector Any circuit designed to identify the 100 volt AC, 20-Hertz signal used to ring a telephone.

Ring network An arrangement of switches in which they are all connected in a series, with the last one connected back to the first (in a circle).

Ring voltage A 100-volt, 20 Hertz signal used to ring the bell or power an electronic ringer on a telephone.

Ring-down A telephone that automatically rings another telephone when it is picked up. A "hotline."

Ringback tone The sound you hear when you're ringing somebody else's phone. This is generated by an oscillator and may be totally unrelated to the rate the phone at the other end rings.

ROM *See read only memory.*

Rotary dial A telephone dial with ten holes in it, which you rotate to dial.

Rotary dial pulse *See dial pulse.*

Routing The technique by which a path is selected for a call.

RS-232 A standard interface for serial data transmission.

RS-449 Another standard serial interface—uses balanced lines.

Run (program run) Starting the program counter and having the computer execute the instructions it finds in a list is called *running* the program.

Satellite link A microwave link that uses a satellite to receive and retransmit signals. Uses a geosynchronous orbit to keep a satellite above a fixed position on the equator.

Scanner *See line finder.*

SCC *See specialized common carrier.*

SCR switching matrix An array of contact points between rows and columns that use solid-state switches instead of moving contacts.

SDM *See space division multiplex.*

Sender A device used to send out the digits of a telephone number. Senders are necessary in a common control system, because the switch needs to know all the digits of the number you are calling before it can select a trunk.

Serial Transmission A technique for transmitting multidigit binary information on a single line, by sending each bit at a separate time.

Series circuit A loop in which all the electrical devices are lined up, one after another, so that the same current must pass through every device, to return to its source.

SF signaling *See single-frequency signaling.*

Shared link A circuit that can be used by multiple users. For example, the trunk that links a call from Chicago to New York is a shared link. It is released when the call is completed, and any other caller in Chicago, wanting to call New York, may be able to use it.

Shielded cable A conductor or group of conductors, surrounded by a grounded foil or jacket of braided wire. This "shield" conductor, grounded to zero volts potential, is used to prevent external "noise" from entering the internal conductor(s) of the cable, or radio-frequency energy from being radiated out of the inner conductor(s).

Shielding (radio frequency interference)   Enclosing electrical devices or conductors in a grounded, conducting container. This is sometimes called a *Faraday Cage*. It prevents the pickup of external radio-frequency "noise," and emission of radio-frequency energy from within the enclosure. *See shielded cable.*

Short circuit   A circuit whose electrical resistance is zero, or close to zero Ohms. Any circuit which has much lower resistance than it is supposed to have is usually referred-to as "shorted."

Sidetone   A characteristic of the telephone handset that enables you to hear, on your earpiece, what you are saying into the mouthpiece. There are various ways to do this, but they are all intended to give you "feedback" that the circuit is supplied with electrical power, even before you are connected to a dial-tone or another telephone.

Signal   A voltage, current or electromagnetic wave (radio wave or light wave) that carries information.

Signaling   Certain types of information carried by a varying signal in a telecomm. system. This information describes the status of circuits in the system (on-hook or off-hook), or identifies the destination of a call (addressing).

Simplex   A transmission in which information travels only in one direction. A one-way communication.

Single-frequency signaling   A frequency used on interoffice trunks, whose absence or presence indicates whether a circuit is in-use or busy. The frequency is present when the circuit is idle, and goes away when the circuit is in-use. It is used for supervision, and, rarely, for inter-register signaling. The frequencies used on the two pairs of a four-wire line is 2600 Hertz, and on two-wire lines, two frequencies are used, 2400 Hertz in one direction and 2600 Hertz in the other.

Single line phone   A telephone with only one voice circuit on it. A common household telephone (having only one telephone number).

Skin effect   Electrical characteristic of wire conductors that causes their resistance to increase with higher frequencies. Skin effect is caused by the electrical current travelling on only the outside surface of the wires. As frequencies become higher, the conducting portion of the wire becomes more and more limited to the outside surface of the conductor, and this reduction in usable current-carrying area appears as an increased resistance in the current path.

Sleeve   One of the three conductors of a phone plug—the tip, ring, and sleeve. The sleeve conductor is often grounded. Many telephone circuits do not have this connection, only the 'tip' and 'ring.'

Smart peripheral   an input or output device in a computer system that has on-board encoding or decoding circuitry of its own, and does not require the computer to perform code-conversion tasks with software. Often, the on-board "smarts" of the peripheral device are provided by a microprocessor and a small program in ROM memory.

Software   The binary information held in digital *hardware*. In a computer, the software consists of the instructions of the computer program, and any data or tables of numbers held in RAM memory along with the program. Program information held in ROM memory (which is almost like a part of the permanent wiring of the system) is sometimes called *firmware*. Software gets its name from the fact that RAM memory contents are easily altered or erased, therefore "soft," by comparison to permanently-wired electrical circuitry (hardware).

Solid state   Devices that use semiconductor characteristics to enable or disable the passage of current are called solid state. Transistors, diodes and integrated circuits are examples of solid-state devices.

Space division multiplex   Transmission of multiple channels of information at the same time, by using a different wire for each channel. In digital terms, this is called *parallel transmission*.

Speaker   *See receiver (handset).*

Speakerphone   A unit, either built-in or attached to a telephone, with an amplifier, microphone and speaker, that allows "hands-free" talking and listening in the room. The microphone and speaker alternate between the two parties as each one speaks.

Specialized common carriers   Are like other common carriers, but specialize in providing specific services such as data or video communications.

Star network   An arrangement of switches in which several "satellite" switches attach to a central "hub" switch, and all traffic passes through the "hub."

Start bit   A digit transmitted before a data word in a serial data transmission. This bit indicates the beginning of a new data word or byte (usually seven or eight bits in length), and is preceded by the *stop bit* at the end of the previous word. *See stop bit.*

Static RAM   A memory device that uses latches as the storage elements for binary digits. The information remains in the latches unless over-written or unless power fails.

Station   Equipment that is used to place a call, or receive a call, in a telecommunications system. It may be a telephone, a data terminal, an electronic workstation, a PC, or some combination of these.

Station feature   An enhancement of the basic capabilities of a piece of station equipment. Speed dialing is an enhancement of the basic dialing capabilities of a telephone, and therefore considered a station feature.

Statistical multiplexing   A form of time-division multiplexing in which channels are allocated to specific time-slots, but the number of channels may exceed the number of time-slots. This depends on the likelihood that some time-slots will be idle, and can be shared with other channels. The number of additional channels depends on how many time-slots are likely to be idle (not in use by their regular users) at a given time.

Step-by-step switching   A method of connecting a call in which each digit dialed rotates contacts in a switch matrix, completing part of the path to the number being dialed. When the last digit is dialed, the final connection is made, and the called phone rings (unless it is busy). This method has largely been replaced by *common control* switching.

Stepping switch   An electro-mechanically operated rotary switch, used for addressing in the earliest dial telephone systems. This device, while several generations behind the state-of-the-art in switching technology, is still in use in many places.

Stop bit   A digit added at the end of a serially-transmitted word or byte to indicate completion of the sequence of bits. It is followed by the start bit of the next character. *See start bit.*

Stored program control   Computer control of switching. The software program that is stored in memory and defines the behavior of the switch. Stored-program control replaces the hard-wired logic of earlier common control switching systems.

Strowger switch   A stepping switch invented by Almon B. Strowger, and first used in a telephone switch in Kansas City in 1892. *See stepping switch.*

Subscriber code   The last four digits of a telephone number dialed in the North American Dialing Plan.

Supervision signals   Signaling on a telephone trunk that identifies, for instance, whether the called party has picked up the phone, and the telco can begin to time the length of the call for billing purposes (this is called answer supervision). If the call just completed has been hung up (this would be called disconnect supervision), the telco equipment signals various switches along the route that the call is over, so that the various parts of the public system can decide what to bill the customer.

Switch (short for *switching system*)   A system for the connection of telephone calls. It links the caller together with the called party, or with a trunk going to other switches that can complete the connection. These calls are usually voice (telephone) but can also be data (computer communications) or digitized voice.

Switch cable   A cable of wires connected between a switch and the main distribution frame in the switch room.

Switch room   The place where the PBX or switch is located. It usually contains the switch cabinets, main distribution frame and various optional or peripheral devices.

Switchboard   A manual panel for connection of calls, equivalent to an automatic PBX, except that the console operator's tasks include making all the electrical connections. Sometimes attendant consoles are still called "switchboards."

Switchhook   *See hookswitch.*

Switching system   A switch or network of switches. *See switch.*

System feature   A feature or ability of the switching system that is generally available to all its users. Examples: Least-cost call routing (Automatic Route Selection) and Queing are system features.

T1 span   A standardized high-speed serial bit stream, for data or digitized voice transmission at a rate of 1.544 million bits per second. T1 is often used for connections between switches.

Tandem switching   Routing of a call through a switch to another switch (the station being called is not connected to this switch) repeated from switch to switch until the connection is completed.

Tape recording   Data or voice recording on magnetic tape.

Tariff A description of services, and the price of those services, that has to be filed by common carriers and approved by regulating agencies such as the Federal Communications Commission and/or State Communications Commissions.

TDM *See time division multiplex.*

Telco Abbreviation for telephone company. A company that provides connection services and the sires needed for those services to their subscribers in the area. Telcos are often Bell Operating Companies.

Telecommunications Transmission of information from one place to another. This information may be a voice signal, digital data or video pictures.

Teleconferencing Combined voice/video communications, used to schedule conferences without the cost of transporting the people to a common location.

Telegraphy Transmission, by wire, of serial alphanumeric information by Morse Code. Nonvoice, Morse code transmissions are usually characterized as telegraphy, and voice as telephony. Wireless transmission of Morse code is called *radiotelegraphy.*

Telephony Transmission of voice signals by wire. Wireless transmission of voice is called *radiotelephony.*

Teleprinter *See teletypewriter.*

Teletypewriter A "robot typewriter," controlled by digital transmission from a remote location, usually via wire. Teletypewriters may be used as terminals in data communications.

Test handset (also called butt set) A telephone handset designed to be connected directly to the telephone lines. It usually has a dial, but no ringer. It also has a "test monitor" switch so that you can listen to a line without drawing loop current and activating the circuit.

Tie trunk (tie line) A dedicated trunk used to connect two locations that routinely need to contact one another. Stations connected to a switch at one end of the tie trunk may dial stations connected to a switch at the other end directly, without using the public network.

Time division multiplex A method of placing many channels of data, or voice, on a single line by switching each channel's information into a different time-slot and scanning all time-slots faster than new information is generated.

Time-slot An interval of time in which a sample is taken from one of many channels in a TDM system, and transmitted.

Tip One of the two conductors (tip and ring) of a line or trunk circuit. (The name originated with the fact that this conductor was originally connected to the tip of a 'phone plug)

Tone generator (sender) An oscillator designed to produce advisory tones that the user of the telephone can interpret by listening—like the dial-tone, busy signal and ringing tone. May also be used to generate signalling like DTMF.

Total internal reflection Principle on which fiber-optics is based. Light is trapped in a transparent medium when it strikes the outer surface of the transparent material at a shallow angle, and 100% of the beam is reflected back into the medium. This allows all the light to continue down the length of a fiber.

Touch tone® *See DTMF.*

Traffic Calls placed on a system or link. Traffic is an important consideration; systems or links can handle some maximum amount of traffic before blocking occurs.

Traffic data statistics Information about trunk usage, common circuits like registers and senders. Statistics include how many calls were placed on a trunk group, the number of times all the trunks were busy, total number of seconds a trunk group was in use, and traffic between stations within the switch as well. Statistics such as how many registers were in use, and how often all the registers were in use and calls were blocked are important in determining if the present system is capable of handling traffic efficiently.

Transfer To connect a call to another station in a switching system.

Transformer A coupling device that can transfer a signal from one electrical circuit to another without a direct electrical contact between the two circuits. This separation of electrical circuits while signals are transferred is called *isolation*.

Transistor switching matrix An array of transistors used to carry out the switching functions of a crossbar or strowger switch without moving parts.

Transmission forms The forms into which information is transformed or converted before transmission. A voice communication, for instance, is transformed into an analog electrical signal called an *audio signal*.

Transmitter (handset) The mouthpiece into which you talk on a telephone. This telephone mouthpiece is usually carbon but other types of microphones are also used.

Trunk A link from one switch to another. *See tie trunk, foreign exchange trunk, etc.*

Trunk group Several trunks arriving at a switch from a common source, or are used to carry calls with a common purpose. For instance, we would group Band 5 Wats trunks from AT&T together because they are used to call a certain long-distance area. A group of MCI trunks would be grouped separately because they are supplied by a different common carrier.

TTY *See teletypewriter.*

Turnkey system A digital system that "knows what to do" when you "turn the key". Programming for computers (like the so-called "bootstrap" program) that gets things started when power is turned on, is usually stored in ROM memory. A computer that "kicks-in" this kind of program at power-up (when the "key" is "turned") is called a turnkey system. The user of such a system can start using it immediately when the power is turned on.

Twisted pair A wire-pair loop in a complete electrical circuit, which has the outgoing-current wire and the incoming-current wire twisted around each other to reduce cross-talk. The pair is twisted together so that the electrostatic and magnetic fields induced in an adjacent pair by one of the wires are cancelled via the oppositely-directed current of the other wire.

Two-wire circuit A telephone line or trunk that has just one current loop (one pair).

UART (Universal Asynchronous Receiver/Transmitter)   A serial interface that permits multidigit binary information to be time-division multiplexed onto a single line. A device that converts parallel data to serial data (Transmitter) and vice-versa (receiver).

Vacuum tube   The first electronic amplifier capable of boosting the power in a voice signal. These are largely obsolete, except for the CRT (picture tube) of video displays and televisions, and for high-power amplifiers in radio transmitters.

Video   The method of picture display used in television. An electron-beam pattern of scan lines, called a *raster,* covers the surface area of a phosphor-coated screen with light. The brightness of this light is varied from point to point on this screen by varying electron-beam current. A picture is formed by timing these variations in electron-beam current to coincide with specific locations on the scan-pattern, and repeating the pattern 60 times a second—with slight variations in pattern—to give the illusion of movement. When used in computer displays or data terminals, the information used to modulate beam brightness can produce alphanumeric characters on the screen, as well.

Virtual circuit   A digital point-to-point connection between one station and another, in which the route travelled by the data may change, but the connection remains intact.

Virtual machine   The machine that you "see" when you operate a computer or terminal through its keyboard and displays. The "rules" this machine follows are defined by software. By putting different software into the machine, the same hardware can control a personal computer, a microwave oven controller, a PBX or a video arcade game.

VMS   *See voice message system.*

Voice   Information that varies an elecrical, optical or data signal in proportion to air pressure variations in the spoken voice of a person. Normally an *analog signal,* voice can be digitized, and transmitted as binary information, intended for digital-to-analog reconstruction at the receiving end.

Voice-grade line   Any telecommunications link capable of responding to analog signals in the frequency range of 300 to 3000 Hertz. Voice information can be bandwidth-limited into this range without losing intelligibility.

Voice message system   An equivalent to the "electronic mailbox" that allows spoken input. You can call a "voice mailbox," and leave a message for a person or a group of people. This differs from an "answering machine," since the message can be duplicated for distribution to many people—even for everyone with access to the system, if necessary. The voice message is typically digitized and stored as hard-disk data, so that it can be accessed and distributed under computer control.

Voltage   Potential energy per unit of electrical charge. It is sometimes called "electromotive force," and controls the flow of electrical charges through a circuit in much the same way that pressure controls the flow of water through a pipe.

Volts   Units of voltage. A *volt* is the amount of electromotive force that will push electrical charge through a one-Ohm resistor at a rate of one Coulomb per second (one Ampere of current).

Wats   *See wide area telephone service.*

Waveguide   A conductor for microwave transmission. Often, a waveguide is a metal pipe with a rectangle cross-section, half a wavelength wide, and a quarter of a wavelength high, which is only useful for one specific frequency of microwaves. Circular or helical waveguides are used for wider ranges of frequencies.

Wavelength   The distance electromagnetic or other forms of energy can travel between successive oscillations of the wave generator. The wavelength is the physical length of the wave as it propagates through space for a conductor. It is usually measured from one node (where the wave has a momentary value of zero) to the next.

Wide area telephone service   A service of AT&T that provides long-distance calling at reduced rates. It is not always cheaper than other services.

Wink start   A method of signalling on trunks, that identifies a busy/ready status to receive digits. The "wink" is sent by the answering switch to indicate that it is ready to receive the dialled digits from the calling switch.

Wireless propagation   Transmission or electrical information without the use of conductors. Usually, this name connotes the use of radio waves, but could be infrared, light, or ultrasonic transmission.

Workstation   *See electronic workstation.*

Index